Engineering, Science, and Sustainability: Advancements in Technology and Techniques

First edition published 2024
by CRC Press
4 Park Square, Milton Park, Abingdon, Oxon, OX14 4RN

and by CRC Press
2385 NW Executive Center Drive, Suite 320, Boca Raton FL 33431

CRC Press is an imprint of Informa UK Limited

ISBN: 978-1-032-48423-5 (pbk)
ISBN: 978-1-003-38898-2 (ebk)

DOI: 10.4324/9781003388982

Typeset in Sabon LT Std
by Ozone Publishing Services

About

ISC 2022 is dedicated to the Niti Aayog policies to promote sustainability through exchange of ideas emerging out of the academia. The ISC is an annual conference that would be held in virtual mode until COVID restrictions on travel exist.

Vision

The vision of the conference is to capacitate Academia with the necessary ideas that will provide insights of the grassroot level development to various stakeholders of the Niti-Aayog policies. Towards this goal, the conference will create a conjunction of various stakeholders of Niti-Aayog policies that include- academic institutions, government bodies, policy makers and industry.

Mission

The ISC organizers will make concerted efforts to promote academic research that would provide technological, scientific, management & business practices, and insights into policy merits & disruptions. The framework of exchange of ideas will be geared towards adoption of deep technologies, fundamental sciences & engineering, energy research, energy policies, advances in medicine & related case studies. This framework will enable the round table discussions between the academia, industry and policy makers through its range of plenary and keynote speakers.

<u>Chairs</u>
Ravindra Pratap Singh, PhD
Indira Gandhi Tribal University, India.

Dimitrios A Karras, PhD
NKUA Greece.

Amarendra Pratap Singh, PhD
Indira Gandhi Tribal University, India.

K.K. JACOB, PhD
Faculty of Applied Sciences and Technology, Universiti Tun Hussein Onn Malaysia.

<u>Organizing Commitee</u>
Mengistu Tulu Balcha, PhD
Ambo University Ambo, Ethiopia.

Kannaiya Raja, PhD
Ambo University Ambo, Ethiopia.

Karthikeyan Kaliyaperumal, PhD
Ambo University Ambo, Ethiopia.

<u>Publishing Committee:</u>
Dimitrios A Karras, PhD
University of Athens (NKUA), Greece.

Karthikeyan Kaliyaperumal, PhD
Ambo University Ambo, Ethiopia.

Sudenshna Ray, PhD
RNTU-AIISECT University Bhopal, India.

Contents

Editor's Biography...

Foreword..*xii*

Preface ....*xiii*

Introduction...*xiv*

Details of Programme Committee ....*xv*

1. Comprehensive Survey on Artificial Intelligence-Based Malware Analysis and
    Detection Methods 1

2. Integrated Artificial Intelligence-Based Traffic Light Detection System for
    Autonomous Vehicle...7

3. Identification and Authentication of SSR Markers for Purity of Hybrids
    in Sesame (Sesamum indicum L.)...12

4. Detection of Specular Reflection on the Smart Colposcopy Images
    Using Fine-Tuned U-Net Model...16

5. Mapout a Crop Protection System Against Forest Animals Using Internet of Things 22

6. Patrolling Robot for Fugitive Emissions in Industries 26

7. Investigation of Antioxidant and Photo Catalysis of Natural Honey and Cow
    Urine-Doped CeO$_2$ Nanoparticles Fabricated by Reflux Method 31

8. Assessment of Reactive Oxygen Species, Antioxidant in Mercury-Induced
    Freshwater Cyprinus Carpio...37

9. (M^{II}Al-CO$_3$) Based Layered Double Hydroxides: Synthesis and
    Photocatalytic Efficiency Against Anionic Dye...44

10. Millimeter Wave Communication in Autonomous Vehicles 49

11. Designing an Enhanced Self-Supported Transition Metal Complexes Based on
    Electrocatalysts for Hydrogen Evolution Reaction 54

12. Study of Self-Resilient Cloud Server Using Machine Learning Techniques for
    Local Area Networks 59

13. Performance Study of Dual Rotor Induction Motor: Simulation Study....................64

14. Hausdorff Measures in Caratheodary Construction, Vital Covering Theorem,
    Steiner Symmetrization & Isodiametric Inequality...69

15. Heuristic for Symmetric Euclidean TSP with 3/2-approximation Ratio 75

16. New Approach to Solve COVID-19 Via Fuzzy Mathematical Model........................79

17. Machine Learning Strategies for Understanding Autistic Neuro Images...................................84

18. Solution Synthesis, Characterization, Photoluminescence and Afterglow
    Studies of Y_2O_3:Eu^{3+}, Ho^{3+} Phosphor 89

19. A Comprehensive Review of the Forensic Significance Toxic Medicinal Plants.......................94

20. Investigation on SIDO Quadratic Buck Converter for Battery Charging and
    LED Driver Applications 99

21. Secure RSA based Text File Encryption and Decryption using LabVIEW...............................104

22. OPAL-RT-Based Implementation of a Power Controller for a PV-Fed
    Grid-Connected Inverter...108

23. Extraction of Essential Oils From Medicinal and Aromatic Plants in
    Oman Using Solar Energy 113

24. A Review on Multi-Modal Classification for Emotional Intelligence 118

25. Factors Regulating the Human Microbiome 123

26. Citrus indica: Characterization and Potential as a Nutraceutical.................................128

27. Evaluating Impact of Clotrimazole on Inhibition of Candida Albicans and
    Surface Roughness in Permanent Soft-Liners..132

28. Identification of Road Surface Deficiencies Using Convolutional Neural Network...................137

29. Analysis of Road Surface Deformation Using Radar Image 141

30. A Deep Dive into Motor Imagery EEG Classification with Transfer Learning
    Approach Using DenseNet-121 Model 146

31. Realistic Method of Domestic Power Management System 151

32. Performance Evaluation of High Voltage Gain Bidirectional DC-DC
    Converter for Modern Electric Vehicles..156

33. Zagreb Index In Various Signed Graphs 162

34. Domination-Chromatic Number in Operations on Fuzzy Graphs 168

35. Special Dio - Triples Involving Primes...173

36. On The Non-isolated Resolving Number of Product Graphs and
    Some Families of Graphs 177

37. Entropy and Distance Measures of Bipolar Pythagorean Neutrosophic Soft Set 182

38. Mastitis Disease Diagnosis in Dairy Cattle Using Artificial Neural Networks.....................189

39. Upper Vertex Square Free Detour Number of a Graph 194

40. AI-Driven Proctoring System 200

41. Preliminary Research for Adhesive Peeling Device 206

42. Guidelines for Material Selection for Brackets to Treat Orthodontic Patients 210

43. An Efficient Prediction of Kidney Disease Using Machine Learning Algorithms 214

44. Prediction of Liver Diseases at Earliest Using Machine Learning Algorithms.......................220

45. Detection of Lung Cancer Using Deep Learning Techniques.................................225

46. Brain-Computer Interface Based Home Automation..229

47. Defect Exposure in Vegetables and Fruits Using Machine Learning Algorithms....................236

48. Design of Remote Controlled Feeding and Smart Monitoring System for Aquaculture..........241

49. Hospital Data Security Using Blockchain and AI..247

50. MIMO Antenna for 5G Communication Using HFSS 253

51. Multiple Biometric Authentication Through Image Assessment
    Using Machine Learning...257

52. Skin Cancer Detection Using CNN..261

Editor's Biography

Proceedings of the 1ˢᵗ International Sustainability Conference (ISC 2022)

Edited by

Dimitrios A Karras

Bio: Dimitrios A. Karras received his Diploma and M.Sc. Degree in Electrical and Electronic Engineering from the National Technical University of Athens (NTUA), Greece in 1985 and the Ph. Degree in Electrical Engineering, from the NTUA, Greece in 1995, with honors. From 1990 and up to 2004 he collaborated as visiting professor and researcher with several universities and research institutes in Greece. Since 2004, after his election, he has been with the Sterea Hellas Institute of Technology, Automation Dept., Greece as associate professor in Intelligent Systems-Decision Making Systems, Digital Systems, Signal Processing, till 12/2018, as well as with the Hellenic Open University, Dept. Informatics as a visiting professor in Communication Systems (the latter since 2002 and up to 2010). Since 1/2019 is Associate Prof. in Intelligent Systems-Decision Making Systems, Digital Systems and Signal Processing, in National & Kapodistrian University of Athens, Greece, School of Science, Dept. General as well as Adjunct Assoc. Prof. Dr. with the EPOKA university, Computer Engineering Dept., Tirana (1/10/2018-25/9/2020). He has published more than 70 research refereed journal papers in various areas of intelligent and distributed/multiagent systems, Decision Making, pattern recognition, image/signal processing and neural networks as well as in bioinformatics and more than 185 research papers in International refereed scientific Conferences. His research interests span the fields of intelligent and distributed systems, Decision Making Systems, multiagent systems, pattern recognition and computational intelligence, image and signal processing and systems, biomedical systems, communications and networking as well as security. He has served as program committee member as well as program chair and general chair in several international workshops and conferences in the fields of intelligent Systems-Decision Making Systems, signal, image, communication and automation systems. He is, also, former editor in chief (2008-2016) of the International Journal in Signal and Imaging Systems Engineering (IJSISE), academic editor in the TWSJ, ISRN Communications and the Applied Mathematics Hindawi journals as well as associate editor in various scientific journals, including CAAI, IET. He has been cited in more than 2220 research papers (https://scholar.google.com/citations?user=IxQurTMAAAAJ&hl=en), (Google Scholar) and 1626 (ResearchGate Index, after recent recounting based only on ResearchGate database, with more than 9105 ResearchGate Index Reads metric) citations, as well as in more than 1101 citations in Scopus peer reviewed research database, with H/G indexes 20/48 (Google Scholar), Scopus-H index 15, RG-index 30.96 and RG-Research Interest index 886.1 (https://www.researchgate.net/profile/Dimitrios_Karras2/).

Sai Kiran Oruganti

Indian Institute of Technology Patna India.

ORCID: https://orcid.org/0000-0003-4601-2907

Areas: Wireless Power Transfer, Wireless technologies, IoT, Radio Science, Electromagnetics & applications.

Profile Summary

Prof. Dr. Sai Kiran Oruganti is with the School of Electrical and Automation Engineering, Jiangxi University of Science and Technology, Ganzhou, People's Republic of China as a full Professor since October 2019. He is responsible for establishing an advanced wireless power transfer technology laboratory as a part of the international specialists team for the Center for Advanced Wirless Technologies. Between 2018-2019, he served as a senior researcher/Research Professor at Ulsan National Institute of Science and Technology. Previously, his PhD thesis at Ulsan National Institute of Science and Technology, South korea, led to the launch of an University incubated enterprise, for which he served as a Principal Engineer and Chief Designer in 2017-2018. After his PhD in 2016, he served Indian Institute of Technology, Tirupati in the capacity of Assistant Professor (Electrical Engineering) between 2016-2017.

Research

Prof. Dr. Oruganti, prime research focus is in the development of Wireless Power Transfer(WPT) for applications- Internet of Things (IoT) device charging, Agriculture, Electric Vehicle Charging, Biomedical device charging, Electromagnetically induced transparancy techniques for military and defence applications, Secured shipping containers, Nano Energy Generators.

Achievements

Prof. Dr. Oruganti has more than 21 patents pending on his credit and with several of those patent applications passing the NoC stage. As of 2021, 16 of 21 patents have been granted. He is credited with the pioneering work in the field of Zenneck Waves based Wireless Power Transfer system. Most notably, he has been regarded as one of the only few researchers in the field of WPT to be able to conduct power and signal transmission across partial Faraday shields. His recent paper accepted by Nature Scientific Reports has generated a lot of interest and excitement in the field. International Union of Radio Science(URSI) recognized his research efforts and awarded him Young Scientist Award in 2016. He is also recipient of IEEE sensors council letters of appreciation.

Sudeshna Ray

Dr. Ray's research at the interface of Chemistry and Material Science is focused on the development of Novel Inorganic based Luminescent materials for the application in white Light Emitting Diodes (LEDs) and as Spectral Converters in Solar Cell. The novel materials can be a single composition multi-centred phosphor or near UV/blue excitable blue, green, yellow and red emitting phosphor. The utilization of 'Green' solution based Synthesis Methodology for the precise control of the composition of the phosphors and achievement of a homogenous distribution of small amounts of activators in the host compounds is the main paradigm of my Research. In addition, to my previous focus on Synthesis, Characterization and Optical studies of size and shape tuned nanocrystalline Y_2O_3, YVO_4 and YPO_4 based phosphors, I am extensively involved into the research for the development of Advanced Luminescence Materials so called Quantum Cutters for the application in Solar Cell. A unifying theme of my research is the compositional tuning of the properties of extended solids through solid solution; sometimes referred to as the game of x and y, as, for example, in $Sr_2(1-x-y/2)Eu_{2x}La_ySi_{1-y}Al_yO_4$. Design of New phosphor for LEDs and fabrication of LEDs using the phosphors is an integral part of my Research. Currently, I am involved in the synthesis of Persistent Phosphors for the fabrication of Glow Bullet for Defense Application. RESEARCH INTERESTS • Exploration of New Phosphors by Mineral Inspired Methodology • Development of water soluble silicon compound by alkoxy group exchange reaction • Synthesis of Eu^{2+} and Ce^{3+} doped silicate phosphors using water soluble silicon compound • Solution Synthesis of 'Size' and 'shape' tuned Nanomaterials • Characterization of Nanocrystalline phosphors by XRD, TEM, FE-SEM and Raman spectroscopic measurement. • Study of 'Up-conversion', 'Down-conversion', 'Down-shifting' phenomena by steady state photoluminescence and lifetime measurement and analysis. • Measurement of 'Quantum Efficiency' and Thermal stability of phosphors. • Development and optical study of 'Quantum Cutting' Materials as Spectral Converter for Solar cell

RESEARCH EXPERIENCE

- Postdoctoral Fellow Phosphor Research Laboratory, Department of Applied Chem. September (2012) –July (2013) National Chiao Tung University, Taiwan Advisor: Prof. Teng Ming Chen
- Study of Energy Transfer from sensitizer to activator by lifetime analysis.
- Measurement of Quantum Efficiency

Foreword

I had the privilege to serve as the convenor for the first ISC2022 (formerly ICTSGA-1) which is dedicated to the realization of Niti-Aayog policies of the government of India. As a chairman for the Technology Innovation Hub IIT Patna whose National Mission on Interdisciplinary Cyber-Physical Systems is dedicated to Analytics, I have always felt the need to make an outreach to the Indian as well as international academia. ISC2022 is a prelude to the larger vision to capacitate Academia with the necessary ideas that will provide insights of the grassroot level development to various stakeholders of the Niti-Aayog policies. Towards this goal, the conference has aimed to create a conjunction of various stakeholders of Niti-Aayog policies that include- academic institutions, government bodies, policy makers and industry.

I hope that the contents of this series will serve as a guide for young researchers and policy makers for their future endeavours towards a holistic growth of the human society.

Trilok Nath Singh, PhD
Professor & Director, Indian Institute of Technology Patna.
Chairman and Board of Directors, Technology Innovation Hub,
Indian Institute of Technology, Patna-INDIA.

Preface

ISC 2022 (Formerly ICTSGA-1) is dedicated to the Niti Aayog policies to promote sustainability through exchange of ideas emerging out of the academia. The ISC is an annual conference that would be held in virtual mode until COVID restrictions on travel exist. The conference featured several plenary and keynote speeches from UN, African Environmental Sustainability, Mahatma Gandhi University, Universidad Politécnica de Valencia, Technical University of San Luis Potosí Mexico, Delhi Technological University, University at Johannesburg. The sessions were divided into two major sections: (a) Sciences & Engineering (b) Management & Humanities. In addition, a dedicated session on women in sciences, and academia was held. The sessions included ~50% representation from Women in academia, and research.

Statistics

The ISC 2022 received 816 abstract submissions of these only 340 were selected.

- Total Plenary Talks: 4
- Total Keynote Talks: 8
- Total Invited Talks: 7
- Total Oral Talks spread across two days: 35.
- Total Women in SDG talks: 26

Conference Chairs

Ravindra Pratap Singh,
IGNTU Amarkantak, India

Dimitrios A Karras
National and Kapodistrian
University of Athens, Greece.

Amarendra Pratap Singh
IGNTU Amarkantak, India

Introduction

Vision

The vision of the conference is to capacitate Academia with the necessary ideas that will provide insights of the grassroot level development to various stakeholders of the Niti-Aayog policies. Towards this goal, the conference will create a conjunction of various stakeholders of Niti-Aayog policies that include- academic institutions, government bodies, policy makers and industry.

Mission

The ISC organizers will make concerted efforts to promote academic research that would provide technological, scientific, management & business practices, and insights into policy merits & disruptions. The framework of exchange of ideas will be geared towards adoption of deep technologies, fundamental sciences & engineering, energy research, energy policies, advances in medicine & related case studies. This framework will enable the round table discussions between the academia, industry and policy makers through its range of plenary and keynote speakers.

Details of Programme Committee

Organizing Committee

T. Sunder Selwyn, PhD
Prince Dr. K. Vasudevan College of Engineering and Technology, Chennai India

K.K. JACOB, PhD
Faculty of Applied Sciences and Technology, Universiti Tun Hussein Onn Malaysia.

Mengistu Tulu Balcha, PhD
Ambo University Ambo, Ethiopia.

Kannaiya Raja, PhD
Ambo University Ambo, Ethiopia.

Karthikeyan Kaliyaperumal, PhD
Ambo University Ambo, Ethiopia.

Publishing Committee:

Dimitrios A Karras, PhD
University of Athens (NKUA), Greece

Karthikeyan Kaliyaperumal, PhD
Ambo University Ambo, Ethiopia.

Sudenshna Ray, PhD
RNTU-AIISECT University Bhopal, India

Comprehensive Survey on Artificial Intelligence-Based Malware Analysis and Detection Methods

M. OmaMageswari[a] and P. Vijayakumar[b]

[a]Research Scholar, Vellore Institute of Technology, Chennai, Tamilnadu, India.
[b]Associate Professor, Vellore Institute of Technology, Chennai, Tamilnadu, India
E-mail: [a]omajazped@gmail.com, [b]vijayrgcet@gmail.com

Abstract

In recent years cybercrime is a worldwide problem that constitutes a threat to individual security and very big imminence on large international companies, banks, and government sectors. To protect the confidential data and Network, Artificial Intelligence (AI) based malware analysis methods are used to monitor and detect any malware in the system. AI-based approach helps to arrive at better accuracy, less false positive rate with minimal training and processing time. This article provides a comprehensive summary and recent literature on Artificial Intelligence for detecting malware, including surveys, new developments, and research challenges pursued by the scientific community to address the problem. This article helps researchers with a thorough understanding of various machine learning models, the performance of various feature extraction methods, malware datasets, and malware detection tools, as well as the opportunity to compare performance analysis with state-of-the-art techniques, strengths, and limitations.

Keywords: Artificial intelligence, Deep Learning, False Positive rate, Feature Extraction, Machine Learning Algorithms, Malware Analysis

Introduction

Cyber-attacks have escalated dramatically in the year 2021. However, certain areas were struck worse than others, with education, research, and healthcare suffering the most. This indicates that cyber threat actors are focusing their efforts on places that are rapidly growing more technologically reliant while also being the least prepared to defend themselves against cyber threats. The quantity and severity of cyber-attacks are only set to increase as cyber-threat actors improve their strategies and employ machine learning and automation. Cyber Security concentrates on the industry's standards of confidentiality, integrity, and availability. Confidentiality means data can be accessed only by authorized people, integrity means the information can be added, modified, or removed only by authorized users; and availability means systems, functions, and data must be available on demand according to collaborative parameters. Cyber security is a set of techniques used to secure the integrity of networks, programs, and data against attack, damage, or unauthorized users. AI will make Cyber security easier, more provident, less costly, and far more successful. Machine learning is used to create patterns by combining information from past attacks and modifying the patterns with algorithms. To construct the pattern, a large amount of malware data from all possible sources is required, and the data must be full, relevant, and rich in context.

The malicious software programs that are designed by cyber criminals are called "malware." Malware is a file or a code that is dangerous to computer, such as worms, viruses, trojans, spyware, and so on. It creates a backdoor entry to steal personal information, confidential data, etc. As attackers use increasingly knowledgeable programs to elude detection, finding malware remains a very big challenge. Malware analysis can help you figure out if a suspicious file is harmful, investigate its origins, processes, and capabilities, and estimate its impact to make detection and prevention easier. Traditional software systems cannot detect huge volumes of advanced malware developed every week, therefore this is an area where artificial intelligence can really help with Antivirus solutions and VPN can help against remote malware and ransomware attacks, they often work based on signatures but signature protection may not be able to protect the new type of malware attack. AI-driven endpoint protection takes a different approach, relying on a repetitive training procedure to establish a baseline of behavior for the endpoint. The main contribution of the author is to (i) Detailed survey of most relevant AI-used malware detection papers- methods, proposals, and results (ii) Advanced methods of malware analysis were presented (iii) Various recent malware Datasets were discussed (iv) Latest tools for malware analysis were given.

The paper is presented in the following way. Section II gives a detailed survey on malware detection based on Static, Dynamic, Hybrid, and Graphical analysis. Section IV gives a summary of various recent datasets for malware detection. Section V presented the simulation results and the evaluation of different models on different datasets. Finally, future scope is discussed and concluded.

Related Works

To understand the behavior and characteristics of malware, Static Analysis, Dynamic Analysis Hybrid method, and Graphical method of Analysis were used. The survey provides a detailed summary of the research done so far to detect malware and discusses the various detection techniques and analysis of the performance of different classifiers.

Static Features-Based Malware Analysis

Static method is used to analyze the malware characteristics without running the program. It uses Signature based approach which compares the code with various signatures using strings and hashing mechanism. It analyzes the file names, hashes, strings, metadata, domains, and header data. Hashing is used to uniquely identify malicious programs and produces a sort of fingerprint. Strings of a program give it functionality, metadata in header provides code libraries and functionality, It has the drawback to detect advanced unknown malware, suffers from code obfuscation technique, encryption, and compression techniques will completely change the features of a binary program so it is failed to classify its classes. Literature survey below discussed the analysis and performance of Static properties with various techniques in Artificial Intelligence for detecting malware. Vasileios et al. (2021) used static analysis data has API and Permissions with machine learning to detect malware using Drebin data set, various ML classifiers and Regression methods were compared for feature selection and found that SVM has achieved accuracy of 99% and Ridge Regression has got best model optimizer to compare to Least Absolute shrinkage and Selection Operator (LASSO) Regression, Naive Bayes, Random Forest (RF), Support Vector Machine (SVM), Artificial Neural Network (ANN). It shows that the class imbalances dataset will affect the performance metrics and also discussed the total no. of features required for each model to give better performance. Guosong Sun et al. (2021) developed RMVC – (Recurrent Neural Network (RNN), Minhash, visualization, and Convolutional Neural Network (CNN)) it combined the locality-sensitive information from static opcodes with RNN then the feature images will be sent to CNN for training and classification. The accuracy increased by 92% for smaller datasets however, for larger datasets, the accuracy increases to 99.5 %. Damin moon et al.

(2022) convert static features of variable length vectors into fixed length by using Feature hashing for malware detection using Machine Learning for reducing memory space. It uses random forests for classification and compares the performance metrics by changing the vector size.

Dynamic Features-Based Malware Analysis

Dynamic analysis allowed running the program in an encapsulated environment to inspect the characteristics of malware. It can be examined by monitoring the function calls, analyzing parameters, tracing the instructions, and autostarting. The properties are API call Traces, network traffic, and Run time traces. API call is used to capture the behavior of malware, network traffic provides uniquely specific insights about malware. Rida Nasir et al. (2021) focused on deep learning-based insider threat detection using behavioral analysis. The data from the .csv file is processed, features are extracted, and the model is trained using the Long Short Term Memory (LSTM) autoencoder. The value is greater than the threshold for "Insider," and it is lower for normal. The Deep Learning (DL)-based LSTM autoencoder is compared with other Machine Learning (ML) classifiers and produces an accuracy of 90.6%, precision of 97%, and F1 Score 94.4%. But MobiTive, an Android malware detection system described by (Ruitao Feng et al., 2020) uses customized deep neural networks. Feature preparation can be accomplished by obtaining raw data, extracting features, and evaluating performance separately to determine feature selection. The behavior-based feature extraction method improves accuracy while also allowing for a thorough comparison of seven distinct neural networks and achieved an accuracy (i.e., 96.78 percent accuracy) with relatively reduced overhead.

Hybrid Method-Based Malware Analysis

Hybrid method combines the advantages of both static and dynamic techniques to detect hidden code, unknown threats, and sophisticated malware. Deepti Vidyarthi et al. (2019) performed hybrid analysis and also compared the features with ML classifiers. SVM achieved better accuracy of 96.39 for static analysis and 85.98 % for dynamic analysis. Nan Zhang et al. (2021) proposed CoDroid in Natural Language Processing (NLP) for text using the static opcode and dynamic system call. LSTM is used for feature extraction and CNN as classifier to categorize Android apps as benign or harmful. Processing speed is 80s, Accuracy-0.976, F1-0.986, precision-0.954, Recall- 0.978. Vinayakumar et al. (2019) proposed a machine learning and deep learning model for a hybrid system pipeline based on static, dynamic, and image processing that is commonly referred to as Windows-Static-Brain Droid (WSBD) used to detect zero-day malware. The Ember dataset was used to test with the machine learning classifiers, in those deep learning architectures, which are independent of packaging and operating systems. The Deep Image MalDetect (DIMD) model uses the CNN-LSTM to extract data, the malware is converted into an image and converted to 1D vectors, then ReLU activation function prevents vanishing and exploding gradient issues, attaining the best accuracy of 96.3 %. The obfuscation techniques make malware detection and categorization extremely difficult.

Graph Method-Based Malware Analysis

Nowadays Graph neural network (GNN) plays a vital role in malware detection, in GNN each node is labeled and we can predict the unlabelled node with the help of a labeled neighbor node. Reverse engineering using dissembler technique generates high-level representations includes, function all graph (FCG) and Control flow graph (CFG). Shanxi Li et al. (2021) presented 3 feature extraction processes, first weight model is extracted from API, then the samples will be mapped on the weight model, the model combines the test sample features with general sample features then the Graph Convolutional Network is trained, tested and achieved an accuracy of 98.32% with less FPR. Zikai Zhang et al. (2021) also used weighted graph method in Spectral-based Directed Graph Network that transforms 3 symmetrical node information combined for graph representation, then Graph convolutional network is used for classification. F-1 score is increased by 1.14% and 1.02% compared with other algorithms.

Malware Analysis Tools and Dataset

To analyze and understand the behavior of malware deeply, there is no. of latest open source and free tools available for reverse engineering the malware samples. Pe Studio: Pe studio is an excellent tool for analyzing malware at initial stage, if binary input is given it will provide hashes of malware, if strings were given it will give useful information such as domain & IP address. The entropy of the malware will identify whether the sample is packed or not, if it is packed the malware is obfuscated and prevented from analysis. For analyzing the packed malware, Pe studio gives the level of entropy from 0 to 8. The high level of entropy denotes that the malware is packed well. Process Hacker: Process hacker monitors the processes running on a device and also displays how the malware is making to hide. This tool is also used to get memory information like IP addresses, domains, and User agents. Process Monitor: Process monitor records the action of file systems like process creation and registry changes.

Malware Dataset Information

Malware attackers use polymorphism to the malicious components to evade detection. Polymorphism means the files belonging to the same malware "family," with same behavior, are instantly modified and/or obfuscated using various methods so that it looks like many different files. As a result, large amounts of files are required for effective analysis and classification. This can be solved by gathering data manually or using customized versions of already available datasets. The dataset is categorized into two types (i) balanced dataset and (ii) unbalanced dataset. If the malware and benign samples are the same proportion then it is said to be a balanced dataset. If the sample is not in the same proportion then it is called an unbalanced dataset. For accurate detection of malware and also to get the best classifiers the dataset should be balanced, to achieve this we have to oversample or undersample the unbalanced data. Undersample

is used to reduce the samples from majority class which makes it equal to minority class so most of the useful data will be lost because of undersampling. This leads to getting a poor model and it will be overcome by Oversampling technique which increases the no. of samples from minority class up to the size of majority class. A popular oversampling technique SMOTE will generate synthetic samples by randomly sampling attributes from occurrences in the minority class. The following are the dataset available in online MNIST dataset, Microsoft BIG 2015, KDD Cup, Drebin data set, MISOT Botnet and Ransomware Detection Datasets, CIC-MalMem-2022, CIC-Evasive-PDFMal2022.

Simulation Results and Their Discussion

This section summarizes the results which are obtained by using Machine Learning and Deep Learning methods as feature selection (Gradient Boosting, LASSO, Extra Trees Classifier) and classifiers (KNN, XGBoost, RNN-LSTM) and compared the performance using three various datasets (DSC-class malware dataset, Drebin dataset, Virus Share). After performing a Comparative Analysis of KNN, XGBoost, RNN-LSTM with three types of datasets shown in Fig. 1 represents accuracy metrics, Fig. 2 represents precision metrics, Fig. 3 represents recall metrics and Fig. 4 represents F1 metrics and Fig. 5 shows the overall comparison of all the three classifiers. The DSC-Class dataset has got better and uniform results with all classifiers while comparing all the metrics involved.

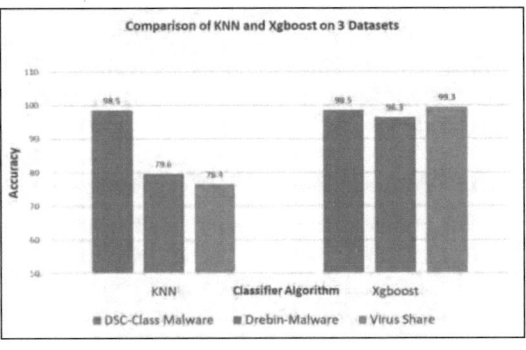

Fig. 1. Comparison of Accuracy in KNN and Xgboost on 3 Datasets

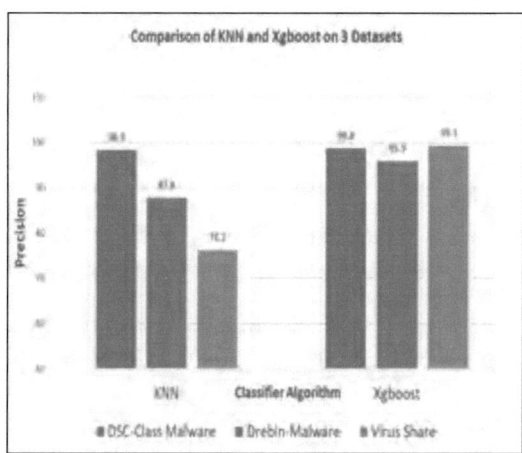

Fig. 2. Comparison of Precision in KNN and Xgboost on 3 Datasets

Fig. 1 shows the comparison of accuracy in KNN and Xgboost with three different datasets in that KNN and Xgboost have achieved 98.5 % in DSC dataset, compare to KNN, Xgboost has achieved above 96% in all the 3 datasets. Figure 2 shows the comparison of Precision in KNN and Xgboost with three different datasets in that Xgboost classifier has achieved higher precision values compare to KNN in all the 3 datasets.

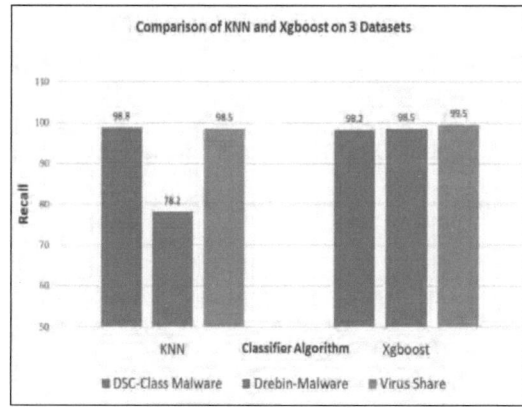

Fig. 3. Comparison of Recall in KNN and Xgboost on 3Datasets

Figure 3 shows the comparison of Recall in KNN and Xgboost with three different datasets in that also Xgboost has achieved higher values of Recall in all the 3 datasets.

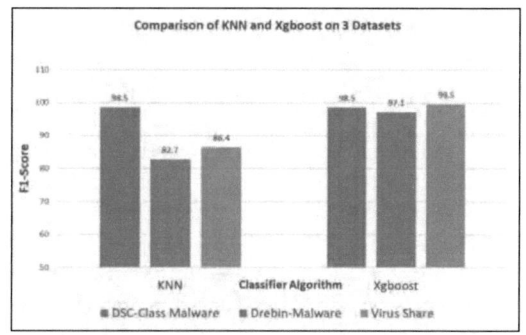

Fig. 4. Comparison of F1-Score in KNN and Xgboost on 3 Datasets

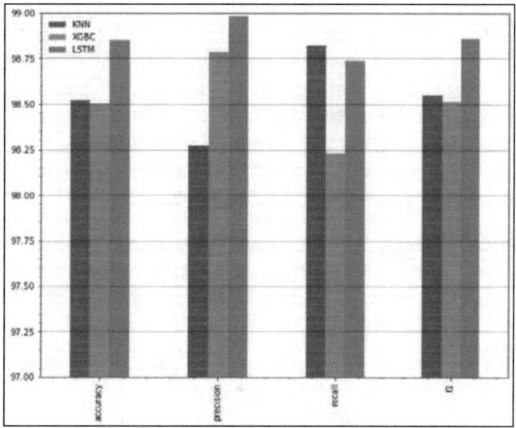

Fig. 5. Comparison of KNN, XgBoost and LSTM

Figure 4 shows the comparison of F1-Score in KNN and Xgboost with three different datasets in that Xgboost has achieved higher F1 Score values compare to KNN in all the 3 datasets. After performing training on the model, Fig. 5 shows that the RNN-LSTM has achieved 98.9% accuracy, compared to other classifiers.

Conclusion

This research study summarizes the literature review to get thorough knowledge of Different analysis techniques and various Machine Learning and Deep Learning detection methods for malware analysis. The survey shows that the graphical analysis with Deep Learning algorithms gives better performance compared to traditional methods. It is also proved that the datasets which are used for malware detection are very important to understand the performance of the model. For accurate prediction of malware the model must be trained with different datasets. In this way, RNN-LSTM has been chosen as best classifier by testing its performance with different datasets. By comparing LSTM with the other classifiers such as KNN and Xgboost, LSTM achieved average 98.9% accuracy with all three different datasets. By same way Extra Trees Classifier has the best feature extraction technique for all the datasets as compared to Lasso Regression and gradient boosting.

References

Syrris, V and Geneiatakis, Dimitris (2021). On Machine Learning Effectiveness for Malware Detection in Android OS Using Static Analysis Data. Journal of Information Security and Applications. 65, 2214–2126. https://doi.org/10.1016/j.jisa.2021.102794.

Sun, Guosong, and Qian, Quan (2021). Deep Learning and Visualization for Identifying Malware Families. IEEE Access. 18(1): 283-295. doi: 10.1109/TDSC.2018.2884928.

Moon, Damin, Lee, JaeKoo, and Yoon, MyungKeun (2022). Compact Feature Hashing For Machine Learning Based Malware Detection. ICT Express. 8(1), 124–129. https://doi.org/10.1016/j.icte.2021.08.005.

Nasir, Rida, Afzal, Mehreen, Latif, Rabia, and Iqbal, Waseem (2021). Behavioral Based Insider Threat Detection Using Deep Learning. IEEE Access. 9, 143266–143274. doi: 10.1109/ACCESS.2021.3118297.

Feng, Ruitao, Chen, Sen, Xie, Xiaofei, Meng, Guozhu, Lin, Shang-Wei, and Liu, Yang (2020). A Performance-Sensitive Malware Detection System Using Deep Learning on Mobile Devices. Journal of IEEE Transactions on Information Forensics and Security. 16, 1563–1578. doi: 10.1109/TIFS.2020.3025436.

Vidyarthi, Deepti, Damri, Gaurav, Rakshit, Subrata, Suthikshn Kumar, C. R., and Chansarkar, Shailesh (2019). Classification Of Malicious Process Using High-Level Activity-Based Dynamic Analysis. Security and Privacy. Wiley. 2(6), 1–14. https://doi.org/10.1002/spy2.86

Zhang, Nan, Xue, Jingfeng, Ma, Yuxi, Zhang, Ruyun, Liang, Tiancai, and Tan, Yu-an, (2021). Hybrid Sequence-Based Android Malware Detection Using Natural Language Processing. International Journal of Intelligent Systems. Wiley, 36(10): 5770–5784. https://doi.org/10.1002/int.22529

Vinayakumar, R., Alazab, Mamoun, Soman, K. P., Poornachandran, Prabaharan and Venkatraman, Sitalakshmi (2019). Robust Intelligent Malware Detection Using Deep Learning. IEEE Access. 7, 46717-46738. doi: 10.1109/ACCESS.2019.2906934.

Li, Shanxi, Zhou, Qingguo, Zhou, Rui, and Lv, Qingquan (2021). Intelligent Malware Detection Based On Graph Convolutional Network. The Journal of Supercomputing. Springer. 78, 4182–4198. https://doi.org/10.1007/s11227-021-04020-y

Zhang, Zikai, Li, Yidong, Dong, Hairong, Gao, Honghao, Jin, Yi, and Wang, Wei (2021). Spectral-based Directed Graph Network for Malware Detection. IEEE Transactions on Network Science and Engineering. 8(2), 957–970. doi: 10.1109/TNSE.2020.3024557

Integrated Artificial Intelligence-Based Traffic Light Detection System for Autonomous Vehicle

Abdullah Ghazi[1,a], Tannu Singh[2,b], Tayib Rizvi[2,c],
Syed Asghar Abbas Jafri[2,d], Devendra Yadav[3,e]

[1]Department of Mechanical Engineering, Jamia Millia Islamia, New Delhi 110025, India
[2]Department of Mechanical Engineering, Galgotias College of Engineering and Technology, Greater Noida 201310, India
[3]School of Chemical Engineering and Technology, Xi'an Jiaotong University, Xi'an, Shaanxi 710049 China
E-mail: [a]abdullahghazi1997@gmail.com, [b]rktmrsingh@gmail.com, [c]tayibrizvi1@gmail.com, [d]syedasgharabbasjafri@gmail.com, [e]ydevendra393@gmail.com

Abstract

Autonomous vehicles (AV) are an important part of intelligent transportation. Intelligent vehicles are capable enough to share the street with human drivers, which is soon going to revolutionize the transport industry. In this paper, we have demonstrated the model by using the concept of machine learning. We have used different components like the Raspberry pi camera module, motor driver L298n, Raspberry module, power switch, and breadboard to build the model. We have used different samples of traffic light images from different angles and zones to train the model and to get the required result. We have taken thousands of images of traffic lights and prepared a dataset, which is used to train our model. This model can easily detect traffic lights and give the desired output. A mini-AV prototype investigation confirms the efficiency of the proposed technology.

Keywords: Autonomous Vehicle, Traffic light detection, Machine Learning, Raspberry pi

Introduction

Traffic light detection and recognition are very important to follow traffic lights by the autonomous vehicle (Iftikhar et al., 2022) and traffic signals are essential driving indicators operating throughout inter-city regions (Lu, et al., 2008). The majority of intersections in the inter-city region are regulated by traffic lights. Furthermore, multi-lane roads may be regulated by traffic signals, with numerous auxiliary signals (Iftikhar, et al., 2022) signifying the correct route for various traffic patterns. AVs require traffic light detection (TLD) and phase classifications to operate. The traffic lights can signal their information to the vehicle (Gong, et al., 2010).

For autonomous vehicles, robust traffic light detection and recognition were developed using a camera-based algorithm. This approach can be used to reduce traffic fatalities caused by drivers' incompetence. Traffic signals recognize in real time using motion cameras and distinguish between traffic signals that are circular in the shape of an arrow traffic light. It can adjust to changes in the quantity of illumination, kinds of lighting, and time spaces in the environment (Bach et al., 2018). There are reduced traffic fatalities and crashes, due to a reduction in the demand for safety gap and the ability to control traffic circulation, greater lane capacities, and lower traffic congestion were achieved. The necessity for traffic cops is reduced and also the cost of automobile insurance is reduced. Along with these benefits, some challenges include high production costs and adapting a vehicle navigation system to many types of weather, like rainy, snowy, or foggy circumstances (Ondrus et al., 2020).

AVs are unable to interpret traffic signals using lidar or radar sensors. To decide when to stop and move, they must depend purely on their recognition system. The cameras are mounted on the vehicles to give essential vehicle info (Lu et al., 2008). The suggested framework may be used with moveable cameras, is responsive to environmental changes, and can be implemented in real-time (Ondrus et al., 2020). It is divided into two segments: detection and classification of traffic lights. Color, region, and boundary characteristics are used in both segments (Gong et al., 2010). To eliminate holes and noise, morphological techniques are used. After that, region labeling is used to identify potential traffic light regions. Border detection is implemented during the classification phase to produce area boundaries, along with region colors, and matching characteristics, both circular and arrow traffic signals can be copied (Bach et al., 2018). Following the previously stated pre-processing, a machine-learning approach was used to continue the effective recognition of traffic lights (Lu et al., 2008).

We proposed a mini-AV prototype for TLD, a camera mounted on the vehicle prototype will capture images, these images then be processed using TLD image-processing Technology, and the autonomous vehicle will be operated based on traffic signal conditions. The goal of this research is to use vehicle-mounted cameras to recognize and classify traffic lights. To accomplish detection, a traffic light detection technique with color fragmentation and machine learning is presented for a complicated built-up setting.

Literature Review

A passive camera-based TLD using vehicle positioning and assuming previous data of traffic light position, he demonstrated how to use navigation, back-projection, and localization to map traffic light positions using recorded video data (Charette et al., 2009). A concept for traffic light recognition and classification, we created a deep convolutional object sensor based on the Faster R-CNN framework and assessed its performance. This will help future research attempts to identify tiny things more reliably and enhance detection distance (Gong et al., 2010). A system is used with flexible cameras, is responsive to environmental changes, and can be implemented in real-time. The RGB color space is first transformed into the HSI color space to reduce holes and noise. Border detection is used to produce area boundaries, together with region colors (Bach et al., 2018). The segmentation threshold was obtained after statistical analysis of traffic lights employing the numerous elements of that color space and color segmentation was completed based on the data. The machine learning approach based on the color-based object tracking algorithm was introduced to increase the system stability and real-time performance (Lu et al., 2008).

Methodology

The purpose of the traffic light detection was to develop an autonomous system that could subsequently be integrated into the autonomous vehicle to recognize calibrated traffic lights. The vehicle should eventually be able to stop at the red signal and go at the green signal (Brugger et al., 2022). TLD using camera-based recognition systems has been studied since last three decades. A traffic signal detection technique based on machine learning is adopted (Lu, et al., 2008).

The traffic light dataset is created by a fixed camera set up at the windshield of the vehicle, (Iftikhar et al., 2022) which captures high-resolution images at 2048 × 1024 pixels, hand-labeled images of traffic lights with over 2,00,000 unique annotations (Fregin et al., 2018). This dataset was developed to capture more than simply training data for machine learning algorithms, to study intersection configurations in relation to traffic lights. We describe the dataset that was utilized for training and assessment, as well as the changes that were made to meet the goals of this project. An overview of the dataset is also available from the journal paper (Jensen et al., 2016). A few papers use learning-based approaches based on Deep Convolutional Neural Networks (DNNs), which establish a compact classification network to better discriminate traffic light statuses and reduce false positives (Gong et al., 2010). Block diagram of traffic light detection and recognition is shown in Fig. 1.

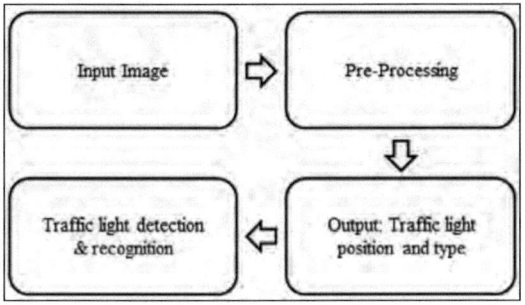

Fig. 1. Block diagram of traffic light detection and recognition

Prototype Modeling

<u>Construction:</u> The raspberry pi (RPI) module, RPI cameras module, motor driver L298n, L-shaped DC gear motor, li-ion batteries (power bank), breadboard, power switch (ON/ OFF), are used

in the construction of the model, so we can detect traffic light and stop signal that could be monitored on the computer display. We can also employ the TLD and stop symbol detection by the vehicle acting like an autonomous vehicle. Currently, raspberry pi 5 MP camera is on-boarded on the prototype for surrounding perception. These cameras are used to detect traffic signals and stop symbols. The RPI 5MP camera is connected to the raspberry pi 3 modules to send real-time information, which takes place through the pre-processing of the captured images, then detection and classification of the traffic light are done as per the model trained and dataset provided. The output is sent to the motor driver to drive the necessary motor according to the raspberry pi output. A breadboard is used to make necessary connections and a power ON/OFF switch is provided to OFF the power supply when not needed. A vehicle constructed in this research is shown in Fig. 2.

<u>Training:</u> First, we prepare a dataset by collecting camera images at a resolution of 1280 × 720 of 2 distinct classes for the objective of traffic light and stop sign detection. We train our model through machine learning concepts by providing the necessary dataset of traffic lights

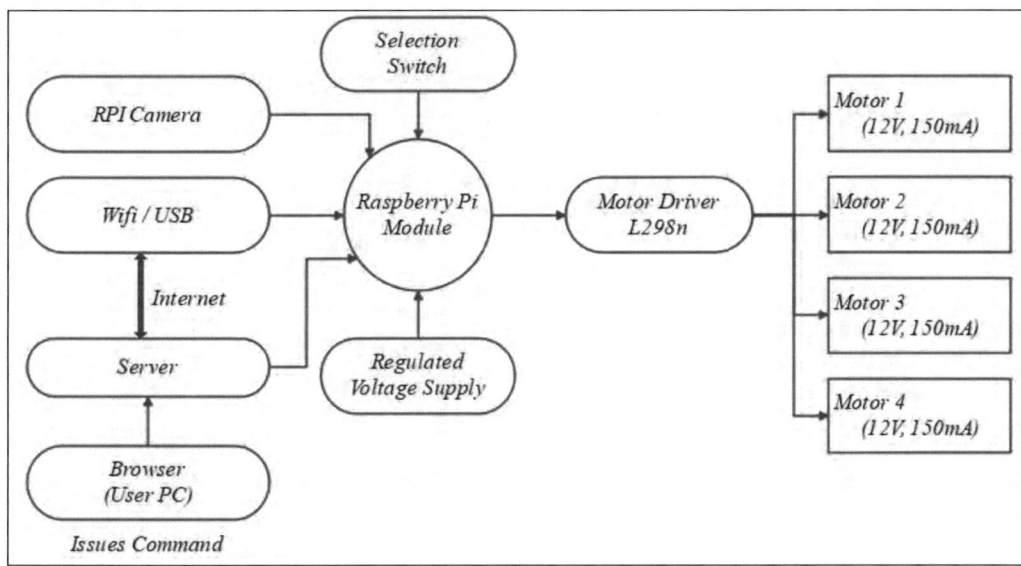

Fig. 2. Block diagram of vehicle prototype

and stop signs. Then, we put our python code into the raspberry pi module, and connect our raspberry pi module with the RPI camera which gets a regulated power supply through a power bank (li-ion battery). When the RPI camera detects the red traffic light or stops signal, it sends the captured image to the Raspberry Pi Module which detects and classifies the images from the dataset images. When system detection is completed, the signal command is sent to the Raspberry Pi which stops the motion of motor drivers after getting the signal. Thus, the autonomous vehicle stops automatically at the red light or the stop sign and again starts moving when green light is active. A Schematic diagram of the proposed model is shown in Fig. 3.

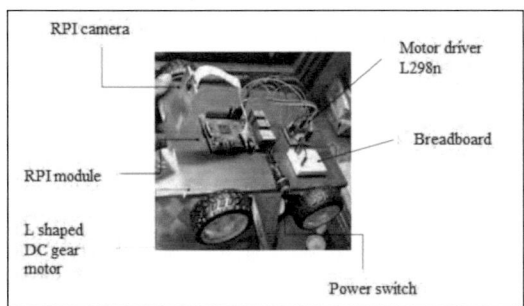

Fig. 3. Mini AV traffic light detection prototype

<u>Working:</u> The Raspberry Pi module is getting regulated voltage supply for its functioning using a power bank (or li-ion battery), the user computer (browser) is connected through a server to which a raspberry pi module is connected through internet broadband or Wi-Fi. When the RPI Camera detects the red traffic light or stop signal, it sends the captured image to the Raspberry Pi Module which detects and classifies the images from the dataset images. After detection, it sends signal to the motor driver to stop the motion of the motors. When the traffic light's green light is active or stop sign is removed from the vision of the raspberry pi camera the motor driver again sends the signal to start the motion of the motor to move vehicle. Figure 4 demonstrates that the model detects the red light of the traffic light and does the required action for it was trained. After capturing the red light, the model shows the 'STOP' command on computer

display and follows the required action to follow the command stop. Figure 5 is showing how the model recognizes the stop signal and takes required action and stops the moving vehicle.

Fig. 4. Traffic light detection of AV model

Fig. 5. Stop sign detection of AV model

Results and Discussion

An AV prototype (model) is created to construct the robustness of the presented method in different scenarios under various lighting and weather situations. We use a set image sequence to train our model through machine learning concepts by providing the necessary data. The outcomes indicate that the suggested approach can identify a traffic signal in a reasonable amount of time and with more confidence than conventional technology. The suggested technology not only saves processing time but also enhances the detection results. Open circular objects are more likely to be spotted in the suggested technique, the traffic signal was accurately detected. If the camera's angle is fixed, we can estimate the positioning of a traffic signal in the picture, considering the position of the detected region will be useful for boosting accuracy. At each phase of detection, additional data is used to enhance the efficiency, precision, and performance of *traffic light detection and recognition*. In addition, the system's performance in real-time detection is excellent. The results reveal that the proposed system has a satisfactory overall performance.

Conclusions

In this paper, we have used a machine learning-based AV model to detect the traffic light and stop signal. We have used basic image-processing algorithm for TLD and stop symbol detection. In this, we have taken two datasets one is the images of traffic lights and another is of stop signals, is from open source to train the model. Our model gave accurate results while identifying the traffic light and stop sign. Although learning and experimenting never stop. So, to make it better we should focus on improving the performance to get the prior result. As it helps in reducing false input saving time and memory. We should also work on improvement on 3D information like height and width to get an accurate result. We should also focus on the Region of interest to reduce time and work on relevant areas. We should develop more robust traffic light detection and recognition methods based on video cameras on the demands in the zone of navigation in autonomous vehicle to improve the intelligent transportation system.

References

Iftikhar, M., Riaz, O., Ali, T., Mumtaz, S., Sharif, W., and Arshad, H. (2022). Traffic Light Detection: A cost effective approach. VFAST Transactions on Software Engineering. doi: http://dx.doi.org/10.21015/vtse.v9i4.836

Lu, K. H., Wang, C. M., and Chen, S. Y. (2008). Traffic Light Recognition. Journal of the Chinese Institute of Engineers, 31(6). doi: 10.1080/02533839.2008.9671460

Gong, J., Jiang, Y., Xiong, G., Guan, C., Tao G., and Chen, H. (2010). The Recognition and Tracking of Traffic Lights Based on Color Segmentation and CAMShift for Intelligent Vehicles. 2010 IEEE Intelligent Vehicles Symposium University of California, San Diego, CA, USA. doi: 10.1109/IVS.2010.5548083

Bach, M., Stumper, D., and Dietmayer, K. (2018). Deep Convolutional Traffic Light Recognition for Automated Driving. 2018 21st International Conference on Intelligent Transportation Systems (ITSC). doi: 10.1109/ITSC.2018.8569522

Ondrus, J., Kolla, E., Vertal, P., and Saric, Z. (2020). How Do Autonomous Car Works? Transportation Research Procedia 44 (2020), 226–233. doi: 10.1016/j.trpro.2020.02.049

Brugger, T., and Albrecht, T. (2022). Traffic Light Detection for an Autonomous Model Vehicle. Technical Reports in Computing Science University of Applied Sciences Kempten. doi: 10.13140/RG.2.2.11710.15686

Charette, R., Nashashibi, F. (2009). Traffic light recognition using image processing compared to learning processes. 2009 IEEE/RSJ International Conference on Intelligent Robots and Systems October 11–15, 2009 St. Louis, USA. doi.org/10.1109/IROS.2009.5353941

Fregin, A., Müller, J., Krebel, U., and Dietmayer, K. (2018). The DriveU Traffic Light Dataset: Introduction and Comparison with Existing Datasets. 2018 IEEE International Conference on Robotics and Automation (ICRA). https://doi.org/10.1109/ICRA.2018.8460737

Jensen, M. B., Philipsen, M. P., Mogelmose, A., Moeslund, T. Baltzer, and Trivedi, M. M. (2016). Vision for Looking at Traffic Lights: Issues, Survey, and Perspectives. IEEE Transactions on Intelligent Transportation Systems. 17(7), July 2016. https://doi.org/10.1109/TITS.2015.2509509

Identification and Authentication of SSR Markers for Purity of Hybrids in Sesame (*Sesamum indicum* L.)

Kanak Saxena,ᵃ and Rajani Bisenᵇ

ᵃDepartment of Genetics and Plant Breeding, Rabindranath Tagore University, Bhopal - 462045
ᵇProject Co-ordinating Unit (Sesame and Niger), Department of Genetics and Plant Breeding, JNKVV, Jabalpur-482004
E-mail: ᵃkanak.saxena@aisectuniversity.ac.in, ᵇrajanitomar20@gmail.com

Abstract

Microsatellite markers are the most suitable markers for fingerprinting hybrids. The present investigation was undertaken for molecular identification of parental lines and to confirm the hybridity of sesame hybrids based on the amplification banding pattern of the molecular markers. Eight sesame hybrids and their parental lines were screened using 18 SSR primers. Five of the primers were found to be polymorphic, resulting in distinct banding patterns between the hybrids and their parents, helping to confirm that the developed F1 plants were genuine. The polymorphic single sequence repeat (SSR) markers (GBssr-sa-08, GBssr-sa-12, GBssr-sa-184, GBssr-sa-07, and GBssr-sa-08) have been found as unique markers that aid in the differentiation and identification of hybrids from their parental lines. The results of our study show that a single microsatellite polymorphic marker was enough to identify the confirmed hybrid.

Keywords: *Sesamum indicum* L., Microsatellite markers, Hybrid Purity, Polymorphic loci, Alleles

Introduction

Sesame (*Sesamum indicum* L.) is a tropical and subtropical oilseed crop. Due to its high oil content (38–54%), high oil quality, and potent antioxidants such as sesamol, lignan, sesamin, and sesamolin, it is also known as the "Queen of oilseed crop." Sesame lowers blood pressure via lowering plasma cholesterol due to its high content of pufas (>80%). Because of its medical benefits to human health and the growing need for vegetable oil around the world, researchers have recently renewed their focus on developing a high-yielding sesame variety. Sesame favors high heterosis because of its high outcrossing rate, so the development of hybrids in sesame is a good option (Pathirana, R., (1994)). Due to their ease, safe, and lack of undesirable effects on yield potential, conventional breeding methods involving emasculation and hybridization are still commonly used today to develop hybrid varieties that have high productivity and quality. Before releasing hybrid varieties at a commercial level, confirmation of hybridity i.e., identifying certified hybrid seeds from non-certified hybrid seeds, is essential for the commercial production of certified hybrid seeds.

Among the polymerase chain reaction (PCR) based DNA markers, single sequence repeat (SSR) markers are widely used in molecular fingerprinting, hybrid purity testing, germplasm identification, and characterization as they are more reliable, highly polymorphic, and co-dominant in nature (Shashibhushan, Muchanthula, and Pradeep, 2021).

Hybrid purity can be easily assessed using SSR markers, as the hybrids' heterozygous status can be easily discerned by the presence of alleles from both male and female parents involved in the hybrid, and generate unique banding patterns for the molecular identification of parental lines and confirm the hybridity of sesame. The main objective of the current work is to identify and confirm the hybridity of F1 plants utilizing SSR markers, which are being used under study.

Materials and Methods

Plant Materials

Eight sesame hybrids were developed using the line x-tester method by crossing six female lines with three male testers. TKG-478 (L1), TKG-503 (L2), RT-373 (L3), AT-253 (L4), TKG-506 (L5), TKG-21 (T1), TKG-22 (T2), and GT-10 (T3) are eight sesame genotypes obtained from diverse sesame growing locations in India. The experiment was conducted under the Project Coordinating Unit (Sesame and Niger) at the Department of Genetics and Plant Breeding, Jawaharlal Nehru Krishi Vishwa Vidhyalaya, Jabalpur, Madhya Pradesh. The crossing was carried out among lines and testers (Table 1) and harvested F_1 seeds.

DNA Isolation and PCR Amplification

Young leaf tissues from both male and female parents and their descendants were used to isolate genomic DNA using the cetyl trimethyl ammonium bromide (CTAB) method with a few alterations [3]. Quantification of DNA was accomplished by a Nanodrop Spectrophotometer at OD260/280nm. Finally, DNA was stored in Tris-EDTA (TE) buffer to a concentration of approximately 25 ng/μL for a polymerase chain reaction. A total of eighteen SSR markers were analyzed to find out polymorphism among the all-parental lines and hybrids. PCR reaction was carried out with a volume mixture of 20μL containing 20ng of template DNA, 2.0μl of 10x PCR buffer, 0.5μl of 15mM MgCl2, 0.5μl of 5mM dNTP mix, 0.5μl of 5pmol of each forward and reverse primer, and 1U of Taq DNA polymerase (5U/μl). The amplification criteria in PCR programming were performed for 35 cycles consisting of pre-denaturation at 95⁰C for 4 minutes (1 cycle), denaturation at 95°C (30 sec), primer annealing at 55°C (1 min), extension at 72°C (1 min), and the final extension at 72°C (7 min) in a thermal cycler. The PCR products were used for electrophoresis on a 2.5% (w/v) agarose gel stained with ethidium bromide at 100–120 V/cm (10mg/ml) using 1X TBE buffer and photographed using a gel documentation unit under UV light. The size of PCR products was estimated by using a wide-range 100 bp ladder for the confirmation of hybridity. For analyzing the hybridity of F_1, the banding patterns of SSR markers were compared and confirmed when they showed the presence of both male and female parent alleles.

Results and Discussion

Identification of Suitable SSR Primers

The current study used SSR markers to identify eight hybrids and their parental lines. A total of 18 SSR markers were used to screen parental genotypes. Five SSR markers (GBssr-sa-06, GBssr-sa-07, GBssr-sa-08, GBssr-sa-12, and GBssr-sa-184) are polymorphic between the parents based on the size of amplicons obtained following PCR between the parental lines. These five polymorphic SSR primers detect highly polymorphic bands in hybrid plants and are used to assess and evaluate progenies' hybridity. These markers amplified alleles with different sizes ranging from 145 bp (GBssr-sa-184) to 340 bp (GBssr-sa-8) as shown in Table 2. The F1 hybrid had alleles from both of its parents. This showed that the hybrid was heterozygous because it had both of its parents' line bands.

Table 1. List of hybrids confirmed by the SSR primers

Hybrid	Cross	Line x Tester	Markers for Hybridity	Band Size (bp)
H1	TKG-478 X TKG-21	L1 x T1	GBssr-sa-07	220
				260
H2	TKG-478 X GT-10	L1 x T3	GBssr-sa-08	290
				340
H3	TKG-503 X TKG-22	L2 x T2	GBssr-sa-184	145
				180

H4	RT-373 X TKG-21	L3 x T1	GBssr-sa-06	230
				260
H5	AT-253 X TKG-22	L4 x T2	GBssr-sa-12	230
				310
H6	AT-253 X GT-10	L4 x T3	GBssr-sa-08	290
				340
H7	TKG-506 X TKG-21	L5 x T1	GBssr-sa-06	230
				260
H8	AT-306 X TKG-22	L6 x T2	GBssr-sa-184	145
				180

When it comes to determining the hybrid purity of developed F_1 seeds, polymorphic SSR markers may be an essential requirement for their commercial use because there is always the possibility of contamination in the hybrid seed production plot due to pollen shedders, out crossings, and physical mixtures in the subsequent handling of the harvested material always pose a risk of contamination in the hybrid seed production plot (Kumar et al., 2021; Saxena and Bisen, 2017).

Table 2. Sequences of polymorphic markers identified

Marker	Foreward Sequence	Reverse Sequence	Tm (°c)	Size of Alleles (bp)
GBssr-sa-06	CCATTG AAAACTG CACACAA	TCCACA CACAGA GACCC	58	230–260
GBssr-sa-07	TCTTGCA ATGGGG ATCAG	CGAACT ATAGAT AATCAC TTGGAA	59	220–260
GBssr-sa-08	GGAGAA ATTTTC AGAGAG AAAAA	ATTGCT CTGCCT ACAAAT AAAA	57	290–340
GBssr-sa-12	GCTGAG GAGTCT TGAAG CAGA	CAAAAT CCCCC AACTC GATA	60	290–340
GBssr-sa-184	TCTTGC AATGGG GATCAG	CGAACT ATAGAT AATCACT TGGAA	55	145–180

Confirmation of Hybridity of F_1

When comparing the banding patterns of the hybrids and their parental lines, researchers discovered that the hybrids gave two different-sized bands corresponding to the bands amplified in their parent lines. Off-type plants or mixtures were shown heterozygous polymorphic banding patterns which was not corresponding to bands observed in parental lines. Therefore, off-type plants were removed from the field. Among all the crosses, F_1 plants namely, TKG-478 X TKG-21, TKG-478 X TKG-503 x TKG-22, RT-373 x TKG-21, AT-253 x TKG-22, AT-253 x GT-10,

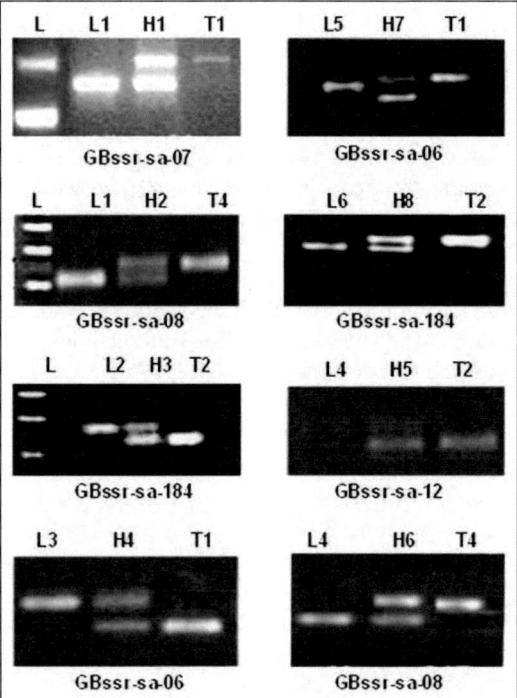

Fig. 1. Polymorphic SSR markers profiles confirming hybridity in sesame, M = 100 bp ladder, L1 – L6 = Lines, T1 – T4 = Tester, H1 – H8 = Hybrids

TKG-506 x TKG-21, and AT-306 x TKG-22 all had both alleles found in both male and female parents. This means that there were three male parents (TKG-21, TKG-22, and GT-10) and six female parents (TKG-478, TKG-503, TKG-506, RT-3. The present study exploited the SSR marker techniques to detect true hybrids of sesame along with their parental lines, demonstrating that identified polymorphic markers may be successfully used to discriminate and recognize the hybrids from their parental lines. The findings of this study are in line with those published in Brassica (Sharma et al., 2018). SSR markers have also been used for genetic purity testing in rice (Wang et al., 2006), soybean (Zhang et al., 2014), sesame (Saxena and Bisen, 2017).

Conclusion: This study found that compared to the time-consuming, resource and energy-intensive, and inaccurate approaches traditionally used, SSR markers are superior for assessing genetic superiority. The results of the study demonstrated the utility of SSR marker technology for the confirmation and early detection of sesame hybrids, paving the way for additional hybrid sesame breeding programs in the country. This study shows that SSR markers could be used to find and confirm hybrid sesame.

Acknowledgments: The authors are grateful to the Department of Genetics and Plant Breeding and Project Coordinating Unit Sesame and Niger, JNKVV, Jabalpur (M.P.) for providing the seeds and all the required facilities for conducting the research work.

References

Pathirana, R. (1994). Natural Cross-Pollination in Sesame (*Sesamum indicum* L.) Plant Breeding. 12: 167–170.

Shashibhushan, D., Muchanthula, A. R., and Pradeep, T. (2021). Molecular finger printing and genetic purity assessment of a rice hybrid using microsatellite markers.

Saghaimaroof, M. A., Soliman, K. M., Jorgensen, R. A., and Allad, R. W. 1984. Ribosomal DNA spacer-length polymorphism in barley: Mendelian inheritance, chromosomal location and population dynamics. Proc. Natl. Acad. Sci, U.S.A. 81: 8014–8019.

Kumar, S. P., Susmita, C., Agarwal, D. K., Pal, G., Rai, A. K., and Simal-Gandara, J. (2021). Assessment of genetic purity in rice using polymorphic SSR markers and its economic analysis with grow-out-test. Food Analytical Methods, 14(5), 856–864.

Saxena, K., and Bisen, R. (2017). Genetic variability studies in sesame (*Sesamum indicum* L.). Annals of Plant and Soil Research. 19(2):210–213

Sharma, M., Dolkar, D. Salgotra R. K., Sharma, Deepika, Punya, Singh, Amrinder., and Gupta, S. K. (2018). Molecular Marker Assisted Confirmation of Hybridity in Indian mustard (*Brassica juncea* L.). Int. J. Curr. Microbiol. App. Sci. 7: 09.

Wang X., Zhao, X., Zhu, J., and Wu, W. (2006). Genome-wide investigation of intron length polymorphisms and their potential as molecular markers in rice (*Oryza sativa* L.). DNA Res. 12: 417–427.

Zhang, C. B., Peng, B., Zhang, W. L, Wang, S. M., Sun, H., Dong Y. S., and Zhao, L. (2014). Application of SSR Markers for Purity Testing of Commercial Hybrid Soybean (Glycine max L.). J. Agr. Sci. Tech. 16: 1389–1396.

Saxena, K., and Bisen, R. (2017). Use of RAPD marker for the assessment of genetic diversity of sesame (*Sesamum indicum* L.) varieties. Int. J. Curr. Microbiol. App. Sci. 6(5): 2523–2530.

Detection of Specular Reflection on the Smart Colposcopy Images Using Fine-Tuned U-Net Model

Jennyfer Susan M.B.,[a] and Subashini P.[b]

[a]Avinashilingam Institute for Home Science and Higher Education for Women, India.
[b]Avinashilingam Institute for Home Science and Higher Education for Women, India
E-mail: [a]jennyfersusan26@avinuty.ac.in, [b]Subashini_cs@avinuty.ac.in

Abstract

Cervical cancer is the most prevalent genital malignant tumor, causing a serious health risk to women. Smart colposcopy is the screening procedure used to identify the early stages of cervical lesions. Due to physiological mucus in the human body, the bright luminous is generated and appears in the high-definition image produced by smart colposcopy. The highlighted area in the colposcopy images resembles similar to the metaplasia epithelium and causes difficulty during visual analysis. This paper presents a deep learning strategy for identification of the luminous spots in colposcopy images using the finetuned U-Net model with customized hyperparameters. The specular reflection region is labeled based on the intensity range and trained with the U-Net model to distinguish the specular region from the non-specular region on the smart colposcopy images. With a loss value of 0.8234, the automated technique correctly recognizes the specular reflection with an accuracy of 97.28%.

Keywords: Smart Colposcopy Images, Specular Reflection, U-Net Model

Introduction

Cervical cancer is a highly prevalent malignancy that affects women around the world. In the year 2020, it is estimated that 604,000 cases are newly reported, and 342,000 deaths are reported by the World health organization (WHO). Due to inadequate knowledge of cervical cancer, persons in rural areas are more likely to be affected by cervical cancer. Initially for screening the traditional pap smear method is used for screening. Due to the difficulty in providing the laboratory structure in developing countries very few screening programs are provided to the rural and tribal people. To provide simple effective screening smart colposcopy is used to identify the abnormalities in the cervical region. Smart colposcopy is the device used to capture high-quality images and videos for the visual

inspection of cervical images. The physician analyzes the color density of the metaplasia epithelium to detect neoplasm and helps analyze the severity of lesion on the cervical images. During the analysis, the specular reflection appears on the cervical images affecting the diagnosis process and significantly decreasing the performance of the smart colposcopy images. Various image processing techniques are applied to detect the specular reflection on the cervical image, but few are automated. Meslouhi et al. (2011) proposed the identification of the reflection region on the colposcopy images by using the chromatic properties of the reflection area which is formulated from the properties of the dichromatic reflection model approach. After identifying the inpainting method is applied to the detected region to remove the glare region on the cervical images. Wei et al. (2019) modified the traditional U-Net segmentation model for the automatic segmentation of seabed mineral images. The model is applied to the grayscale microscopy and seabed mineral images dataset. The experimental method achieves 0.916 for the EM dataset and provides higher quality in the segmentation of the mineral images. Attard et al. (2020) proposed the semantic segmentation deep learning model for the localization of the specular highlights. The proposed method is trained and tested with the reflected affected image dataset. The model achieved a frequency weight of 0.98 and a mean iou of 0.80. Liu et al. (2021) suggested a novel deep learning framework to fix the highlighted region on the endoscopy images using a surgical fix deep neural network (SFDNN). Reflective region and the anti-reflection regions are labeled before training through the deep learning model. It identifies the defects and is further used for refilling the missing portion of the detected region. Jus et al. (2018) presented the U-Net deep learning model for the segmentation of the iris, and a modified U-Net is proposed by Ayalew, et al. (2021) for the segmentation of the ultrasonic lung's images. Based on the analysis of the related paper the U-Net is suitable for the segmentation process in many medical images. The next section discusses the segmentation of the specular reflection on the smart colposcopy images for automatic identification.

So, in this paper, proposed the detection of specular reflection on the smart colposcopy images using the finetuned U-Net model. In section 2 discuss the methodology and in section 3 Results and analysis and finally the conclusion of the paper.

Methodology

To detect the specular reflection on the smart colposcopy images initially the glare regions are labeled using binary masking. The labeled images are trained using a simple convolutional neural network model and Fine-tuned U-Net models for the identification of the glare region of the specular reflection on the smart colposcopy images.

Proposed Binary Masking

Smart colposcopy is a mobile-based digital device for the screening of cervical regions. During the analysis, colposcopy images are affected by the white bright region pixel causing luminosity on the surface of the images. The proposed masking method is formulated from the intensity value of the colposcopy images. The RGB images are initially converted to grayscale images. The intensity range of the pixel ranges from 0 to 255 where 0 represents the darkest pixel value of the image and 255 represents the whitest pixel value of the images. On analyzing the intensity value, the pixel value ranges from 191–255 falls in the specular reflection region and pixels from 0–190 are labeled as the non-specular region. The specular reflection pixel is rescaled and set as 0 and another region is set as 1 for creating masking images on the smart colposcopy images as in equation 1. The $f(x)$ represents the rescaled pixel value of the cervical images.

$$f(x) = \begin{cases} 1 = Non\ specular\ Reflection \\ 0 = Specular\ Reflection \end{cases} \quad (1)$$

Network Architecture for the Detection of Specular Reflection:

The rescaled input images and the original images are trained using the simple convolutional neural network and finetune U-Net model for the detection of specular reflection on cervical images.

Simple Convolutional Neural Network

The labeled images and original images are trained using a convolutional neural network (CNN) to predict the specular reflection pixel on the cervical images (Susan et al., 2022). The max-pooling layer reduces the feature map dimension by half of the size of the input image. The maximum value of the convolutional process is the max-pooling of 3x3 with stride values of 1, 2, and 3. The thirteen convolutional layers with max-pooling are applied to each convolutional layer. The average pooling is concatenated to create the original images in the output images. The convolutional transpose is applied along with the convolutional and concatenate layer to segment the desired pixel on the cervical images. The batch normalization layer is used to normalize the value of the cervical images, and the dropout layer is utilized to reduce overfitting of the training model. In the hidden layers, the sigmoid activation function is employed for binary segmentation of pixel-like 1 and 0 from the smart colposcopy images. The workflow for the detection of specular reflection using simple convolutional neural network is shown in Fig. 1. The model constructed is trained and tested for the identification of specular reflection on the colposcopy images.

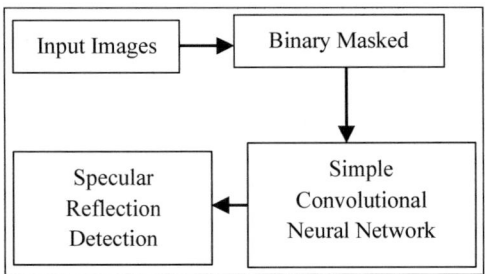

Fig. 1. Flowchart for the Detection of specular Reflection using Simple CNN mode

Fine Tuned U-Net Convolutional Neural Network

Fine Tuned U Net Model

The U-Net model is built for the segmentation of medical images (Ronneberger, et al., 2015). The main advantage of this method is attaining higher segmentation accuracy with minimum trained dataset. It consists of a contracting path also known as the encoder and an expansion path representing the decoder part of the images. The contracting and the expansion section consist of four convolutional blocks and each convolutional block has two convolutional layers. The convolutional layer of the U-Net model is represented in equation.2. Each convolutional kernel is set as (3×3) with the max polling layer size of (2x2).

$$F_{i,j} = f\left(\sum_{m=0}^{2}\sum_{m=0}^{2} W_{m,n} I_{x+m,y+n} + W_b\right) \quad (2)$$

Where $I_{x,\,y}$ represent the row and column of the images, $W_{m,\,n}$ represent the weight of the (m, n), and $F(i,j)$ represents the feature map extracted from the smart colposcopy image. The rectified linear unit (ReLu) is applied to each layer of the convolutional layer in the contracting path and the batch normalization is enabled as true for each layer. It consists of the transposed convolutional layer in the expansion path. It also consists of the four convolutional layers where the padding is set as the same to get the actual size of the original images. The model is tuned for the detection of smart colposcopy images:

- The stride value of the fine-tuned U-Net model is set as (1×1) to consider each pixel of the image.
- The Dropout layer is attached to each block layer of the contraction path to reduce overfitting during the feature extraction process.
- The batch normalization is added in the contraction path to normalize the pixel value and to improvising the computation accuracy of the model.
- The neuron size or filter dimension is set as 8, 16, 32, 64, and 128 for the contraction path and 128, 64, 32, 16, and 8 for the expansion path of the finetuned U-Net model. The finetuned uNet model is shown in Fig. 2.

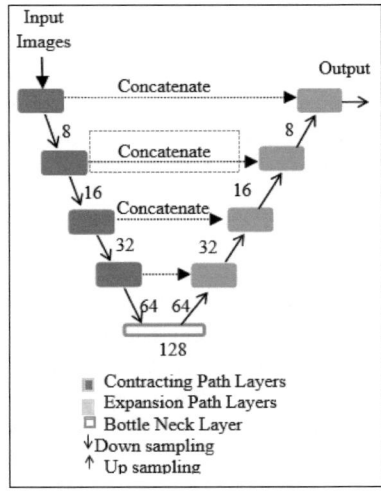

Fig. 2. The Finetune U-Net Architecture

Training and Testing for Simple CNN and Finetuned uNet Model

The input size of the images is resized as 256×256. The model is trained with the Adam optimizer for faster computation and the binary cross-entropy loss calculation is used for the binary segmentation of the cervical images. The model is trained with the epoch value of 10 and with a learning rate of le^{-4}. The batch size is set as 100 and trained the model with 1063 images. For the validation 200 images are selected that are not used in the training which is collected from the Kaggle dataset (Kaggle, 2017). The workflow of the fine-tuned U Net model is shown in Fig.3.

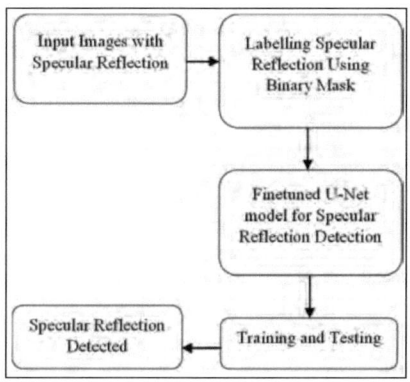

Fig. 3.

Experiment Result and Analysis

Smart colposcopy images are taken from the Kaggle dataset to train the detection of specular reflection on the cervical images. The dataset contains three types of the cervix images such as CIN1, CIN2, and CIN3 based on the severity of cancer affected on the tissue region. CIN1 is the initial stage of cancer in which one-third of the tissue region is affected by the cancer cell and these cancer cells will be higher in CIN 3 which means that the acetowhite region is higher for the CIN3 cervical cancer compared to the CIN1 and CIN2. It will be a challenging task in the CIN3 to exactly extract the specular reflection from the acetowhite region. The binary accuracy, Intersection over Union (IoU), and the dice coefficient are calculated to evaluate the performance of the model on the cervical images.

Qualitative Analysis

The qualitative analysis has been carried out on the cervical images which are collected from the Kaggle dataset.

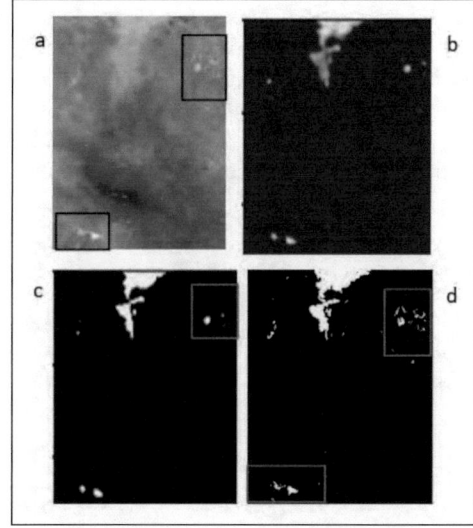

Fig. 4. (a) Original Images (b) Masked Images (c) Predicted Images using Simple CNN method (d) Predicted images using finetuned UNet model

Fig. 4a represents the original image with the specular reflection. The proposed binary masking is applied on the original to label the specular reflection on the cervical as shown in Fig. 4b. The yellow portion on the masked is labeled as the specular region and the other portion is the non-specular reflection region. The masked region was trained using the simple CNN model and fine-tuned U-Net model. Figure 4c detects the specular reflection on the cervical images but some of the small, dotted reflections are not identified as the reflection. Based on the comparison of the fine-tuned U-Net model identifies the small dots of the reflection on the cervical images as shown in Fig. 3d. Based on the qualitative analysis the fine-tuned U-Net model gives the correct prediction of the specular reflection including the small dots of the reflection.

Quantitative Analysis

For the quantitative analysis, the binary accuracy, intersection of union (IoU), Dice-Coefficient, and binary cross entropy loss value are calculated. The binary accuracy predicts the total number of correctly predicted reflection regions divided by the total number of predictions on the smart colposcopy images. The intersection of union (IoU) specifies the amount of overlap that happened between the predicted mask and the original masked images of the smart colposcopy images. The dice coefficient calculates twice the multiple of the area of overlap in the predicted and the original images divided by the total number of pixels in both images.

Table 1. Quantitative Metrics for the detection of specular reflection on the smart colposcopy images

	Binary Accuracy (%)	IoU	Dice-Coefficient	Loss value
Simple CNN[]	0.924	0.915	0.921	1.269
U-Net[]	0.972	0.967	0.977	–0.984
Fine tuned U-Net	0.989	0.965	0.981	–0.980

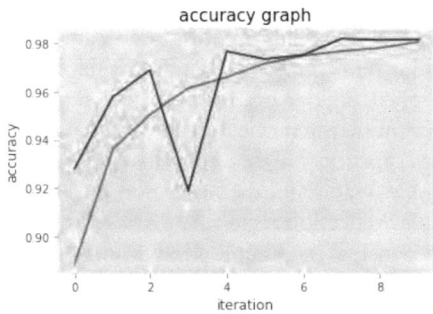

accuracy graph

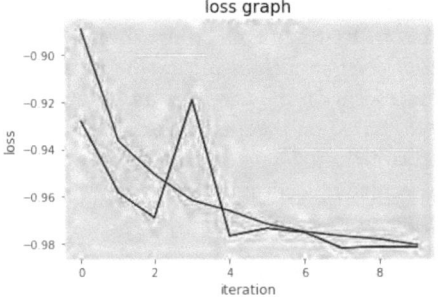

loss graph

Fig. 5. The Accuracy Graph for the fine Tune Model (a) The Loss Graph of the fine Tune Model(b)

A binary cross entropy loss value is calculated to determine the loss of the pixel in the predicted output. These metrics are calculated for the simple convolutional neural network for the segmentation of the specular reflection. Based on the analysis the Finetune U Net model attains the binary accuracy of 98.96% for 10 epochs as in Table 1. The loss value for each epoch is visualized in Fig. 5b and accuracy value for each epoch is visualized in Fig. 5a. The time and consumption are also important to challenge in medical images. The U-Net model took five hours and 26 minutes with a memory usage of 874 MB to train 1063 cervical images with 10 epochs. The time and memory usage are shown in Table 2. The process is computed using the GPU processor and the time

Table 2. Computational time and memory usage for Simple CNN and U-Net model

Methods	Time	Memory Usage
Simple CNN	2 hrs. 03 mins	741 MB
U-Net	5 hrs. 30 mins	874 MB
Finetuned U-Net	5 hrs. 26 mins	874 MB

Conclusion

The reflection identification on the specular reflection is a challenging task and no automated method is applied to the smart colposcopy images. Initially, the simple CNN model for the segmentation is applied but for this model, a greater number of cervical images are required to obtain higher accuracy. The U-Net model advantage is obtaining higher segmentation accuracy with minimum trained images. So, the U-Net model is fine-tuned based on the smart colposcopy images and helps predicting the specular reflection with promising accuracy with few trained images.

Acknowledgments

The author expresses deep gratitude to the "Centre for Machine Learning and Intelligence" for supporting and sharing the resource for this research work. It is supported by the Department of Science and Technology under the scheme of DST CURIE (AI) sanctioned for the "Core Research grant for Artificial Intelligence (AI)," in the period 2021–2023.

References

Attard, L., Debono, C. J., Valentino, G., and Castro, M. D. (2020). Specular Highlights Detection Using a U-Net Based Deep Learning Architecture, *Fourth International Conference on Multimedia Computing, Networking and Applications,* Valencia, Spain.

Ayalew, Y. A., Fante, K. A., and Mohammed, M. (2021). Modified U-Net for liver cancer segmentation from computed tomography images with a new class balancing method, BMC Biomedical Engineering, 3(4).

Jus, L., Meden, B., Struc, V., and Peer, P. (2018). End-to-End Iris Segmentation Using U-Net, *IEEE International Work Conference on Bioinspired Intelligence*, San Carlos, Costa Rica

Kaggle (2017). Intel & MobileODT Cervical Cancer Screening. Retrived from https://www.kaggle.com/c/intel-mobileodt-cervical-cancer-screening

Liu, T., Wang, C., Chang, J., and Yang, L. (2021). Specular Reflections Detection and Removal Based on Deep Neural Network for Endoscope Images. Seventh International Conference on Image Processing Theory, Tools and Applications, Montreal, Canada.

Meslouhi, O. E., Kardouchi, M., Allali, H., and Gadi, T. (2011). Automatic detection and inpainting of specular reflections for colposcopic images, Central European Journal of Computer Science. 1(3), 341–345.

Ronneberger, O., Fischer, P., Brox, T. (2005). U-Net: Convolutional Networks for Biomedical Image Segmentation, Paper presented at International Conference on Medical Image Computing and Computer-Assisted Intervention, Munich, Germany.

Susan, M. B. J., Subashin, P., and Krishnaveni, M. (2022). Comparison of various deep learning inpainting methods in smart colposcopy images. International Journal of Computational Intelligence Studies, 11(1), 53–72.

Wei, S., Zang, S., Liu, X., and Qiu, L. (2019) An Improved U-Net Convolutional Networks for Seabed Mineral Image Segmentation, IEEE Access, 7(1), 82744–82752.

Mapout a Crop Protection System Against Forest Animals Using Internet of Things

Mali Sujith Kumar Reddy,[a] Janapati Krishna Chaithanya,[b] and Katroju Rajapriyudu[c]

[a,b,c]Vardhaman College of Engineering (Affiliated to JNTUH), Hyderabad, India.
E-mail: [a]sujithkumarreddy147@gmail.com

Abstract

India is having the backbone of farming culture of various crops in various seasons but at night these are damaged by forest animals such as buffaloes, cows, goats, birds, and wild elephants many times. It leads to major losses to the farmers in terms of wealth and health too. The proposed model addresses the issue of crop protection systems using "Internet of Things (IoT)" like farmers cannot stay in the farm field 24×7 to protect and also reduce the losses further. The proposed design includes an animal detection system using night vision camera interfaced with Yolov3 algorithm (Image Processing Algorithms) and processes to Raspberry Pi by interfacing high-intensity buzzer when animals are detected. Buzzer will work at different intensities respective to the type of animals, whole system is having better efficient than the existing systems and ensure the complete protection of crops.

Keywords: Raspberry pi, Camera, Internet of Things, Image Processing Algorithms

Introduction

In recent years, wild creatures such as bears, elephants, tigers, and monkeys have posed a huge threat to farmers all over the world. Animals such as wild bears, elephants, tigers, and monkeys, among others, have wreaked havoc on crops by trampling them. As a result, farmers are facing financial challenges.

In this project, we provide a solution to this problem. The Raspberry Pi is used in this project to safeguard agriculture. To detect and analyze the animals Rasberry Pi is employed in this research. Farmers can keep an eye on their farms without having to leave their houses. The Rasberry Pi recognizes an animal entering the farmland and sounds a buzzer with varied intensities of noise depending on the animal's classification. This effort decreases the circuit's complexity.

Compared to the other projects, this one has a better level of precision. Sensor malfunctions will become less common. The components used in this project will make the work easier for the user.

Literature

Smart intruder detection system which uses motion sensor and infrared sensor to detect motion and to save crops from the objects without damage. Arduino is used in to interface all the sensors and hardware components (Srushti Yadahalli, 2020). An "IoT" based smart security system takes over "sensor panel grid made up of PIR sensor. Passive Infrared Sensor (PIR)' is used to detect the motion of an object. This system uses pattern recognition algorithm for classifying object into human or animal (Tanmay Baranwal, 2016). YOLOv3 based neural network for de-identification technology is an algorithm for object detection and identification of objects and categorize them into humans or animals which speeds up the farms for rescuing farms (Ji-Hun Won, 2019). An Intelligent crop monitoring and protection system which uses buzzer as an alarm sound and an alarming system where it is notified with mobile. This system uses wifi, cloud with data provided to the cloud periodically through wifi (Ramprasad, 2019). Automation of automobiles system used "Pulse width modulated (PWM)" signals to control the speed of the motor using Raspberry pi 3B+ coupled to Arduino nano. Proposed method uses Raspberry pi 4 which is more modern than earlier ones "(V. S. Maruthi Revanth)." An "IoT" solution for crop protection system used an integrative approach in the field of "IoT" for smart agriculture which used open-source system for protection of crops against wild animals (Giordano, 2018). Smart irrigation with field protection system has been designed to protect crops from fire and animals. Therefore, they have used "PIR" sensors for detection of animals, and photoelectric sensors are used to detect fire (Gobhinath, 2019).

and then sent to the pi in which these pictures are processed and identify the object. In second stage classification of animals into wild and domestic is done using image processing algorithms. After the classification of animals buzzer will emit noise with different intensities according to the type of animal it is. So, after classification displaying of animals photos and its name will be visible on the screen at home so that farmer can alert forest officers in case of wild animals.

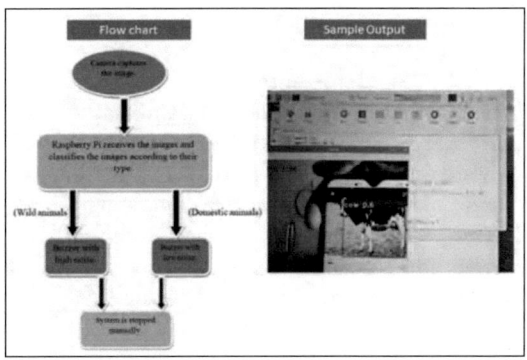

Fig. 2. Flow chart and Output of the proposed system

Raspberry pi

The Raspberry Pi is a tiny computer that can be used for many different purposes, including military and surveillance applications. The Raspberry Pi 4 model B is the most recent addition to the foundation's lineup. Total number of general-purpose input and output pins on the Raspberry Pi is 40. On the board, there are seven zero-volt pins, two 5V pins, two 3V3 pins, and 26 General purpose input output pins (0V). Processor present in Pi 4 Model B uses 20–25 percent less power and performs more than 90 percent more efficient than the older version. It includes two tiny "High-Definition Multimedia Interface (HDMI)" ports with a resolution of 4Kp60. Raspberry pi-4 consists of 4-pole stereo and audio port. It consists of SD card slot which is useful for loading operating system, files, and images needed in it. It has "Universal serial Bus (USB)'-C connector. Since pi-4 has 64-bit architecture Operating should also be 64 bit which was released by the foundation in recent

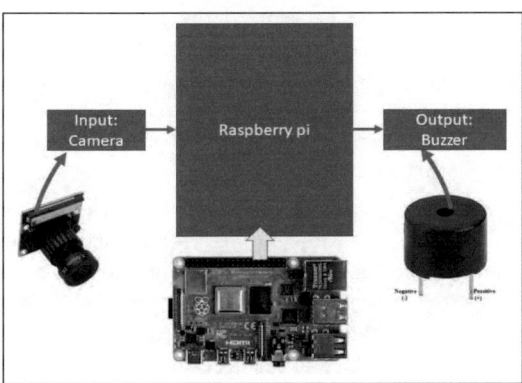

Fig. 1. Block Diagram and all the components used in the hardware realization

Design

In the proposed method entire process is done by Raspberry pi. It consists of three stages for detection and classification in the entire process. We are using camera for capturing the pictures. At first stage, pictures are captured using Camera

times. The PI-4 B's "Graphics processing unit GPU (GPU)" supports Blue Ray video playback and 3.0 graphics. It features four "USB" 3.0 and 2.0 connectors. They can be connected to external devices such as mouse, keyboards, and other peripherals. True Gigabit Ethernet is included in the Pi-4-B, allowing it to send Ethernet frames at a rate of 1 billion bits per second.

Camera

A camera is practical both at night and during the day. These cameras are used to capture images or used for surveillance in presence of low light conditions. The fundamental feature that these work upon is "Infrared radiation (IR)" illumination without which is impossible of capturing and surveillance. The following are the features of a night vision camera: It should have strong infrared illumination that is preferably adaptive. It can be used at any time of day or night. It is of High light sensitivity to the sensor. It consists of ICR filter. It is capable of surrounding 10 meters.

Buzzer

Many industries employ piezo element buzzers, including telecommunications, home appliances, cars, alarm systems, etc. A Piezoelectric buzzer is such that uses piezoelectric effect to produce a tone. The initial motion of buzzer is created by applying a voltage to a piezoelectric material, which is converted to sound. The frequency of the Piezo buzzer is 31Hz-655Hz. Buzzer needs voltage up to 1.5v to 12v. It generates a square wave of the specific frequency with a 50 percent duty cycle. It functions between -20°C and +60°C.

Implementation

The above picture displays the project's execution sequence in three levels.

Initially, boxes will be generated around the object, then the boxes will be suppressed with the command "Non-maximum suppression (NMS)," and then confidence score will be calculated. If the confidence score is greater than 0.2 then it is said that object is detected. When an object is detected, it compares the object with

a list present in the file (where trained images are present with their names), and if the object is detected and it is present in the file, then the object name is displayed.

Results and Discussion

Above figure represents the result of prototype model of particular case. It can be seen that the name of the animal and confidence score is displayed on the top of the box.

Discussion

The protection of crops against animals is the major goal of this initiative. For this application, the project's "Frames per Second (FPS)" image capture rate is far faster and more advanced. In comparison to earlier designs, it quickly takes photos and classifies them. This is a result of the design's use of the Yolov3 algorithm. Consider, for instance, a scenario in which a wild animal joined a farm.

Previous Designs: In earlier systems, "PIR" and "IR" sensors were employed to detect object movements. A buzzer then emits a sound and a notification will be sent to the farmer, but the farmer was unaware of the type of animal that had been entered. For the forest authorities to visit the farm, the farmer must alert them when he arrives. Therefore, the entire crop would suffer during this gap.

Proposed Project Design: With the help of this project, we can quickly determine the sort of animal that has entered the farmland and may notify the forest officials while at home. so that a portion of the farm can be preserved without being harmed. In this design capturing the image detection and classification is done very fast which helps in time-saving and crop-saving. Yolov3 has not yet been used for this application by anyone. Advantages of the proposed design are: Power consumption is less. Less manual work. Less complexity in circuit. More efficient and accurate.

Conclusion and Future Scope

Wild animals destroying crops has become a serious social problem in recent years. It demands prompt attention and a workable solution. Therefore, this project has a lot of societal value because it tries to solve this issue. The suggested Raspberry Pi-based system is discovered to be smaller, more approachable, and less sophisticated, and it may be used to complete tasks with ease. The detection, classification, and display of the animal's name in this project all take less time. For the farmers, this helps them save time and crops. By the aforementioned benefits, we can conclude that this warning system for crop protection is more efficient than earlier approaches.

To further develop this device, one can work on one of the following suggestions. We used image detection and noise emission, but it might be expanded in the future to track the location of animals using "Global positioning system (GPS)." It can be modified to generate a smoke repellant that aids in the eradication of animals. It may be developed so that fire can be detected using the system's cameras and Raspberry Pi.

References

Yadahalli, S., Parmar, A., and Deshpande, A. (2020). "Smart Intrusion Detection System for Crop Protection by using Arduino," 2020 Second International Conference on Inventive Research in Computing Applications (ICIRCA), Coimbatore, India, pp. 405–408, doi: 10.1109/ICIRCA48905.2020.9182868.

Baranwal, T., Nitika, and Pateriya, P. K. (2016). "Development of IoT based smart security and monitoring devices for agriculture," 2016 6th International Conference - Cloud System and Big Data Engineering (Confluence), Uttar Pradesh, India, pp. 597–602, doi: 10.1109/CONFLUENCE.2016.7508189.

Won, J. -H., Lee, D. -H., Lee, K. -M., and Lin, C. -H. (2019). "An Improved YOLOv3-based Neural Network for De-identification Technology," 2019 34th International Technical Conference on Circuits/Systems, Computers and Communications (ITC-CSCC), 2019, Okinawa, Japan, pp. 1–2, doi: 10.1109/ITC-CSCC.2019.8793382.

Ramaprasad, S. S., Sunil Kumar, B. S., Lebaka, S., Prasad, P. R., Sunil Kumar, K. N., and Manohar, G. N. (2019). "Intelligent Crop Monitoring and Protection System in Agricultural fields Using IoT," 4th International Conference on Recent Trends on Electronics, Information, Communication & Technology (RTEICT), 2019, Nagoya, Japan, pp. 1527–1531, doi: 10.1109/RTEICT46194.2019.9016770.

Revanth Pasupuleti, V. S. M., Jollu, B. M., Chaithanya Janapati, K., Naseeha, A., and Chandana, P. (2020). "Partial Automation of Automobiles using Embedded Systems," 2020 4th International Conference on Trends in Electronics and Informatics (ICOEI) (48184), Tirunelveli, India, pp. 191–195, doi: 10.1109/ICOEI48184.2020.9142924.

Giordano, S., Seitanidis, I., Ojo, M., Adami, D., and Vignoli, F. (2018). "IoT solutions for crop protection against wild animal attacks," 2018 IEEE International Conference on Environmental Engineering (EE), Milan, Italy, pp. 1–5, doi: 10.1109/EE1.2018.8385275.

Gobhinath, S., Darshini, M. D., Durga, K., and Priyanga, R. H. (2019). "Smart Irrigation with Field Protection and Crop Health Monitoring system using Autonomous Rover," 2019 5th International Conference on Advanced Computing & Communication Systems (ICACCS), Coimbatore, India, pp. 198–203, doi: 10.1109/ICACCS.2019.8728468.

Patrolling Robot for Fugitive Emissions in Industries

Mannepally Preethi,[a] Janapati Krishna Chaithanya,[b] Bhanavath Bhaskar,[c] and Ravula Prudvi[d]

[a,b,c,d]Vardhaman College of Engineering (Affiliated to JNTUH), Hyderabad, India
E-mail: [a]mannepallypreethi@gmail.com

Abstract

Industries with gas leaks suffer terrible consequences in terms of business, human lives, and the environment. From the factory unit to the household unit, robotics and automation have been vital players in fulfilling and addressing the evolving requirements of humans. Using the Internet of Things, the proposed solution uses a patrolling robot to handle the issue of gas leaks. The gas sensor and flame detector of this system identify leak, detect fire respectively and alert the control room. The robot follows a predetermined path and checks leakage for every few meters at each node, recording the data and alerting the user if the value exceeds the threshold.

Keywords: NodeMCU, Gas sensors, Infrared sensors, Flame detector

Introduction

Uncertainty exists regarding the frequency of gas leaks in industrial operations. The majority of these leaks, whether or not they are acknowledged, get unreported since they do not directly cause physical harm. Industrial gases are extremely hazardous to all living organisms. Industrial accidents not only cause human loss but also leave a huge impact on economic growth of industries along with decrease in country's gross domestic product (GDP)" rate. Moreover, the people working in the industries get affected seriously both physically and financially. It also affects the people and environment in the vicinity. Highly toxic gases also cause global warming.

The main source of inspiration for our idea is the scenarios and challenges that arise in industries as a result of gas leaks. Leakages and industrial fires have become a big problem. People working in the industries, as well as those in the immediate vicinity, face significant danger. Gas leakages in industries have a significant impact on industry's economic growth, affecting the country's shares. To avoid disasters to occur, strict quality standards should be built into the safety system. When it comes to designing any industry, the most important criterion is safety. Internet of things (IoT)" which does not require human or computer intervention is intended to automate daily operations, but its advantages can also be used to improve existing safety regulations.

Literature

A gas leakage detection system that uses MQ gas sensors to detect Liquified petroleum gas (LPG)', Methane and Benzene and uses an ESp-32 wireless fidelity (Wi-Fi)' module to send notifications and also use different types of light-emitting diode (LED)' to indicate different gas leakage was implemented (Kodali, 2018). A patrolling robot with a camera, Global positioning system (GPS)' module which removes human errors such as limitation of sight

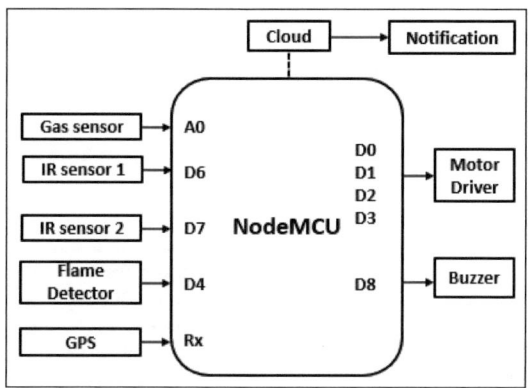

Fig. 1. Block diagram of the proposed model

at night was implemented. It avoids obstacles in its way and moves around the place without human intervention and detects abnormalities and send message to control room (Khalid, 2021). A gas leakage detection system with a field-programmable gate array (FPGA)' and Global System for Mobile Communication (GSM)' module was implemented. "FPGA" detects any gas leakage and sends warning to "GSM" module which further sends a text notification the number registered (Arpitha, 2016). A robot with Infrared (IR)' sensors for moving robot autonomously without human intervention was implemented. It is used in domestic and industrial applications to do a repetitive task such as mopping the floor (Tayal, 2020). A wireless gas leakage detection system that has gas sensors placed around a region in the industrial plant to detect various gas leakage was implemented. Here the data travels in a single mesh network towards a particular point where algorithms are applied to detect gas leakages (Chraim, 2016). Liquefied petroleum gas (LPG)" detection system using "GSM" module which detects "LPG" and sends Short Message Service (SMS)" notifications to mobile was implemented. It is generally used for home safety "(Amsaveni, 2015)." A robot that moves in a pre-defined path set by the user is designed. It uses "IR" sensors which are used for movement along a black line. It is used either for domestic or industrial applications (Pakdaman, 2009).

Design

NodeMCU digital pins D6, D7, and D4 are connected as inputs to "IR" sensor 1, "IR" sensor 2, as well as Flame detector, respectively. The analog pin A0 is connected to the Gas sensor as an input. The Rx pin is also connected to the "GPS" module. The digital pins D0, D1, D2, and D3 of the NodeMCU are connected to the motor driver, while D8 is connected to the buzzer as an output. NodeMCU is used to connect all of the components to the cloud.

NodeMCU

NodeMCU hardware is open-source, thus anyone can modify or design it. The ESP8266 "Wi-Fi" capable chip is used in the NodeMCU Development kit. The ESP8266 is a TCP/IP-based chip that was created by Espressif Systems. The analog (A0), and digital (D0-D8) pins on the board of the NodeMCU Dev Kit are identical to those on an Arduino. Various communication protocols are supported. Lua scripts are commonly used to program the NodeMCU with ESPlorer IDE. Arduino programming environment can be used to create NodeMCU applications.

MQ135 Gas Sensor

The MQ-135 Gas Sensors are efficient in ammonia, NOx, alcohol, smoke, and carbondioxide detection and measurement in air quality management. When only one gas needs to be detected, the digital Pin on the MQ-135 sensor module makes it possible for it to operate independently of a microcontroller. A broad variety of microcontrollers are compatible with the TTL-driven analog pin, which runs at 5V.

Flame Detector

A flame sensor is a type of detector made specifically to find and react to fire or flame. The flame detector's performance may be impacted by how it is mounted. High photosensitivity, fast response time, and adjustable sensitivity are its features. It is simple to use and both analog and digital output switches are available to use.

"IR" Sensor

An electronic device known as an infrared sensor emits infrared light to determine various conditions. An infrared sensor can measure the heat of an item and detect movements. There are two types of "IR" sensors: active and passive "IR" sensors.

Neo6MV2 "GPS" Module

A "GPS"-enabled navigation module is called the NEO-6MV2. Its position on the planet is easily determined by the module, which also supplies its longitude and latitude. Due to their compact architecture and limited power and memory options, NEO-6 modules are well-suited for battery-operated mobile devices with strict cost and space restrictions.

Motor Driver L293D

A motor driver is the L293d integrated circuit (IC). It requires a small voltage to work. While other "IC's might carry out the same tasks as the L293d, they cannot produce the high voltage needed by the motor. The L293d supplies a constant, bidirectional Direct Current to the motor. Without affecting the entire "IC" and any other component in the circuit, current polarity can change at any time.

Implementation

The robot moves on its own, without the need for human assistance. It employs the line-following robot technique, in which "IR" sensors are utilized to detect movement. The robot follows the designer's predetermined path. If a black line is identified, it travels ahead; if a white line is spotted, it stops. It makes use of two infrared sensors. The robot can also move in a curved path by employing the right and left "IR" sensors. If both of the "IR" sensors read high, the robot continues forward; if any of the "IR" sensors reads low, the robot comes to a halt. If both the left and right "IR" sensors read low, it will come to a halt. Motor driver L293D is in charge of the motor movement.

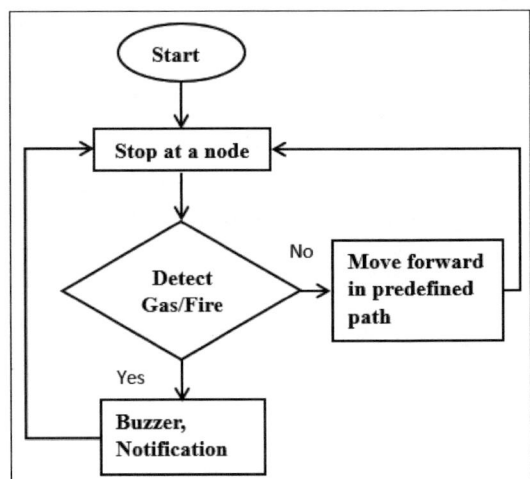

Fig. 2. Flow chart of the proposed model

The robot advances and waits 5 milliseconds before sensing any leakage. The detector module displays a high output and sends a signal to the controller unit if leakage of gas or flame detection occurs. The buzzer sounds louder and provides an alert signal to those nearby. If the robot detects no leakage, it continues on its predetermined path.

The Blynk app which interfaces with the sensors and store, and saves data, and displays it in the app continuously shows the readings such as Alert! Gas leakage is detected or Alert! Fire is detected. The "GPS" module sends the exact location where the leakage is found and its location is shown in the Blynk app.

Results and Discussion

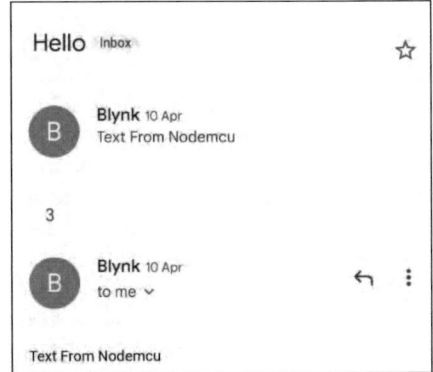

Fig. 3. Initial message notification

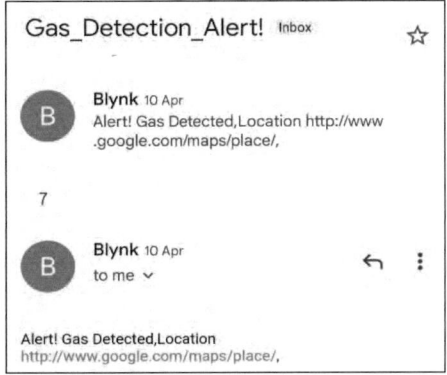

Fig. 4. Gas detection alert notification

Fig. 5. Fire detection alert notification

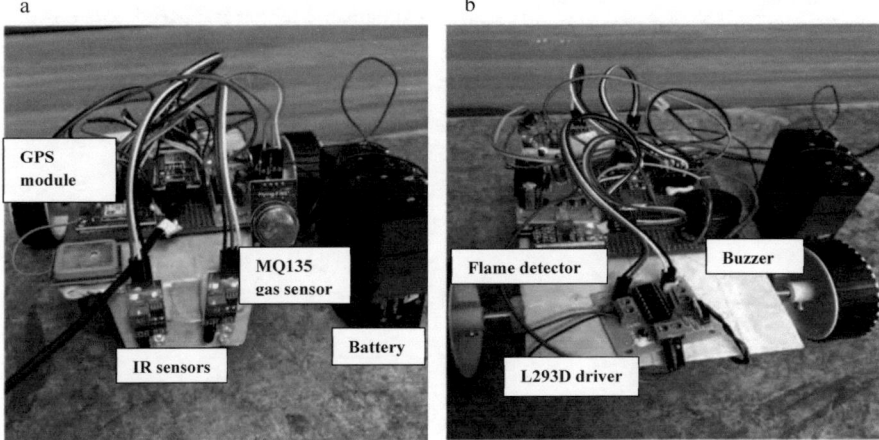

Fig. 6. Prototype of the proposed model (a) Front view of the proposed model (b) Back view of the proposed model

Conclusion

The suggested project employs a technique where the robot moves without human involvement. The suggested project uses gas sensors to detect leakage of various gases at designated nodes and alerts the control room. It also detects and alerts the control room if a fire breaks out in the industry at a predetermined node. If there is an irregularity, such as a gas leak or a fire, a siren will sound to inform residents. If no irregularity is found at a predetermined node, the robot proceeds along the course as planned. This project aids in the prevention of hazardous situations in the workplace and the occurrence of significant financial loss to the business owner.

Future Scope

Future work includes the usage of different "LED" to represent various gas leakage detection. This helps to easily identify the type of gas leakage and can further prevent it in less time. Also, multiple gas absorbers can be interfaced to prevent great damage due to gas leakage. As the proposed model uses a line following a robot it may cause a disturbance in the movement of robot if there is any obstacle present across a predefined path. Therefore, an ultrasonic sensor can be used to detect the obstacle in the predefined path.

References

Kodali, Ravi Kishore, Greeshma, R. N. V., Nimmanapalli, Kusuma Priya, and Yogi Borra, Yatish Krishna (2018). IoT Based Industrial Plant Safety Gas Leakage Detection System. 4th International Conference on Computing Communication and Automation. Greater Noida, India, pp. 1–5. doi: 10.1109/CCAA.2018.8777463.

Farabee Khalid, Itmamul Haque Albab, Dipto Roy, Azad Prince Asif, and Kawshik Shikder. "Night Patrolling Robot." 2nd International Conference on Robotics, Electrical and Signal Processing Techniques (ICREST). 2021, American International University-Bangladesh, pp. 377–382. doi: 10.1109/ICREST51555.2021.9331198.

Arpitha, T., Kiran, Divya, Sitaram Gupta, V. S. N., and Duraiswamy, Punithavathi (2016). "FPGA-GSM based gas leakage detection system." IEEE Annual India Conference. Bangalore, India, pp. 1–4. doi: 10.1109/INDICON.2016.7838952.

Tayal, Satyam, Govind Rao, Harsh Pallav, Bhardwaj, Suryansh, and Aggarwal, Harsh (2020). Line Follower Robot: Design and Hardware Application. 8th International Conference on Reliability, Infocom Technologies and Optimization. 2020, Noida, India, pp. 10–13. doi:10.1109/ICRITO48877.2020.9197968.

Chraim, Fabien, Erol, Yusuf Bugra, and Pister, Kris (2016). Wireless Gas Leak Detection and Localization. Transactions on Industrial Informatics 12.2, pp. 768–779. doi: 10.1109/TII.2015.2397879.

Amsaveni, M., Anurupa, A., Gunasekaran, M. (2015). Efficient Gas Leakage Detection and Control System using GSM Module. International Journal of Engineering Research Technology. pp. 1–5. doi: 10.17577/IJERTCONV3IS15053.

Pakdaman, Mehran and Mehdi Sanaatiyan, M. (2009). Design and Implementation of Line Follower Robot. Second International Conference on Computer and Electrical Engineering. vol. 2. 2009, Dubai, UAE, pp. 585–590. doi: 10.1109/ICCEE.2009.43.

Investigation of Antioxidant and Photo Catalysis of Natural Honey and Cow Urine-Doped CeO$_2$ Nanoparticles Fabricated by Reflux Method

S. Chandruvasan,[a] Harish Madival,[b] M. Mylarappa,[,a] N. D. Naik,[c] S. Kantharaju,[a] and S. Bharath[a]*

[a]Department of Chemistry, SJR College of Science, Arts and Commerce, Bengaluru-560009, India.
[b]Department of Physics, JAIN (Deemed to be University), Bengaluru - 560027, India.
[c]Department of Physics, Dr. A. V. Baliga College of Arts and Science, Kumta-581343, India
E-mail: *mylu4mkallihatti@gmail.com

Abstract

Nanoparticles of natural honey and cow urine doped with CeO$_2$ were synthesized using reflux synthesis method. The prepared samples were characterized using UV-visible Spectroscopy, X-Ray Diffraction (XRD), and Fourier Transform Infrared (FTIR) spectroscopy. Honey and cow urine have been studied for their influence on the crystalline size and functional group characteristics of CeO$_2$ Nanoparticles. Photo catalytic activity of methyl orange has been studied using Cerium oxide Nanoparticles doped with natural honey and cow urine and exposed to ultraviolet light irradiation. Photo catalysis studies on methyl orange degradation revealed that natural honey and cow urine-doped CeO$_2$ Nanoparticles had a greater impact. It was shown that the Nanoparticles containing honey and cow urine were able to scavenge free radicals using the DPPH scavenging method.

Keywords: Honey, Cow urine, CeO$_2$, Antioxidant, Photo catalysis activity.

Introduction

Natural nanoparticles in the size range of 1–100 nm are used in bio-based nanocomposites, which are composite materials based on renewable natural resources. New polymer production could be achieved by using hemicellulose as a raw material. It has also been used to make nanoparticle chitosan (dextran), gelatin (alginate), albumin (albumin), and starch (gelatin) (Isabelle, 2006; Irene Bravo-Osuna, 2006).

For thousands of years, Hindus have revered cows as sacred beings, referring to them as "KAMADHENU" the "mother of all entities" in the universe. An in-depth examination of cow urine in the lab has revealed its high concentration of iron and copper in addition to other minerals and vitamins such as uric acid (creatinine), nitrogen (nitrogen), Sulphur (sulfuric), manganese (manganese salts), carbolic acid (carbolic acid), silicon (silicon chloride), magnesium (magnesium chloride. To maintain our physical and mental well-being, Gomutra's above-mentioned components are extremely beneficial. Honey-mediated nanoparticle synthesis, a green synthesis example, is receiving attention in the scientific community (Eranga, 2017; Maria, 2018).

Tissue engineering and cytotoxicity studies are just two examples of their many uses in biological research (Martinotti, 2014). While honey was the first sweetener, it was also used for medicinal purposes for quite some time. Honey's reducing properties have also been linked to glucose and fructose, according to some researchers. It has been suggested that honey's naturally occurring sugars are responsible for these properties. When cerium is tetravalent (Ce^{4+}) or trivalent (Ce^{3+}), it can exist in one of two oxidation states. Cerium oxide can be found in two distinct forms, Ce_2O_3 (Ce^{3+}) and CeO_2 (Ce^{4+}), depending on the characteristics of the materials (Beaudoux, 2016).

One of cerium dioxide's primary advantages is its ability to create an oxygen vacancy in the lattice. The differences in particle shape, size, and agglomeration tendency characterize the various methods used to make CeO_x nanoparticles. Thermal breakdown, precipitation, hydro/solvothermal or sol-gel processes, surfactant-assisted procedures, and green synthesis using natural stabilizers are some of the methods used in this field (Cerchier and Dabalà, 2017).

Carcinogenic dyes such as methylene blue, direct acid dyes, azo dyes, and reactive black have been photo mineralized using this method. There has been considerable interest in the field of nanomedicine in CeO_2 nanoparticles (NPs), which have shown promise for medical applications in the areas of catalysis, biosensing, and drug administration. It is inexpensive, environmentally friendly, has a low toxicity level, is highly bio-compatible, and has a low rate of instability. Using the reflux approach, the current study demonstrated the fabrication and characterization of honey and cow urine nanoparticles. The DPPH approach was applied to examine the Nanoparticles' antioxidant properties and photo catalytic studies have been done with the prepared Nanomaterials.

Experimental

Materials and Methods

All the chemicals used in the experiment are purchased from Merck, India

Table 1. Details about the chemicals and raw materials used

S. No	Common Name	Specification	Suppliers
1	Ajwain honey	Apis Mellifera	Tamilnadu India
2	Ammonium Ceric Sulphate	632.555 g/mol	Merck
3	Sodium Hydroxide	39.997 g/mol	Merck
4	Cow urine	-	Bengaluru, India

Synthesis of Natural Honey-Doped CeO_2 Nanoparticle

Cerium Nanoparticles containing honey and cow urine have been successfully fabricated using the reflux method. 6.32 g Ammonium ceric sulfate (ACS) prepared for 50 ml was added to a round bottom flask, followed by 10 ml urine. 50 ml of sodium hydroxide solution was produced with 1.2 g of sodium hydroxide. 50 ml of (ACS) and 10 ml of Cow urine were added to a round bottom flask. The mixture was mixed for 35 minutes at 150 °C at 250 rpm, then 40 ml sodium hydroxide was added. The apparatus should be left on at the same temperature and level of heat for approximately 90 minutes. After that, the solution is filtered through Whatman filter paper, and the compound is rinsed with distilled water. After that, the compound is held in an oven for one hour at a temperature of 200 °C.

Results and Discussion

UV-vis Spectroscopy Studies

UV-visible absorption spectroscopy (ELICO UV 1800 Spectrometer) was used to investigate the optical characteristics of cerium oxide nanoparticles in the 260–540 nm wavelength range (Honey and Cow Urine doped cerium oxide nanoparticles) have an absorbance peak around 349 nm. According to a study, many factors affect absorbance, including the band gap, oxygen deficit, particle size, and impurity centers, lattice strain, and surface roughness (Agrawal, 2018).

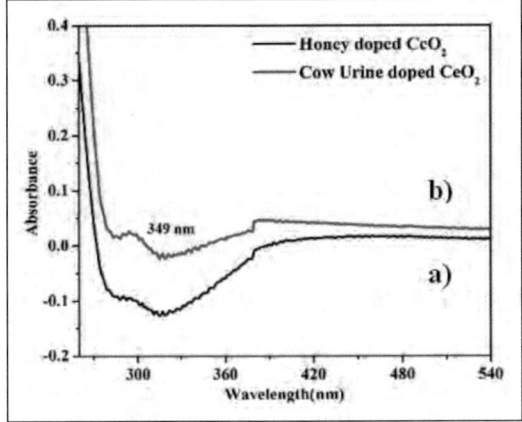

Fig. 1. Absorbance spectra of a) Honey/CeO_2 b) Cow urine/ CeO_2

X-Ray Diffraction Studies

The X-ray diffractometer PROTO XRD LPD used monochromatized Cu-Ka radiation to determine the powder's phase composition and crystallinity. The XRD of Honey/CeO_2 and cow urine/CeO_2 were shown in Fig. 2a and b. According to Scherer's equation. The average particle size was determined. The mean crystallite diameters of Honey/CeO_2 and cow urine/CeO_2 were reported to be 37.65 nm and 29.51 nm respectively. Crystallinity improves with the sharper, more intense, and longer-lasting peaks in the XRD pattern of cow urine/CeO_2. Impurity peaks could not be detected, suggesting high purity. There is a well-crystalline form of a particle in this product, as evidenced by the strong and narrow peak.

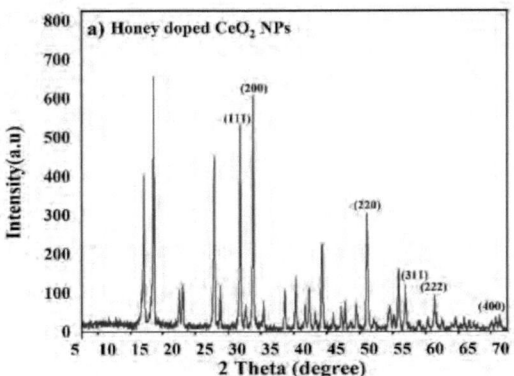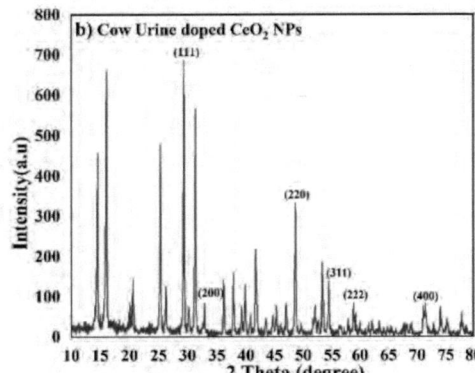

Fig. 2. XRD spectrum of (a) honey-doped CeO_2 (b) Cow Urine doped CeO_2

Fourier Transform Infrared (FTIR) spectroscopy:

Figure 3(a) and (b) depicts the FTIR of honey and cow urine-doped CeO– nanoparticles. In Fig. 3a, we can find broad stretching at 3502 cm^{-1}, 1058 cm^{-1}, and 1360 cm^{-1} corresponds to

the functional group of (O-H, N-H), C-O-C, and C-H respectively. The absorption band at 745 and 540 cm^{-1} corresponds to CeO_2 stretching vibration. The broad adsorption band of about 3502 cm^{-1} corresponds to hydrated and physically adsorbed water in the sample.

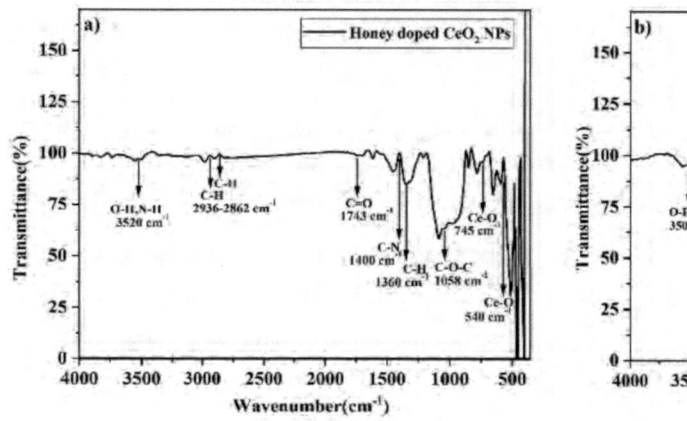 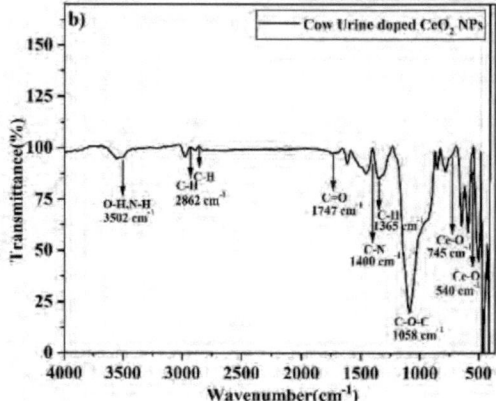

Fig. 3. FTIR spectra of a) Honey doped CeO$_2$ b) Cow Urine doped CeO$_2$

Antioxidant Studies

The antioxidant activity of CeO$_2$ Nanoparticles doped with cow urine/honey was determined using a DPPH test. DPPH in methanol was dissolved and the resultant product was applied to samples at various quantities. An absorbance reading at 517 nm was taken using an Elico UV 1800 spectrophotometer after the tubes had been incubated for an hour at 25 °C. The concentrations of each fraction were determined as follows in order to remove 50 % of the free radicals in a specific amount of time. To determine how a substance reacts to various concentrations of a solution, divide the solution among the five test tubes and number them one through five (100 mL, 200 mL, 300 mL, 400 mL, and 500 mL). Methanol should be added into the test tubes by way of micropipette, and the mixture should be thoroughly dispersed. To keep the DPPH solution from drying out, add 2 cc to each test tube and then cover it with Aluminium foil. For about an hour, it should be left alone in the dark. A UV-Visible spectrophotometer should be used to measure the absorbance of the solution. As a result, the following percentage of activity was calculated (Mylarappa, 2022).

$$\% \text{ Activity} = \frac{A_C - A_S}{A_C}$$

Where As the absorbance of the sample, Ac is the absorbance of the control.

Table 2. Antioxidant activity of honey-doped CeO$_2$ nanoparticle

Volume of Sample (mL)	Volume of methanol (mL)	Concentration	Absorbance	% Activity	IC50 (g/mL)
100	500	1000	1.080	37.60	
200	400	2000	0.938	45.84	
300	300	3000	0.870	49.78	
400	200	4000	0.782	57.98	326.797
500	100	5000	0.637	63.22	
CONTROL			1.733		

Table 3. Antioxidant activity of Cow Urine doped CeO_2 nanoparticle

Volume of Sample (mL)	Volume of methanol (mL)	Concentration	Absorbance	% Activity	IC50 (g/mL)
100	500	1000	1.169	32.54	
200	400	2000	1.113	35.74	
300	300	3000	0.980	43.45	366.032
400	200	4000	0.948	45.28	
500	100	5000	0.484	72.06	
CONTROL			1.733		

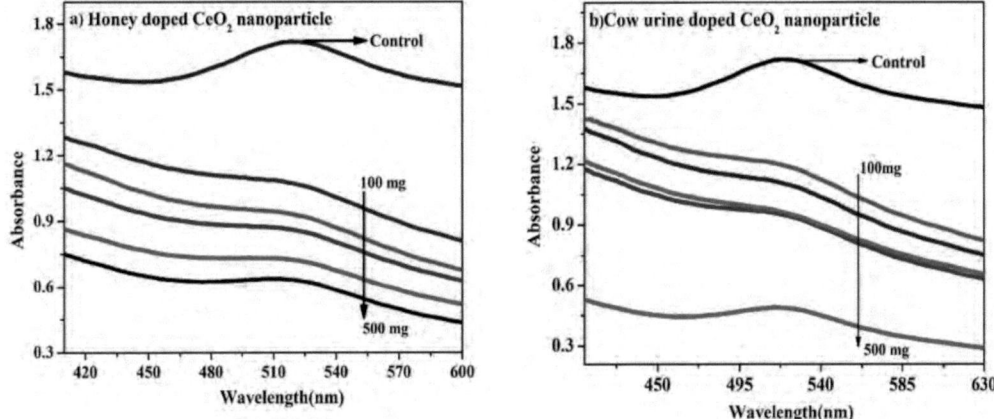

Fig. 4. UV absorption spectrum of (a) Honey doped CeO_2 b) Cow Urine doped CeO_2

Figures 4 (a) and (b) illustrate, respectively, the UV absorption spectra of honey-doped CeO_2 nanoparticles as well as the antioxidant capabilities of cow urine-doped nanoparticles. Tables 2 and 3 demonstrate the methodologies used to calculate an antioxidant activity and the IC50 value. Results discovered that the radical scavenging activity of honey and cow urine-doped CeO_2 nanoparticles was 326.797 mg/mL and 366.032 mg/mL respectively. Scavenging 50% of free radicals corresponds to an IC50 value, which is the sample concentration needed to achieve that desired outcome. Because of their DPPH free radical scavenging action, doped honey, and Cow Urine CeO_2 NPs by donating an electron to an oxygen atom, CeO_2 NPs turn the DPPH solution colorless and produce a stable DPPH molecule.

Photocatalytic studies:

The UV-Vis absorption spectra of produced photocatalysts for Methyl Orange dye were measured as shown in Figs. 5 (a) and b). CeO_2 nanocomposite photocatalysts (20 mg) were investigated with Methyl Orange (10 ppm) and UV light for 60 minutes under UV light irradiation for the photocatalytic studies. Finally, the absorbance of these dye samples was measured in the 464 nm region. An improved photocatalytic activity (92.98%) was observed for Cow urine-doped CeO_2 Nanocomposite composites under UV-Visible light irradiation compared to Honey-doped CeO_2 nanoparticles (91.20%). Performance of photodegradation was influenced by the particle size, shape, and surface characteristics of photocatalysts. Honey and cow urine CeO_2 nanocomposite photocatalysts generated by the Refluxing process have crystallite sizes of 37.65 and 29.51 nm, respectively, based on XRD results (Mylarappa, 2022).

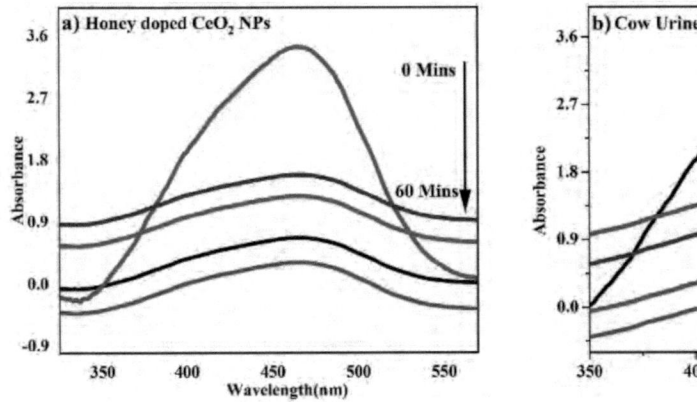
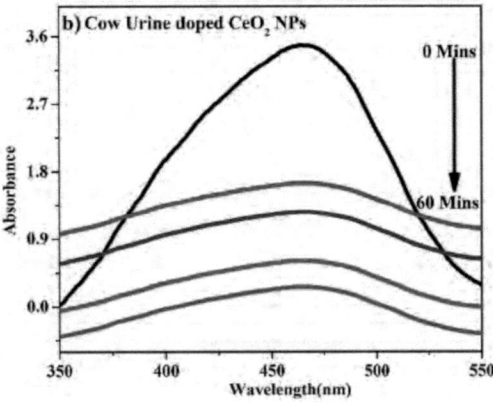

Fig. 5. Photocatalytic activity of a) Honey doped CeO_2 b) Cow Urine doped CeO_2

Conclusion

Honey and Cow urine-doped CeO_2 nanoparticles were successfully synthesized using the refluxing method. Particles with average crystalline sizes of 37.65 nm and 29.51 nm were found. Honey-doped CeO_2 NPs had higher antioxidant activity against the DPPH molecule than CeO_2 nanoparticles doped with cow urine, as determined by antioxidant properties. In Photo catalytic studies, the degradation of methyl orange by synthesized Nanomaterials and shows a greater degradation in honey-doped CeO_2 than in cow-doped CeO_2 nanomaterial.

References

Bertholon, Isabelle and Hommel, Hubert (2006). Properties of Polysaccharides Grafted on Nanoparticles Investigated by EPR. Langmuir, 22(12), 5485–5490.

Bravo-Osuna, Irene and Schmitz, Thierry (2006). Elaboration and characterization of thiolated chitosan-coated acrylic nanoparticles. Int. J. Pharm. 316(2), 170–175.

Balasooriya, Eranga Roshan and Jayasinghe, Chanika Dilumi (2017). Honey Mediated Green Synthesis of Nanoparticles: New Era of Safe Nanotechnology, 1–10.

Bonifacio, Maria A., and Cometa, Stefania (2018). Antibacterial effectiveness meets improved mechanical properties: Manuka honey/gellan gum composite hydrogels for cartilage repair. Carbohydrate Poly. 198, 462–472.

Martinotti, S., and Ranzato, E. (2014). In Cellular and Molecular Mechanisms of Honey Wound Healing. Nova Publishers Inc. ISBN: 978-1-63117-251.

Beaudoux, Xavier, Virot, Matthieu, Chave, Tony, Durand, Grégory, Leturcq, Gilles and Nikitenkoa, Sergey I. (2016). Vitamin C boosts ceria-based catalyst recycling. Green Chem. 18(12), 3656–3668.

Cerchier, Pietrogiovanni, Dabalà, Manuele and Brunelli, Katya (2017). Green synthesis of copper nanoparticles with ultrasound assistance. Green Processes Synth. 6(3), 311–316.

Agrawal, S., Parveen, A., Azam, A. (2018). Microstructural and optical properties of Co-doped NiO nanoparticles synthesized by auto combustion using NaOH as fuel, AIP Conference Proceedings, 1953. 030231.

Mylarappa, M., and Rekha, S. (2022). Synthesis and Characterization of ZnO and MgO Nanoparticles through Green Approach and Their Antioxidant Properties. ECS Trans. 107(1), 689–695.

Mylarappa, M., and Raghavendra, N. (2022). Electrochemical, photocatalytic, and sensor studies of clay/MgO nanoparticles. Applied Surface Science Adv.10, 100268.

Assessment of Reactive Oxygen Species, Antioxidant in Mercury-Induced Freshwater *Cyprinus Carpio*

Sherin Mathew[a] and Baby Joseph[b]

[a]Department of Biotechnology, Hindustan Institute of Technology and Science,
Chennai, Tamil Nadu India.
[b]Dr. Baby Joseph, Department of Biotechnology, Hindustan Institute of Technology and Science,
Padur, Kalambakkam, Chennai – 603103, Tamil Nadu, India
E-mail: [a]sherinmary18@gmail.com; [b]josephbaby2022@gmail.com

Abstract

One of the most important scientific disciplines is determining the toxicity of heavy metals in freshwater, and there is a rising concern about the development of methodologies for identifying toxic effects in aquatic creatures. To test the efficacy of this approach, a laboratory study was conducted on the fish *Cyprinus carpio* from a pond in Bengaluru, Karnataka. As a function of mercury and toxicity indices using a reactive oxygen species (ROS), oxidative stress biomarkers such as catalase (CAT), superoxide dismutase (SOD), and glutathione reductase (GR) in the brain, liver, and kidney. The ROS analysis showed that the higher concentration was in the brain. The CAT, SOD, and GR analysis revealed that mercury had higher activity in the liver compared to the brain, kidney, and controls. This study has elicited a strong response from freshwater aquaculture to implement toxicity control measures and avoid future infiltration into the food chain.

Keywords: Antioxidant analysis, *Cyprinus carpio*, Mercury, ROS

Introduction

Heavy metal contamination in marine ecosystems is increasing at an alarming rate and has become a major global issue due to the variety of anthropogenic activities sewerage drainage systems, discharging of hospital and other wastes, idol immersion, and recreational activities. The identification of Hg as either a priority hazardous substance in the aquatic environment has prompted several study results in the last few decades to evaluate the consequences of Hg pollution for humans and aquatic organisms (Garai et al., 2021). Even though the precise mechanisms of mercury toxicity are unknown, research suggests that it can cause the formation of reactive oxygen species (ROS) and limit antioxidant enzyme activity in fish tissue (Zheng, 2019[the] public have become increasingly aware of the negative health effects caused by mercury pollution in the ocean. Consequently, there has been significant interest in the health of humans eating fish exposed to mercury (Hg). ROS can effect cellular lipid, carbohydrate, protein, and DNA damage. The main purpose of this study is to better understand the toxic effects of mercury exposure on the *Cyprinus carpio*, to determine the acute toxicity of mercury, LC50, ROS, and antioxidant analysis such as CAT, SOD, and GR for the brain, liver, and kidney in common carp (*Cyprinus carpio*) from a pond in Bengaluru,

Karnataka. The effects of mercury on oxidative stress induction and immune response were also investigated. This study aimed to provide references for mercury toxicity research, aquatic environmental protection, and the stable development of aquaculture.

Materials and Methods

Cyprinus carpio weighing approximately (182g) was obtained from a freshwater pond in Bengaluru, Karnataka. The criteria for selecting healthy fish were based on morphological characteristics. Fishes were acclimatized at 32 °C *with natural photoperiod and fed commercial feed once a day before the actual experiment began. The fish*'s body weight and length were initially measured. Standard protocols were used to check for the presence of total dissolved solids and other physicochemical characteristics in the water (APHA, 1995 and 2005).

The tissue samples of 500 gm each from the liver, brain, and kidney of *C. carpio* were extracted and partially defatted protein powders were obtained by acetone extraction/ microwave digestion. The physicochemical properties of the freshwater used to grow the *C. carpio* in laboratory conditions were tested to confirm the absence of mercury (Hg).

Acute Bioassay Test*:* Toxicity tests were conducted under standard methods (APHA, 1995). Short-term toxicity (or) range-finding tests were carried out by exposing test organisms to a wide range of concentrations of mercury (Hg). Within the concentration limits ascertained from the above, 8 to 10 narrow ranges at equal intervals were selected. Experiments were conducted in 20 liters of glass aquaria (30 × 30 × 60 cm). Ten fishes of the same size were exposed to different concentrations with three replicates. Appropriate controls were maintained. Mortality was recorded after 24, 48, 72, and 96 hours respectively. Percentage of mortality was calculated and values were transformed into the Probit scale. Probit analysis was carried out as suggested by (Finney, 1971). A regression line of Probit logarithmic transformations of concentrations was made. Slope function (S)

and confidential limits (Upper and lower) of the regression line with the Chi-square test (UNEP/FAO/LAE, 1987) were calculated as follows:

$$S = \frac{\dfrac{LC84}{LC50} + \dfrac{LC50}{LC16}}{2}$$

$$f = \frac{(2.77 \log S)}{\sqrt{N}} = S\,2.77\sqrt{N}$$

where the is the number of animals tested whose expected effects are between 16 and 84 percent mortality, upper confidence limit = LC_{50} x f, lower confidence limit = LC_{50}/f.

Chronic Toxicity Test: Chronic toxicity is referred to as a long-duration test. In acute bioassay, mortality was observed within a short period whereas, in chronic studies, the effect could be seen only on prolonged duration and observation. Sub-lethal or safe or safe level concentrations were derived from 96 h LC_{50} as per the procedure given by APHA (1995) to observe the various responses of the test fishes to prolonged exposure to mercury. In the present study, 1/2, 1/5, 1/10, and 1/15 of the 96 h LC_{50} were selected as sub-lethal concentrations and the fishes were exposed to each concentration for a period of 7, 14, 21, and 28 days. A control batch corresponding to each test group was simultaneously experimented with. The experiments were repeated 5 times for each test species. Fresh concentrations were supplied daily to maintain a constant toxic media.

Detection of Reactive Oxygen Species (ROS): To detect and assess the intensity of ROS, tissue samples (Liver, Brain, and Kidney) were washed with PBS, digested with 0.5% trypsin and 0.1% collagenase for 40 min, centrifuged at 600 rpm for 5 min and re-suspended in Hank's buffer. Approximately 1 × 10⁶ cells were washed with PBS and re-suspended in Hank's buffer. Forty microliters of dichlorofluorescein diacetate (DCFH-DA; 2.5 mmol/L) were added to each sample and incubated for 30 min at 37°C in the dark. The cells were then completely washed to remove surface fluorescence, and the relative concentrations of H_2O_2 produced were measured with an inverted fluorescent microscope (Olympus CKX 41) in which the background was calibrated to zero using a xenon lamp (100 W) and fluorescein isothiocyanate filter set and measured using Image analysis software.

Antioxidant Analysis: Catalase was measured by the method by Macehly and Chance (Maehly and Chance, 1984). Estimation was done spectrophotometrically following the clearance in the absorbance at 230 nm. The reaction mixture contains 10 mM phosphate buffer (pH-7.0), 2 mM H_2O_2 and tissue extract. Specific activity was expressed in units/mg/protein.

Superoxide dismutase was determined by Kakar et al. (1984). Assay mixture contained 1.2 ml sodium pyrophosphate solution (0.052M, pH 8.3) 0.1 ml 186mM phenazine methosulphate (PMS) 0.3 ml 300 μm nitro blue tetrazolium NBT. 0.2 ml NADH, tissue extract and water in the total volume of 3ml reaction started by the addition of NADH. After incubation at 30°C for 150s, the reaction was started by the addition of 1ml of glacial acetic acid. The reaction mixture was vigorously shaken with 4 ml of n-butanol. The mixture was allowed to stand for 10 minutes, n-butanol layer was taken out. The color intensity of chromogen in the n-butanol was measured at 560 nm against n-butanol. A system devoid of enzymes served as control. The enzyme concentration inhibited the absorbance at 560nm of chromogen produced by 50% in one minute under the assay condition and expressed as specific activity in milli units/mg protein.

Glutathione reductase was determined by the procedure (David and Richard, 1983). Assay system contained 2–6 ml of 0.1 M potassium phosphate buffer (pH - 7.2, 0.1 ml of 15 mM EDTA, 0.1 ml of 10 mM sodium azide, 0.1 ml 6.3 mM oxidized glutathione, and 0.1 ml of tissue extract. Keep for 3 minutes then 0.1 ml of NADPH (9.6 mM/l) was added. The absorbance at 340nm was recorded at intervals of 15 s for 2–3 min for each series of measurement blanks done in the system devoid of tissue extract. The enzyme activity was defined as moles of NADPH oxidized/min/mg.

Results

The physical and chemical properties of the freshwater in which C. carpio was found and analyzed for the presence of dissolved salts and toxic metals. The dissolved oxygen content was found to be 6.2 0.4 mg/l, with a neutral pH of 7.3 0.01. The total hardness of the water was determined to be 345±99 mg/l, whereas the free CO_2 concentration was calculated to be 2.1 0.12 mg/l. In the tested water, there were no residues of mercury or cadmium, though there were traces of calcium (81 88 mg/l) and magnesium (34 0.0 mg/l). In addition, the water sample contained high levels of sulfates and chlorides (Table 1). This test showed that C. carpio had not been exposed to mercury before the trial began.

Table 1. Physicochemical characteristics of water

Parameters	Values
Dissolved Oxygen	6.2 ± 0.4 mg/l
pH	7.3 ± 0.01 m
Temperature	28 ± 2°C
Total hardness	345 ± 99 mg/l
Free CO_2	2.1 ± 0.12 mg/l
Ca	81 ± 88 mg/l
Mg	34 ± 0.0 mg/l
Hg	Nil
Sulfates	112 ± 0.9 mg/l
Chlorides	234 ± 22 mg/l
Pb	Nil
Specific conductance	2340 (Micro siemens/cm) at 2°C

Acute toxicity test of mercury: The lethal concentrations of mercury exposure were checked for a time duration ranging from 24h, 48h, 72h, and 96h. It was observed that there was a significant decrease in the LC values for all the LC ranging from LC5, LC10, LC16, LC50, LC84, LC90, and LC96 ppm with a significance level of <0.05. In the case of LC50 concentration of mercury, the values gradually decreased from 8.665, 7.411, 6.312, and 5.177 ppm on the 24th, 48th, 72nd, and 96th hour exposure period respectively (Fig. 1).

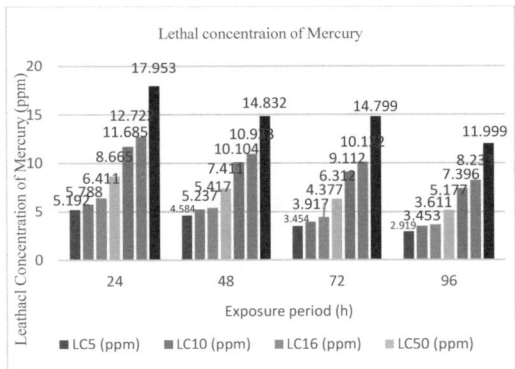

Fig. 1. The lethal concentration of mercury at different exposure times

Acute bioassay test of mercury: The lethal concentration of mercury exposure on the fish was determined using Probit Regression Analysis. As the exposure period increased, the correlation coefficient value decreased. It may be concluded that the dependent variable slope function (S) and the Chi-square values grow as the exposure duration increases (Table 2).

Table 2. Acute Toxicity test showing stress tolerance of *Cyprinus carpio* to Mercury stress.

Exposure Period (h)	Probit Regression Equation	Slope function (S)	Confidence Limit		Chi-Square Analysis
			Upper	Lower	
24	Y= −4.186 + 7.428 X	1.329	9.204	7.821	4.189
48	Y= −2.915 + 7.291 X	1.342	8.107	7.109	6.634
72	Y= −1.768 + 6.280 X	1.424	6.519	6.104	6.367
96	Y= −0.404 + 6.212 X	1.544	5.259	5.191	6.298

ROS analysis of mercury-induced *C. Carpio*.

ROS analysis of mercury revealed that the intensity of dye fluorescence in cells is directly proportional to the number of free radicals released. The mercury-induced *C. carpio* on the 28th day of exposure was observed in the liver, brain, and kidney. The ROS analysis of mercury-induced liver showed that the free radicals generated were quite less in comparison with the

control (Fig. 2a). Further, the mercury-induced brain had a higher number of free radicals compared to the control (Fig. 2b). Also, the mercury-induced kidney had lesser free radicals in comparison with the control (Fig. 2c).

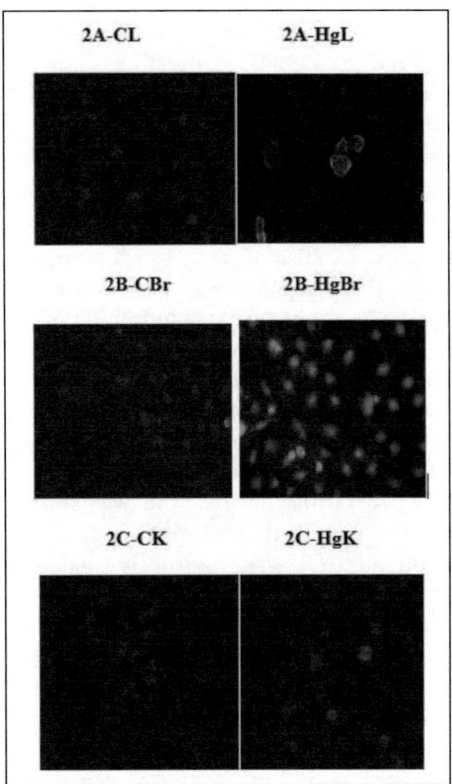

Fig. 2a. ROS analysis in *C. carpio* on the 28th day of exposure (a) In control (CL) and mercury-induced Liver (HgL) (b) In control (CBr) and mercury-induced brain (HgBr) (c) In control and mercury-induced Kidney

Estimation of Catalase Activity: The effects of mercury on catalase activity in the liver, brain and kidney of the fish *C. Carpio* are shown in Table 3. The CAT activity in the liver varied from 9.52±0.89 to 12.72±3.76, for the brain it varied from 6.18±0.52 to 7.93±5.35 and for the kidney, it varied from 8.42±0.81 to 9.78±5.63. In comparison to control fish, catalase activity increased significantly in the liver, brain, and kidney in this study. The CAT peaked in the liver and kidney after 7 days, except for the brain, which peaked after 14 days.

Table 3. Catalase activity in liver, brain, and kidney of *C. Carpio* treated with Mercury

Duration/ Time interval (Days)	Catalase (u/mg protein)					
	Liver		Brain		Kidney	
	Control	Hg	Control	Hg	Control	Hg
0	09.52±0.89	09.52±0.89	06.18±0.52	06.18±0.52	08.42±0.81	08.42±0.81
7	09.52±0.89	10.97±4.76	06.18±0.52	06.82±5.83	08.42±0.81	08.93±2.24
14	09.52±0.89	11.51±5.29	06.18±0.52	07.12±3.59	08.42±0.81	08.96±2.89
21	09.52±0.89	11.98±5.54	06.18±0.52	07.31±6.54	08.42±0.81	09.44±3.71
28	09.52±0.89	12.72±3.76	06.18±0.52	07.93±5.35	08.42±0.81	09.78±5.63

Estimation of Superoxide Dismutase activity: SOD was also measured in the fish's liver, brain and kidney to confirm the existence of oxidative stress (Table 4). The SOD activity in the liver varied from 1.41±0.14 to 2.86±0.27, for the brain it varied from 0.99±0.35 to 1.73±0.23 and for the kidney, it varied from 1.09±0.27 to 1.51±0.33. In the mercury-treated fishes, superoxide dismutase activity increased from the 7th to the 28th day in liver and brain organs, whereas SOD activity was recorded as less than normal on the 7th day for the kidney and thereafter gradually increased when compared with the control.

Table 4. Superoxide dismutase activity in liver, brain, and kidney of *C. carpio* treated with Mercury.

Duration/ Time Interval (Days)	Superoxide Dismutase (u/mg protein)					
	Liver		Brain		Kidney	
	Control	Hg	Control	Hg	Control	Hg
0	1.41±0.14	1.41±0.14	0.99±0.35	0.99±0.35	1.09±0.27	1.09±0.27
7	1.41±0.14	1.89±0.16	0.99±0.35	1.11±0.37	1.09±0.27	1.02±0.27
14	1.41±0.14	2.37±0.19	0.99±0.35	1.28±0.31	1.09±0.27	1.21±0.39
21	1.41±0.14	2.56±0.24	0.99±0.35	1.44±0.28	1.09±0.27	1.38±0.47
28	1.41±0.14	2.86±0.27	0.99±0.35	1.73±0.23	1.09±0.27	1.51±0.33

Activity of Glutathione Reductase

The effects of mercury exposure on glutathione reductase activity in fish liver, brain, and kidney were recorded from the 0th to 28th day (Table 5). The Glutathione Reductase (GR) activity in the liver varied from 1.31±0.54 to 1.62±0.69, for the brain it varied from 0.72±0.16 to 0.96±0.26 and for the kidney, it varied from 0.97±0.89 to 1.35±0.84. The GR activity increased gradually in all three organs in comparison with the control.

Table 5. Glutathione Reductase activity in liver, brain, and kidney of *Cyprinus carpio* treated with Mercury.

Duration/ Time Interval (Days)	Glutathione Reductase (u/mg protein)					
	Liver		Brain		Kidney	
	Control	Hg	Control	Hg	Control	Hg
0	1.31±0.54	1.31±0.54	0.72±0.16	0.72±0.16	0.97±0.89	0.97±0.89
7	1.31±0.54	1.44±0.75	0.72±0.16	0.78±0.32	0.97±0.89	1.14±0.76
14	1.31±0.54	1.53±0.60	0.72±0.16	0.80±0.21	0.97±0.89	1.22±0.91
21	1.31±0.54	1.55±0.61	0.72±0.16	0.81±0.37	0.97±0.89	1.33±0.89
28	1.31±0.54	1.62±0.69	0.72±0.16	0.96±0.26	0.97±0.89	1.35±0.84

Discussion

Mercury induces considerable alterations in *C. carpio*, according to the conclusions of the study. This research will help establish the negative consequences of mercury on fish, which is a rare occurrence. As ROS levels rise, the biological system adjusts the activity of antioxidants including catalase, superoxide dismutase, and glutathione-related enzymes to produce the first line of defense. Mercury can cause oxidative stress by increasing the production of a large number of reactive oxygen species in the body (ROS). In the liver, heavy metal exposure stimulates antioxidant enzymes like SOD and CAT (Awasthi et al., 2018). The activity of these enzymes is also reduced after exposure to heavy metals (Zhao et al., 2020). Antioxidant and detoxifying activities are also seen in GSH. From the 7th to the 28th day of exposure in this study, the CAT, SOD, and GR levels began to rise. In addition, the liver was ranked higher than the brain and kidney in CAT analysis. SOD activity was also shown to be higher in the liver than in the brain and kidney. In addition, the GR activity revealed that the liver had higher amounts than the rest of the body. Reduced glutathione and GSH-related enzymes play a key role in free radicals as well as other cellular processes. These could also be implicated in mercury detoxification, with glutathione being functional in many situations. GSH may be oxidized to GSSG during mercury reduction, which would typically be regenerated back to GSH by glutathione reductase activity using NADPH as an electron donor. Cell biomarkers may be used to detect early signs of biological changes caused by chemical toxins, as well as to anticipate or diagnose long-term toxicological or environmental effects. As a result, CAT, SOD, and GR activity could be used as biomarkers of metal pollution and integrated into a biomonitoring program to assess ecological health in metal-polluted areas.

Metal deposition in fish tissues indicates that metals with high absorption but low clearance rates in fish tissues are likely to be acquired at greater levels. Higher levels of mercury could even cause neurodegenerative disorders like Parkinson's disease (PD) as it blocks the P-450 enzymatic process. The levels of trace components of mercury in the brain, kidney, and liver of C carpio were measured using inductively coupled plasma-mass spectrometry (ICP-MS). As a result, mercury can enter the food chain and concentrate in fish, which are at the top of the pecking order and have a proclivity for accumulating mercury. Mercury contamination in marine products has become a major global problem. To determine the synergistic effects of metals at the molecular level, a comprehensive investigation with vast data sets, including genomes and proteomics data, is necessary.

Conclusion

This study shows that mercury has a significant impact on *C. carpio*'s liver, brain, and kidneys. This is the first study to look at many endpoints in a single fish species after mercury exposure, including LC50, enzymatic and non-enzymatic antioxidants, and oxidative damage in different organs. However, there are significant differences across teleports due to several physiological and ecological characteristics. Both acute and chronic toxicity studies revealed significant flections, ROS generation, antioxidant enzymes (CAT, SOD, and GR), as well as morphological abnormalities. This study could also be used to assess pollution stress in aquatic environments and their inhabitants, enabling the formulation of policies and activities to reduce pollution flow.

References

Garai, P., Banerjee, P., Mondal, P., and Saha, N. C. (2021). Effect of Heavy Metals on Fishes: Toxicity and Bioaccumulation Journal of Clinical Toxicology Effect of Heavy Metals on Fishes: Toxicity and Bioaccumulation, Artic. J. Clin. Toxicol., vol. 11, no. June, p. 1, [Online]. Available: https://www.researchgate.net/publication/353848075.

Zheng, N. A., Wang, S., Dong, W. U., Hua, X., Li, Y., Song, X., and Li, Y. (2019). The toxicological effects of mercury exposure in marine fish. Bulletin of environmental contamination and toxicology, 102(5), 714–720.

Federation, W. E. (2012). APHA, AWWA, WEF. "Standard Methods for examination of water and wastewater," An. Hidrol. Médica, vol. 5, no. 2, pp. 185–186.

Probit Analysis, 3rd ed. By D. J. Finney, Cambridge University Press, 32 E. 57th St., New York, NY 10022, 1971. xv + 333 pp. 14.5 × 22 cm. Price \$18.50," J. Pharm. Sci., vol. 60, no. 9, p. 1432, 1971, doi: https://doi.org/10.1002/jps.2600600940

Maehly, C., and Chance, B. (1954). The assay of catalases and peroxidases., Methods Biochem. Anal., vol. 1, pp. 357–424, doi: 10.1002/9780470110171.ch14

Kakkar, P., Das, B., and Viswanathan, P. N. (1984). A modified spectrophotometric assay of superoxide dismutase. Indian J. Biochem. Biophys., vol. 21, no. 2, pp. 130–132, Apr. 1984.

Awasthi, Y., Ratn, A., Prasad, R., Kumar, M., and Trivedi, S. P. (2018). An in vivo analysis of Cr6+ induced biochemical, genotoxicological and transcriptional profiling of genes related to oxidative stress, DNA damage and apoptosis in liver of fish, Channa punctatus (Bloch, 1793), Aquat. Toxicol., vol. 200, pp. 158167, doi: https://doi.org/10.1016/j.aquatox.2018.05.001.

Zhao, Y., Chen, Y., Zheng, Y., Ma, Q., and Jiang, Y. (2020). Quantifying the heavy metal risks from anthropogenic contributions in Sichuan panda (Ailuropoda melanoleuca melanoleuca) habitat. Sci. Total Environ., 745, p. 140941, doi: https://doi.org/10.1016/j.scitotenv.2020.140941

$(M^{II}Al\text{-}CO_3)$ Based Layered Double Hydroxides: Synthesis and Photocatalytic Efficiency Against Anionic Dye

Savita Soni,[a,b] Sonika Kumari,[a,b] Ajay Sharma,[*a,b] and Anil Kumar Sharma[*,c]

[a]Department of Chemistry, Career Point University, Tikker-Kharwarian, Hamirpur, Himachal Pradesh, India
[b]Centre for Nano-Science and Technology, Career Point University, Tikker-Kharwarian, Hamirpur, Himachal Pradesh, India
[c]Department of Biotechnology, Maharishi Markandeshwar (Deemed to be University), Mullana, Ambala, Haryana, India
E-mail: [*]ajayansu1@gmail.com; anibiotech18@gmail.com

Abstract

The use of dyes in different sectors is rapidly becoming a more common concern due to these molecules in wastewater, which continues to be a critical public health issue as well as environmental impact. This presents a substantial challenge to standard water treatment systems that are currently in place. Layered double hydroxides (LDHs) belong to the group of anionic clays with large anion exchangeability, surface area, and regeneration ability, that have been utilized as an adsorbent to remove a wide range of contaminants. These materials are important from a green chemistry point of view because of having numerous advantages, including cost-effective and simple synthesis processes, facile separation from the reaction mixture, easily available precursors, mild reaction conditions of synthesis, and recycling potential. In this study, M^{II} metals composed $M^{II}/Al\text{-}CO3$ LDHs were synthesized by the co-precipitation method applied for photocatalytic degradation of anionic [Methylene Orange (MO)] dye under UV light exposure.

Keywords: Layered Double Hydroxides, Photocatalyst, Absorption, Green Chemistry

Introduction

Owing to rapid increase in use of dye materials and generation of the industrial waste, various adsorbents like metal oxides, carbon-based materials, polymer-based materials, bio-adsorbents, and metal-organic frameworks are used for the absorption/degradation of these pollutants from environmental segments. However, the invention of clay-based hybrid materials has a wide range of practical uses that can effectively remove contaminants/dyes from wastewater. LDHs/HTs (Hydrotalcite- Type anionic clays) are a type of anionic clays with a large surface area, anion exchangeability, and regeneration ability (Kumari et al., 2022a). LDHs materials have been efficiently employed as an adsorbent to remove a wide range of contaminants from wastewater. These are depicted by following the general formula

$$[MII{+}1{-}x\ MIII{+}(OH)(An^-)]\cdot yH\ O$$

Their structure is made up of brucite-like layers of divalent (M^{II}) and trivalent (M^{III}) metal cations (positively charged) as well as charge-balancing anions (A^{n-}) and water (H_2O) molecules in the interlayer space. The main advantages of these materials are the possibility to synthesize the different combinations of divalent (M^{II}= Mg, Zn, Co, Li, Cd, and Ni) and trivalent (M^{III}=Al, Cr, and Fe) metals as well as interlayer anions (CO_3, NO_3, Cl) (Benicio et al., 2018; Kumari et al., 2022b). Their utility becomes broad-spectrum due to the numerous structural features such as Lewis- Bronsted basic and acidic sites, self-healing variable basal spacing, high ion mobility, and low anion selectivity Besides, these materials also have a uniform distribution of MII and MIII metal cations in the brucite layer, different type of anions intercalation ability, surface hydroxyl groups, sustained/controlled release capability of intercalated anions, flexible tunability, high thermal and chemical stability. (Wang et al., 2012). This study investigated the performance of photodegradation of anionic dye [Methylene Orange (MO)] using M^{II}/Al- CO_3 LDHs composed of different M^{II} metals (Mg, Zn, Co, Li, Cd, and Ni; M^{II}/Al atomic ratio of 3:1) at different concentrations.

Materials

Cadmium Nitrate Hexahydrate, Cobalt Nitrate Hexahydrate, Lithium Nitrate Magnesium Nitrate Hexahydrate, Nickel Nitrate Hexahydrates, Zinc Nitrate Hexahydrate, Sodium Carbonate, Aluminium Nitrate Nonahydrate, Sodium Hydroxide were obtained from LOBA Chem. Methyl Orange (MO) and NaOH were purchased from Merck life science.

Methods Preparation of LDHs

M^{II} Al -CO_3 LDHs with (M^{II}: Al molar ratio = 3:1) were prepared through the co-precipitation method. Dropwise addition of MII metal salt solution (0.3M) M and Al(NO_3) 3·9 H_2O solution (0.1 M) into the mixture of NaOH solution (1 M) and Na_2CO_3 solution (0.5 M) carried out by maintaining pH to a range of 10-11 with constant stirring without heating for 2 h. After that, the reaction mixture was further stirred for 10 h under heating at 65 ± 5°C. After cooling, the precipitates were filtered followed by washing with distilled water till pH 7. Then the wet solid was dried at 110 ± 2°C for 24 h. The obtained material is known as MIIAl-CO_3 LDH (Bharali et al., 2015). In the same way, other samples were prepared with the substitution of the divalent metal ion by using different metal salts and the prepared samples were named Mg/Zn/Ni/Co/Li/CdAl-CO_3 LDHs.

Materials Characterization

The characterization of synthesized LDH samples was done on basis of X-Ray diffraction (XRD) and Scanning Electron microscopy (SEM) analysis. XRD diffractogram of different LDHs was recorded on "Rigaku corporation smart lab 9 kW rotating anode X-ray diffractometer Philips" by using 2θ values from 10 to 90° with Copper K-α radiation [λ (wavelength)= 1.54 Å]. The surface morphology of LDHs samples was determined by SEM, model: "Nova nano SEM-450". JFEI USA (S.E.A.) PTE LTD.

Photocatalytic Activity

The photodegradation of MO dye solution (0.1 L of 10 ppm) was studied by using LDHs (15, 30 & 45 mg) under UV light exposure for 150 min. The photocatalytic degradation of anionic (MO) dye was evaluated by measuring the change in absorbance using UV-Visible spectrophotometer (MO) along with the rate of degradation.

Results and discussion XRD

The XRD patterns of the LDH samples (Mg/Zn/Ni/Co/Li/Cd Al-LDHs) synthesized with MII: Al molar ratio=3:1, are similar to those shown by LDHs intercalated with carbonate anions as published earlier (Kumari et al., 2022b). The diffraction peaks were found to be more intensive and sharper reflections at (003) and (006) planes with 2θ = 11-23° (Fig 1) which indicates a structure of hexagonal lattice system of rhombohedral 3R symmetry (Radjii et al., 2018). The interplanar distance/basal spacing of the plane (003) and is equal to the sum thickness of inter-lamellar space (~2.8 Å for CO3 anions) and brucite layer (~ 4.8 Å). First peak of doublet

2θ value ~ 60° of is owing to diffraction of (110) planes that co-relates the brucite-like layers nearest Metal- Metal distance. The basal spacing (d) and crystallite size (D) have been calculated by using Bragg's and Scherrer's equations, respectively as shown in Table 1 (Mahjoubi et al., 2017; Kumari et al., 2022b).

Table 1. Basal spacing (d) and Crystallite size (D) of the different LDHs

Divalent Metal (MAl-CO3) LDHs	2θ	d (Å)	D (nm)
Ni	11.4	7.71	9.59
Zn	11.7	7.52	37.72
Mg	11.2	7.84	18.52
Co	11.6	7.56	28.97
Li	11.5	7.66	6.67
Cd	11.7	7.54	44.29

SEM-EDX

The average particle size and morphology of different LDHs (3:1) examined through FESEM. FESEM images of various CO3 intercalated Table 2: Average Particle Size (nm) of the different LDHs

Table 2: Average Particle Size (nm) of the different LDHs

Divalent Metal (MAl-CO$_3$) LDHs	Average Particle Size (nm)	Shapes
Ni	40.31	Spheroidal
Zn	103.09	Rod
Mg	34.92	Spheroidal
Co	63.31	Spheroidal
Li	81.64	Plate
Cd	94.55	Spheroidal

LDHs and these sheets are composed of particles with different average particle sizes of 34.92–103.09 nm with spheroidal, rod, and plate shapes as shown in Table 2.

Photocatalytic activity

The photocatalytic degradation of anionic (MO) dye was performed with a change in concentration of LHDs (as a catalyst) with time. The adsorption characteristics of MO absorbed at 464nm were reported to be decreased continuously over time for all the samples and almost complete degradation of dye within 150min was observed by using catalyst of different concentrations. From Table 3, it has been observed that the increase in the concentration of LDHs increases the % degradation of dye. The % anionic (MO) dye degradation has been calculated through the following formula

$$\% \text{ degradation} = [1 - C/C_0] \times 100\%$$

C_0 = absorbance of anionic pure dye solution,

C = absorbance of the reaction mixture (dye and catalyst) at a time, t.

Table 3. % degradation of MO dye at different concentrations of the LDHs LiAlCO$_3$ showed the highest (94%) degradation at 45mg concentration of the catalyst while an average degradation was shown by NiAlCO$_3$ LDH.

Divalent Metal (MAl-CO$_3$) LDHs	Concentrations (mg)	Degradation (%)
Ni	15,30,45	75,83,93
Zn	15,30,45	11,36,58
Mg	15,30,45	26,56,87
Co	15,30,45	25,48,69
Li	15,30,45	29,64,94
Cd	15,30,45	24,52,89

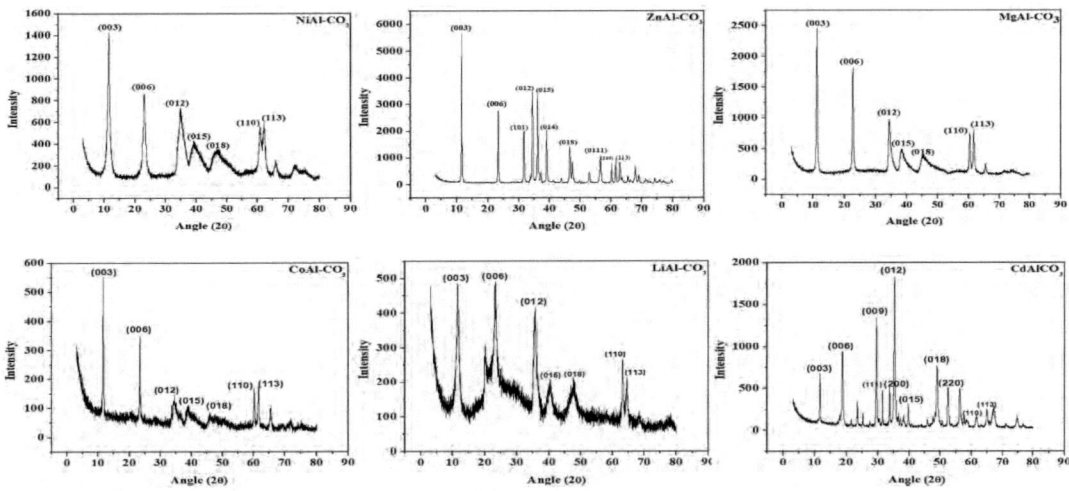

Fig. 1. The patterns of different M^{II}/Al-CO$_3$ LDHs in XRD diffractogram

Fig. 2. Scanning Electron Microscopy images of different M^{II}/Al-CO$_3$ LDHs

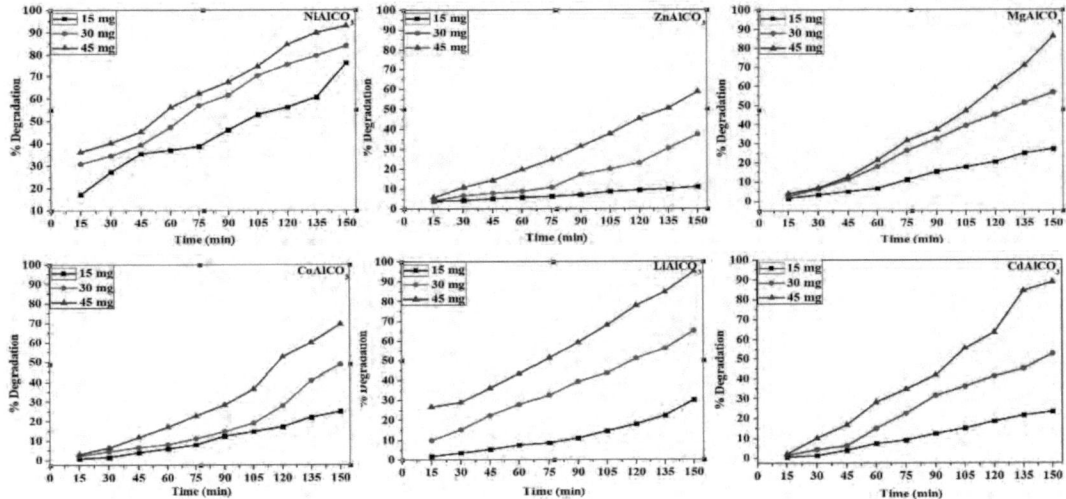

Fig. 3. % degradation of dye by different M^{II}/Al-CO$_3$ LDHs

Conclusions

The crystallite size and average particle size of prepared M^{II}/Al- CO$_3$ LDHs were observed between the range of 6.67 to 44.29 nm and 34.92 to 103.09 nm, respectively. The photocatalytic activity was observed up to 94% in the presence of UV light. NiAl-CO$_3$ LDH revealed the highest average % degradation (83%). Furthermore, the increase in LDHs concentration increases the rate of photodegradation.

Acknowledgments

The authors would like to gratefully acknowledge the Centre for Nano-Science and Technology, Career Point University, Tikker-Kharwarian, Hamirpur, Himachal Pradesh for providing the facility for this study.

References

Benício, L. P. F., Eulálio, D., Guimarães, L. D. M., Pinto, F. G., Costa, L. M. D., and Tronto, J. (2018). Layered double hydroxides as hosting matrices for storage and slow release of phosphate analyzed by stirred-flow method. Mater. Res. 21(6), 1–13.

Bharali, D., Devi, R., Bharali, P., and Deka, R. C. (2015). Synthesis of high surface area mixed metal oxide from the NiMgAl LDH precursor for nitro-aldol condensation reaction. New J. Chem. 39(1), 172–178.

Kumari, S., Sharma, A., Kumar, S., Thakur, A., Thakur, R., Bhatia, S. K. and Sharma, A. K. (2022a). Multifaceted potential applicability of hydrotalcite- type anionic clays from green chemistry to environmental sustainability. Chemosphere. 306:135464.

Kumari, S., Thakur, N., Kumar, R., Thakur, R. C., and Sharma, A. (2022b). Effect of Synthetic Parameters on Crystallinity of Hydrotalcite-Like Anionic Clays with Elucidation and Identification through X-Ray Diffraction Analysis. ECS Trans. 107, 18903–18921.

Mahjoubi, F. Z., Khalidi, A., Abdennouri, M., and Barka, N. (2017). Zn-Al layered double hydroxides intercalated with carbonate, nitrate, chloride and sulphate ions: Synthesis, characterisation and dye removal properties. J. Taibah Univ. Sci. 11(1), 90–100.

Radji, G., Bahmani, A., Ezziane, K., Bettahar, N., and Sellami, M. (2018). Synthesis, Characterization, and Applications of New Hydrotalcite-Like Nano and Innovative Materials. Der Pharma. Chemica. 10(7), 60–66.

Wang, Q., and O'Hare, D. (2012). Recent advances in the synthesis and application of layered double hydroxide (LDH) nanosheets. Chem. Rev. 112(7), 4124–4155.

Millimeter Wave Communication in Autonomous Vehicles

M. Shravan, B. Sneha, and P. Vijayakumar*

Vellore Institute of Technology, Chennai, Tamil Nadu, India
E-mail: *shravanm2k@gmail.com, snehabalasubramanian2001@gmail.com, vijayrgcet@gmail.com

Abstract

Driving has become more automated and vehicles are being embedded with a variety of sensors. A lot of sensors are deployed to enhance autonomous drivability and safety. Technologies for vehicles, such as on-board hardware and sensors, are also evolving rapidly. These sensors will provide a lot of information to the vehicle and can also be used to guide the neighboring autonomous vehicles as they drive along. However, the high data transfer rates required by the connected vehicles for data transmission are above the capabilities of our current techniques, such as DSRC and 4G. This led to mm wave communications, which will support the communication between vehicles using high data transmission rates. Also, Vehicle's sensors might fail or can't sense during the night times, mm waves can be used to overcome such sensor failures. The expected outcome of this project is to observe the mm wave connection of autonomous vehicles.

Keywords: Mm wave, sensors, autonomous vehicles

Introduction

Autonomous vehicle technology is increasing rapidly in terms are drivability and safety. The primary technology that is employed in autonomous vehicles are radars, Lidars, ultrasonic sensors, and visual cameras. These technologies can be limited to sensing range and information about things that are not in the line of sight, this can limit the autonomous capabilities of the vehicle for improved safety. As to improve the safety of the vehicle, vehicle connectivity (V2V) can be introduced which has two possible benefits: first, cars can interact in non-line-of-sight settings if a suitable carrier frequency is chosen. Second, if higher bandwidth is implemented, the data rate transmission can be increased, resulting in a "fully intelligent connected autonomous vehicle." Raw sensor data collected by an individual vehicle can be processed and shared individually with another vehicle. By sending this information, adaptive platooning of automated vehicles can be implemented, to ease drivability and improve safety. The primary goal of this project is to establish a faster way of short-range communication between autonomous vehicles which is Mm Wave communication.

The current technologies such as 4G are restricted to very high data transmission rates (Gayatri Chittimoju, et al., 2021). Mm wave can operate from 30-300 GHz which is 10–100 times higher than the radio frequencies used in current 4G technologies, wavelength of the mm wave ranges between 1-10mm and a date rate of 10 Gigabits per second as compared to 1 Gigabit per second of the radio wave. 5G consists of mm-Wave and sub-6GHz. Sub-6GHz are the mid-bands that have an operating frequency of less than 6 GHz. Our focus is mainly on the mm wave aspects. Considering the 1-10mm wavelength, smaller antenna dimensions can be achieved to implement MIMO.

Beamforming (Roh et al., 2014) is a technology that helps in deciding the direction of the transmission and reception signals (it can target a particular direction towards a reception device). It is one of the main phenomena that makes the 5G connection more focused which results in massive MIMO. Beamforming uses a phased array antenna system to focus the direction and to make a reliable connection. The reason why beamforming is used in millimeter technology is that millimeter waves cannot penetrate huge walls of buildings and other tough barriers the way mid-bands of 4G did. Phased array antennas are employed to improve the signal strength and are designed in such a way that the radiation patterns from each element combine constructively with those from the neighboring elements forming an effective radiation pattern. This radiation pattern creates a main lode and side lobes. The main lobe decides the direction of the desired signals. The undesired directional signals create the side lobes. The overall antenna array system is designed to maximize the energy radiated in the main lobe eliminating the energy in the side lobes. The aforementioned massive MIMO improves the throughput and connection density of 5G network.

Mm Wave Radar: In recent years, researchers have become increasingly interested in human motion behavior monitoring, especially in surveillance, tracking, and patient monitoring. Currently, cameras and infrared sensors are the only methods available. Even though cameras, including thermal cameras, are accurate and reliable for surveillance, there are disadvantages to their use - they leak private information and are dependent on lighting conditions. As we need an accurate and comprehensive sensor to monitor human behavior, we are using micro-Doppler signatures resulting from mm-Wave radar to detect people's movement.

Autonomous vehicle and ADAS technology: Thinking of future technology in automobiles strike us with the autonomous function.

Autonomous cars are smart cars anticipated to be driverless, efficient, and crash-avoiding ideal urban vehicles. The translation of conventional cars to autonomous cars presents a challenge. Meeting the challenges leads to tremendous technological advancement.

An autonomous vehicle can perceive its environment, choose its route to its destination, and drive itself, but the technology does not stop there. Sensor-based features such as blind spot detection, cruise control, automatic emergency braking, and lane-keep assist, assist the driver in driving more safely. In other words, autonomous vehicles can be advert to smart cars or robots that rely on various technologies.

A wide range of sensors, computer processors, and cloud-based services such as maps, are used to automate some or all of the driving functions currently performed by humans. Crash rates, energy consumption, and pollution will likely be reduced. Several OEMs have announced their plans to sell vehicles with ADAS (Advanced driver assistance system) in the future years.

ADAS is the term used for autonomous vehicles since the technology is more focused on the driver. ADAS is sub-divided into five levels of autonomy. Level 1 deals with based assistance such as cruise control and lane assistance. Drivers need to be fully engaged and responsible for the control of the vehicle.

The level 2 automation system extends to adaptive cruise control and lane switching where some small work can be done by the car itself. Level 3 integrates significantly high computational sensors and cameras to enable parking assistance and traffic jam assistance. Level 4 is when the car can drive itself up to a certain distance. Level 5 refers to a fully autonomous system in which the car will drive independently of its driver and all functions are controlled by the system. This level of autonomy is currently in development.

Related Works

The millimeter-wave spectrum is capable of supporting multi-gigabit transmission, according to Linghe Kong et al. (2017). With this system, vehicles can share sensory data, which helps them fight blind spots and bad weather. Cloud computing is being used to recognize human gestures and small objects using HD video communication between V2I devices. Marco Giordani et al. (2017) summarized the current state of mm-Wave MAC protocols in the context of vehicular applications by illustrating the challenges that arise in an automobile environment. MM-Wave automotive communication systems must include mechanisms to compensate for the large isotropic path loss at higher frequencies caused by higher frequencies. Certain automotive-specific variables, including the vehicle's speed, beam tracking frequency, node density, and MIMO antenna layout, must also be taken into account.

In their study, Choi et al. (2016) stated that existing vehicular communications technologies, such as DSRC and 4G cellular systems, will not be enough for future connected vehicles that want to communicate raw sensor data on a wide scale. mm-Wave can be used in future vehicular networks in three ways: 5G cellular, a modified version of IEEE 802.11ad, or a dedicated new standard. Multiple mm-Wave transceivers are required in automotive contexts to avoid blockage and achieve greater spatial packing. Deluka Tibljas et al. (2018) stated that, based on microsimulation tools, illustrations were made on how different levels of autonomous vehicles contribute around 10–50% of the total vehicle population. He has concluded that the different autonomous vehicles will contribute to a significant change in traffic and safety measures at a single-lane roundabout. He performed the simulation using VISSIM software for two suburban roundabouts (Daniel et al., 2015) feels that the upcoming automation technology has the potential to reduce accidents, and hence it can also enhance the fuel efficiency of the vehicle. Martin-Vega et al. (2018) MmWave vehicular communications are the only viable approach for enabling multi-gigabit connectivity for autonomous and non-autonomous vehicles. Exchanging raw sensor data with surrounding vehicles, giving cloud computing capabilities to vehicles, and having a bird's eye view are some of its key applications. They looked at cutting-edge technology at the physical and MAC levels that can

meet autonomous driving's rigorous requirements, such as low latency, high dependability, security, and broadcasting capabilities. Prediction of the combination of analog/hybrid beamforming with a location-based beam search protocol, using full-duplex radios and physical layer key generation, is capable of meeting autonomous driving requirements (Tassi, A. et al., 2017).

The efficacy of an mm-Wave network deployed along a highway segment was investigated and determined to be satisfactory. The author developed a novel theoretical framework for analyzing the SINR outage probability and rate coverage probability of a user who is surrounded by heavy cars using the other highway lanes. According to (Rojas-Rueda, et al., 2020), autonomous vehicles are a cutting-edge form of transportation that will affect the public's health.

AVs could present significant prospects for public health when they are implemented as entirely electric, in a ridesharing format, and integrated with public and active transportation modes, in addition to the benefits projected to be related to traffic safety. Increase sedentarism and the destruction of the urban environment, as well as worsen traffic congestion and public space utilization. Aiming to priorities more disadvantaged communities and advance urban and transportation planning in the direction of a better urban environment, the installation of AVs should encourage public and active transportation.

Cao Siyang et al. (2018) said this research introduces a mm-Wave radar-based system for tracking human mobility behavior in real time. It can function as a ward in an area of interest to gather data on human motion behavior. Future research may use this technology to classify vehicles, pedestrians, and cyclists using automobile radar. For a better understanding of the surroundings and detection, the network can also be expanded to include more human motion behavior.

According to Rajasekhar et al. (2015), autonomous cars appear to have a lot of potentials. However, the secret is in the safe application of this technology, which may be accomplished by using strict guidelines and regulations. On the other hand, by significantly lowering travel stress, this technology will be highly helpful in resolving many traffic-related issues such as parking and traffic congestion, and accidents. For military purposes, it can significantly cut down on casualties and play a crucial role in minesweeping, search operations, and troop supply. If the link between the vehicle and the controller or the grid is lost, there should be adequate fail-safe plans.

Objectives of this Project:

- To create a wireless mm wave communication between two autonomous vehicles to monitor the various communication conditions between the vehicles.
- To improvise high data transfer between the vehicles.
- To compare the efficiency between radio frequency and mm wave communication frequency.
- To establish an mm wave communication using NS-3 software in ubuntu using a tool called "millicar ()."
- To instruct the nodes to act according to the server's instructions i.e., observe the communication.
- To calculate the data transferred, throughput, and other required necessary network parameters.
- To show the various parameters of the field in real-time.

Mm Wave air interfaces can be seen to the left, while FPGA processing can be seen on the right (Fig. 1). A baseband element typically consists of high-speed Analog-to-digital and digital-to-Analog converters between the up and down converters.

Proposed mm Wave Communication System for Autonomous Vehicle

In this project, we have taken five scenarios to explain the mm wave communication between the autonomous vehicles using NS-3 simulation software.

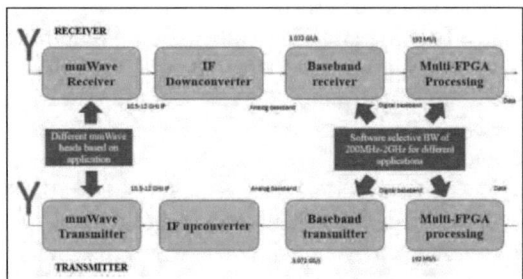

Fig. 1. Block diagram representation of mm wave

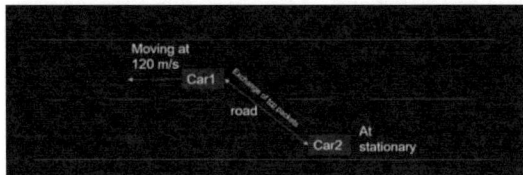

Fig. 2. Block diagram representation of scenario 1

1. Scenario 1

The Car's carrier frequencies are taken as 70 GHz Speed of car is 120m/s, Speed of car 2 is 0 (at stationary), Initial distance between the cars is 1000 m and the final distance = 0 as show is Fig. 2.

2. Scenario 2

In this scenario, we have taken two nodes. Node 1 moving at 20m/s and Node 2 is also moving at 20m/s. Two nodes are moving in the same direction at a distance 10m between the nodes. Carrier frequency is taken as 70 GHz as show is the Fig. 3.

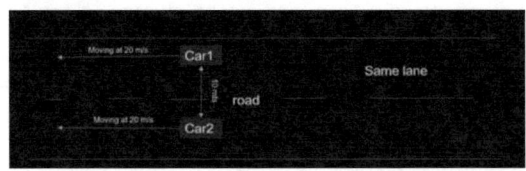

Fig. 3. Block diagram representation of scenario 2

3. Scenario 3

In this scenario, we group two nodes each group having two nodes moving in different lanes, in the same direction.

Speed parameters of each node are set at 20m/s and the Carrier frequency is clocked at 70 GHz. We check the interference between the nodes in the model as shown in Fig. 4.

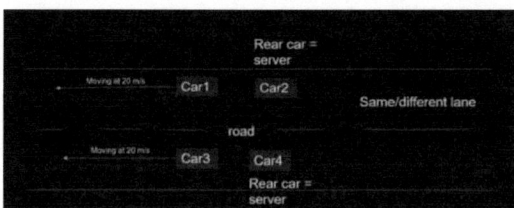

Fig. 4. Block diagram representation of scenario 3

4. Scenario 4

Take the previous scenario 3 but the cars operate at 28 GHz and the Bandwidth is increased to 100 MHz Here the rear car acts as the server which transmits the packets to the car in front of it. Each node moves at a speed of 20m/s as show is Fig. 5.

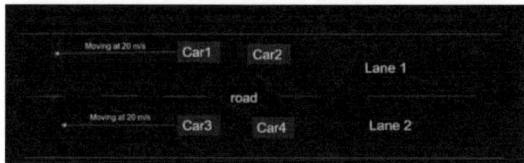

Fig. 5. Block diagram representation of scenario 4

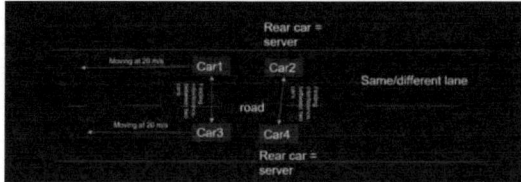

Fig. 6. Block diagram representation of scenario 5

5. Scenario 5

Extending the case scenario, we have tried to find the interference between the nodes in the groups. Cars operate at 28 GHz with a Bandwidth of 100 MHz as shown in Fig. 6.

Conclusion: 5G mm waves have a great scope in the automobile sector as compared to 4G cellular technology will not be equipollent to the current trend. In this study, we examine 5G millimeter wave communication with the demonstration of five scenarios in autonomous vehicles. The mm wave range, which includes 30 to 300 GHz, includes 70 GHz. Thus, we have established a relationship and the output is based on the scenarios. The following were the observation.

References

Chittimoju, Gayatri and Yalavarthi, Usha Devi (2021). A Comprehensive Review on Millimeter Waves Applications and Antennas. 1804(1), 21–36.

Roh, W. (2014). Millimeter-wave beamforming as an enabling technology for 5G cellular communications: theoretical feasibility and prototype results. 52(2), 106–113.

Kong, Linghe, Khurram Khan, Muhannad, Wu, Fan, Chen, Guihai, and Zeng, Peng (2017). Millimeter-Wave Wireless Communication for IoT-Cloud supported Autonomous vehicles: Overview, Design, and Challenges. 55(1), 62–68.

Giordani, M. Zanella, A., and Zorzi, M. (2017). Millimeter wave communication in vehicular networks: Challenges and opportunities. 1(1), 1–5.

Choi, V., Va, N., Gonzalez-Prelcic, R., Daniels, C., Bhat, R., and Heath, R. W. (2016). Millimeter-Wave Vehicular Communication to Support Massive Automotive Sensing. 54(12), 160–167.

Deluka Tibljas, A., Giuffre, T., Surdonja, S., and Trubia, S. (2018). Introduction of Autonomous Vehicles: Roundabouts Design and Safety Performance Evaluation. 1(1), 1060–1070.

Fragrant, Daniel J., and Kockelman, Kara (2015). Preparing a nation for autonomous vehicles: opportunities, barriers and policy recommendations, Transportation Research. 77(1), 167–181.

Martin-Vega, F. J., Aguayo-Torres, M. C., Gomez, G., Entrambasaguas, J. T., and Duong, T. Q. (2018). Key Technologies, Modeling Approaches, and Challenges for Millimeter-Wave Vehicular Communications. 56(10), 28–35.

Tassi, A., Egan, M., Piechocki, R. J., and Nix, A. (2017). Modeling and Design of Millimeter-Wave Networks for Highway Vehicular Communication. 66(12), 10676–10691.

Rojas-Rueda, D., Nieuwenhuijsen, M. J., Khreis, H., and Frumkin, H. (2020). Autonomous Vehicles and Public Health. Annual Review of Public Health. 41(1), 1–5.

Cao, S. (2018). Real-time Human Motion Behavior Detection via CNN using mmWave Radar. 1(1), 1–5.

Rajasekhar, M. V., and Jaswal, A. K. (2015) Autonomous vehicles: The future of automobiles. 1(1), 1–6.

Designing an Enhanced Self-Supported Transition Metal Complexes Based on Electrocatalysts for Hydrogen Evolution Reaction

M. Swathika[a] and N. Arunadevi[b,*]

[a]Research Scholar, Department of Chemistry, PSGR Krishnammal College for Women, Coimbatore, India
[b]Assistant Professor, Department of Chemistry, PSGR Krishnammal College for Women, Coimbatore, India
E-mail: *arunadevi@psgrkcw.ac.in

Abstract

Progressing efficient and low-cost catalysts to reduce overpotential within the hydrogen evolution reaction (HER) in large-scale production is highly desirable in the present era. The catalysts are accustomed to initiating an excellent approach to optimizing the electronic characteristics employed to boost their electrocatalytic production. In this article, we described the activity of transition-metal complexes with the nitrogenous analog of carbonic acid and naphthalene ring containing monocarboxylic acid to study hydrogen evolution reactions. The HER activity in transition metal complexes was varied through physicochemical and electrochemical properties. Density function theory (DFT) was calculated using the B3LYP hybrid functional and the basis set of 6-311 + G (d,p) to deliver the relevant frontier orbitals.

Keywords: HER, naphthoic acid, guanidium, DFT, TG-DTA, Electrocatalyst

Introduction

The development of current technologies for clean and sustainable hydrogen energy has drawn increasing attention over the past few years; hydrogen is hailed as a promising energy source to reduce our necessity for fossil fuels and benefit the atmosphere by reducing the emissions of greenhouse and other toxic gases also. To this end, an effective and promising method is used for the electrolysis of hydrogen production. Hydrogen evolution reaction (HER) of electrocatalytic preferably determined by solar energy produces a desirable methodology for these requirements, making it efficient, low-cost, and environmentally friendly (Du and Eisenberg, 2012; Natarajan Arunadevi et al., 2021; Jain et al., 2018 we report the electrocatalytic behavior of the neutral, monomeric Ni(II).

In this HER reaction, the significant role was played by the H^+ ion, the nature of the catalyst, and the electron produced by the catalyst. The catalyst not only adsorbs the free H^+ ion in acidic electrolytes but breaks the H-O-H bonding in the electrolytes before adsorbing the H^+ ion. So, the acidic electrolytes are preferred in hydrogen production due to their enough H^+ ions in the electrolyte to react with the electrode surface. Hydroxide conducting polymeric material is also used as an electrolyte still in its initial technological stages: due to its unstable and less conductive nature, it produces greater overpotentials in HER. Therefore, it has been appreciated that a metal-based acid electrocatalyst is efficient for hydrogen evolution reactions (Jain et al., 2018) we report the electrocatalytic behavior of the neutral, monomeric Ni(II. Currently, Pt-based electrocatalysts efficiently catalyze in HER, while their extensive uses have been limited due to their low abundance and high cost; later, the progress of HER catalysts that are composed of low-cost and earth-abundant elements is the main target in renewable energy in the recent years. Molybdenum and transition-metal-based compounds are used in HER catalysts, including Mo_2C, MoS_2, $MoSe_2$, MoP, etc., which show excellent activity (Ai et al., 2019; Queyriaux et al., 2015).

However, the transition metal complex was occupied with half-filled 3d orbitals and ligands with hetero atoms, which produce electrons in excited states that are active in both the ground and the excited states. In the excited state of the metal, complexes have a charge transfer reaction from metal-to-ligand or ligand-to-metal, which is the key step for electrocatalytic action. The exact mechanism of HER is as follows; the first step, known as the Volmer step, is H^+ adsorption on the catalyst's surface, followed by Heyrovsky. The proton from the solution was transferred to the metal complex catalyst surface, which absorbed hydrogen atoms to generate H_2.

This research is focused on synthesizing and characterizing metal complexes with electron-rich hetero atoms as electrocatalysts.

Materials and Methods

Chemicals and reagents were obtained commercially from Aldrich. Infrared spectra in the range 4000–400 cm^{-1} were recorded in the FT-IR 8000 Shimadzu spectrometer. The UV-visible spectral range at 180–400 nm was recorded in a Varian-Cary 5000 analyzer. The thermogravimetric analysis was recorded in EXSTAR/63000. The powder XRD for the metal complex was verified in Siemens D-500. Metrohm Autolab M204 was used to analyze the cyclic voltammetry. A three-electrode system is used, with glassy carbon as the working electrode and Ag/Ag^+ electrode as the reference electrode. Tetrabutylammonium hexafluorophosphate was used as the conductive salt. The resulting solution was purged with nitrogen for 10 minutes under a nitrogen atmosphere at a scan rate of -0.15 to +1.0 V until a constant current was obtained. Then acid was added until the current was saturated, a variation in the scan rate (Swathika and Natarajan, 2022).

Synthesis of Selected Electrocatalyst

Figure 1 shows the procedure for the preparation of the ligand in which the acid and base are mixed in a 1:2 ratio by adding 1-naphthoic acid (0.172 g, one mmol) and guanidine (0.0900 mg, two mmol) to the 40 ml of double distilled water and heated in a water bath to dissolve completely. The resulting solution was left undisturbed for 24 hours to attain precipitation and then washed with ethanol to remove impurities and dried. The ligand prepared in the previous step was added dropwise to 10 mL of an aqueous solution containing the Ni (II) nitrate (0.29 g, one mmol) with continuous stirring. The microcrystalline product was obtained immediately, and the solution was stored undisturbed for one day. The product was filtered off, washed with water to remove impurities, and dried at room temperature (Arunadevi et al., 2021).

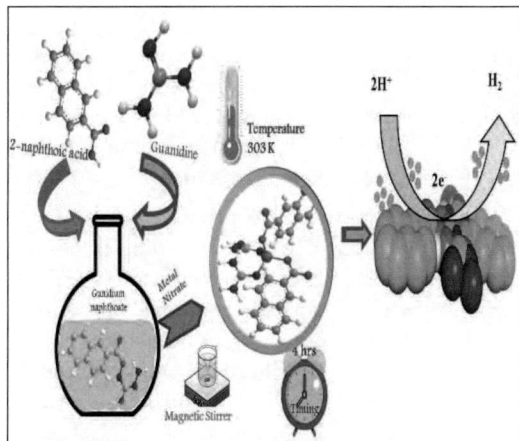

Fig..1. Synthesis of electrocatalyst

Result and Discussion

Elemental Analysis

The proposed molecular formula of the metal complexes agrees with the elemental analysis and their theoretical values. Anal. Calcd. for the $[Ni(CH_5N_3)_2 \{C_{10}H_7(1COO)\}_2.2H_2O]$, was Found:Carbon 52.11;Hydrogen 4.74; Nitrogen 15.19; Oxygen 17.35;Hydrazine 11.3; Metal 10.4,Calcd Value: Carbon 50.09; Hydrogen 4.70; Nitrogen 15.12; Oxygen 17.30; Hydrazine 11.09; Metal 10.3; Colour light green; M.pt/ 216 ℃ (Pervaiz et al. 2019)

Vibrational Spectra

The peak in the region 3600–3000cm⁻¹ is due to the stretching vibrations of NH and OH groups. The asymmetric stretching frequency occurs from 1326–1352 cm⁻¹ due to the carboxylate group presenting strong- to medium-intensity, while the other symmetric stretching was recorded between 1329 and 1355 cm⁻¹. The presence of guanidinium structures was recognized with strong intensity bands around 1700–1655 cm⁻¹, allocated to the δ NH and δ CNH vibrations. N-N stretching frequency of ligand arose between 1111 and 1086 cm⁻¹, but in the synthesized complex, the rage of –NH₂ group was shifted to higher wavenumbers, 1144 and 11734 cm⁻¹ (Arunadevi and Vairam 2009; Mohan et al. 2020).

Electronic Spectra and Optical Band

The nickel metal complexes show the strong, intense band at 291 nm representing the π→π*, and 324 nm due to n→π*, which were assigned by $^3A_{2g}(F)$®$^3T_{2g}(F)$ and $^3A_{2g}(F)$®$^3T_{1g}(P)$ and confirms the distorted of octahedral geometry.

The optical bandgap has been predictable using Tuac"s equations In which hv produces photon energy, α refers to absorption coefficient, A relates to constant relative to the material, and n may be a direct or indirect transition. The optical band gap E_g of absorption peak can be found by extrapolating the linear to the (αhv)ⁿ –hv arc to zero. The attained band gap is 4.8 eV (Enhessari et al., 2013).

Thermal Analysis

The TG-DTA thermogram data for the nickel complex undergoes decomposition in a single step with endotherms at 115–382 °C. The final precipitate formed in the complexes was metal oxides. The last mass loss is attributed to the complete decomposition of the organic moiety to develop the unstable intermediate.

X-ray Powder Diffraction

The significant peaks have been computed and analyzed with the help of a computer program. It is represented by the miller indices *hkl*, and the indexing was calculated by 2θ values. The crystal nature of the metal complexes confirmed the distorted octahedral geometry. The average particle sizes were, determined using Debye-Scherrer equation shown below, ranging from 25–80 nm.

Electrocatalytic Performance

The electrochemical properties were measured by cyclic voltammetry in the range of –0.15 to +2.0 V vs. RHE. The complex exhibits a reversible Ni (III)/Ni (II) oxidation–reduction peak shown in Table 1 and Fig. 2, demonstrating the coordination environment of the Ni (II) ion and Ni (III) ion. The anodic reaction of $H_2O \rightarrow O_2$ and cathodic reaction at $H_2O \rightarrow H_2$ improved the catalysts suitable for HER response.

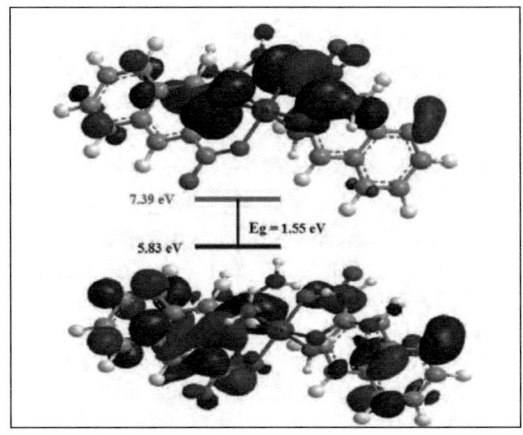

Fig. 2. Cyclic Voltammetry

LSV test under three-electrode exposed the fundamental electrochemical performance of the materials as expected. In Figure 2b, slight initial overpotential produces 98 mV and current density -3.0 V vs. RHE, respectively. The Tafel plot obtained from the LSV curve is suitable for the rapid catalytic kinetics of HER evolution reaction; they depend on the Volmer-Heyrovsky equation (Zhang et al., 2020).

Computational Analysis (DFT)

DFT calculated the computational study for Ni (II) metal complex. The optimized metal complexes structure runs under the Gaussian-09 software in B3LYP/6-311G(d,p). The energy-optimized frames have been displayed in Fig. 3. The HOMO and their LUMO band gap energy range from 5.83–7.39 eV during π-π^* electron due to the interaction between ligand and center metal ions. The calculated values of the HOMO–LUMO gap with the presence of the hetero atoms can be ascribed to the steadiness imparted to the LUMO and slight dislocation of the HOMO. The band gap energy implies negative values due to the metal complexes' stability. ΔE energy gap due to the transition state from HOMO→LUMO is 1.55.

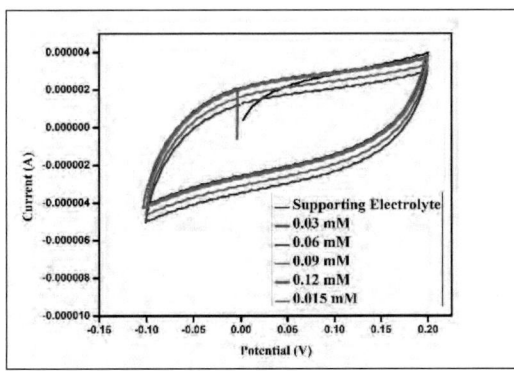

Fig. 3. DFT for optimized molecular structure

Conclusion

In conclusion, we have summarized a new group of metal complexes as an electrocatalyst with high activity in HER. The formulation of the metal complex was determined by using elemental analysis. Vibrational spectroscopy is used to specify the coordination mode of the metal complexes through nitrogen and oxygen atoms. The Uv-visible spectroscopy confirms the distortion of octahedral geometry. The optical band gap E_g is represented by 4.8 eV. The powder XRD is used to measure the diffraction pattern of the crystalline nature of the compound. The average particle sizes ranged from 25-80 nm. The thermal stability of the metal complex was evident from TG-DTA analysis, and the degradation region is from 115- 382 °C. The final precipitate formed from thermal degradation was metal oxides. The redox nature of the electrocatalysts ranged from –0.15 to +2.0 V vs. RHE. It exhibits a reversible Ni (III)/Ni (II) oxidation–reduction peak. LSV test produces 98 mV and current density –3.0 V vs. RHE, respectively. The DFT data exposes an apparent synergistic effect on the compound, creating activated complexes and promoting electron allocation in hydrogen evolution. The metal complex is an effective, low-cost catalyst in which experimental and theoretical data are used to learn the structure and its catalytic property relationships.

Acknowledgments

The authors wish to acknowledge GRG Trust, Coimbatore, for the financial support provided throughout the work.

Table 1. CV current density ranges from -0.15 to +2.0V vs. RHE

Concentration	ΔEp
Supporting electrolyte	0.009
0.03 mM	0.021
0.06 mM	0.032
0.09 mM	0.057
0.15 mM	0.079

References

Ai, Wenying, Zhong, Rui, Liu, Xufang, and Liu, Qiang (2019). Hydride Transfer Reactions Catalyzed by Cobalt Complexes. Chemical Reviews, 119(4), 2876–2953.

Arunadevi, N. (2021). Structural, Optical, Thermal, Biological and Molecular Docking Studies of Guanidine Based Naphthoate Metal Complexes. Surfaces and Interfaces 24, 101094.

Arunadevi, Natarajan. (2021). Synthesis and Crystal Growth of Cadmium Naphthoate Crystal for Second Order Non-Linear Optics and Cytotoxic Activity.

Du, Pingwu, and Eisenberg, Richard (2012). Catalysts Made of Earth-Abundant Elements (Co, Ni, Fe) for Water Splitting: Recent Progress and Future Challenges. Energy & Environmental Science, 5(3), 6012–21.

Enhessari, Morteza, Maryam Shaterian, Mohammad Javad Esfahani, and Mohammad Nasser Motaharian (2013). Synthesis, Characterization and Optical Band Gap of La2CuO4 Nanoparticles. Materials Science in Semiconductor Processing 16(6), 1517–20.

Jain, Rahul (2018). Ligand-Assisted Metal-Centered Electrocatalytic Hydrogen Evolution upon Reduction of a Bis(Thiosemicarbazonato)Ni(II) Complex. Inorganic Chemistry, 57(21), 13486–93.

Swathika, M., and Natarajan, Arunadevi. (2022). Recent Advances in Hydrogen Evolution Reaction Using First-Row Transition Metal Complexes as Catalyst. ECS Transactions 107(1), 5763.

Mohan, Bharti (2020). Synthesis, Characterizations, Crystal Structures, and Theoretical Studies of Copper(II) and Nickel(II) Coordination Complexes. 761961, 73(8), 1256–79.

Pervaiz, Muhammad (2019). Synthesis, Spectral and Antimicrobial Studies of Amino Acid Derivative Schiff Base Metal (Co, Mn, Cu, and Cd) Complexes. Spectrochimica Acta Part A: Molecular and Biomolecular Spectroscopy, 206, 642–49.

Queyriaux, Nicolas (2015). Recent Developments in Hydrogen Evolving MolecularCobalt(II)–Polypyridyl Catalysts. Coordination Chemistry Reviews 304–305, 3–19.

Zhang, Ya Ping. (2020). Cobalt(II) and Nickel(II) Complexes of a PNN Type Ligand as Photoenhanced Electrocatalysts for the Hydrogen Evolution Reaction. Inorganic Chemistry 59(2), 1038–45.

Study of Self-Resilient Cloud Server Using Machine Learning Techniques for Local Area Networks

Antony Joseph,[a] Kukatlapalli Pradeep Kumar,*,[a] Angel Joy,[a] Meghna Madhu,[a] and Cherukuri Ravindranath Chowdary[a]

[a]Department of Computer Science Engineering, Christ University, Bangalore, India
E-mail: *kukatlapalli.kumar@christuniversity.in

Abstract

Servers play a key role in providing required information to clients over the web. Issues associated with server are observed in their capacity, connectivity, and delivery. These parameters are exploited by cybercriminals to compromise the systems and associated networks by performing various malicious activities. In this regard, we propose structure of a server system that is capable of defending against cyber-attacks. Post-attack the server rolls back to a normal state for carrying out its regular operations in the network. A cloud environment is chosen and machine learning techniques are employed to bring in the desired behavior in the server. Architecture of the LAN and the connected server is simulated through CISCO packet tracer environment. Results are discussed and appropriate conclusions are made at the terminal of the paper throwing a window to the readers in extending their knowledge to perform research in the area of self-resilient server systems.

Keywords: Cloud server, Resilience, Machine Learning, Local area networks

Introduction

The computer server is an important component of cloud architecture. As a result, one must protect the mainframe server from failures such as compromises, denial-of-service attacks, and malware threats. The data center (a collection of computer servers) is now transitioning from a conventional to a virtualized age. Since the server is virtualized, Virtual Machine Introspection (VMI) is a common approach for monitoring its health. In-VMI and out-VMI are the two forms of VMI. The performance of a resilient server can be constructed by integrating virtualization tools with a self-repair system model (Winarno et al., 2019). As servers are so important in data processing and transmission for serving a large number of clients, server failures affect not just the server's performance but also all the computers linked to it. When computer structures and networks reach such massive scales as seen in information centers and cloud computing, resilient systems that can self-recognize faults, self-repair failures, and self-replace failing parts are necessary (Winarno et al., 2015). In this regard, we provide an insight into the study of self-resilient servers backed by machine learning techniques, which benefits the application processes to complete the tasks in cloud environment.

Next section is followed by an exhaustive literature review of the concerned fields of server resilience. Results are put across in three phases which are normal zone, attack zone, and recovery zone for a local area network. This is followed by conclusion of the work carried out.

Literature Analysis

This segment deals with the literature review in the area of self-repair and self-resilient computer networks. It also complements the allied studies associated to cloud and the security of servers.

Self-Repair Network Analysis

Synchrophasor technology is critical to the development of the subsequent group of smart grid-wide-area observation, security, and regulation. However as tools and technology-related artifacts advance, so do the attacks. Exploitable facades in the synchrophasor system; attackers use advanced persistent threats (APTs) to influence power system dependability and stability. Platform for the creation of a self-healing and attack-resistant unit can be done by putting cutting-edge intrusion detection system (IDS). Along with an IDS, an intrusion mitigation system (IMS), as well as a notification management system (AMS), are also employed. Noteworthy is that during cyberattacks on computers, the suggested platform identifies irregularities. The produced warnings are subsequently broadcasted to the IMS in the IDS. The proposed IMS takes automatic remedial actions to combat assaults by reorganizing the synchrophasor network to isolate the compromised phasor data concentrators (PDC) and orchestrating new PDCs to prevent attacks from spreading in the future. Furthermore, by attaching the additional PDCs, the IMS restores the system's observability, making the grid attack-resistant (Singh et al., 2020).

Encounters in Intrusion Detection Systems for Internet of Things

The Internet of Things (IoT) is quickly gaining attention. Due to the diversity of expedients and procedures in use, the thoughtfulness of the data stored inside with legal and privacy concerns, IoT security is becoming a rising research importance and commercial concern. Many security approaches are inadequate for IoT networks owing to their resource-intensive nature, hence it's critical to integrate second-layer defenses. These systems will also need to be tested in a range of system categories and conventions to see how effective they function (Pradeep Kumar et al., 2021).

The developments in IoT intrusion detection procedures are the subject of the study offering a thorough examination of contemporary Intrusion Detection Systems for IoT technologies; with a focus on architectural types. Suggestive IoT-based IDS directions are presented and reviewed. This demonstrates how standard approaches are inadequate for the IoT domain due to their inherent traits providing low coverage.

To produce a safe, reliable, and optimized solution for these networks, current IoT intrusion detection research must take a new path (Kumar et al., 2014). An instance is provided to demonstrate how malicious nodes might be discovered passively. It is a new prototype connected with structuring a persistent atmosphere of smart procedures looking for enhanced daily life via universal connectivity (Benkhelifa et al., 2018).

Machine Learning (ML) and Computer Networks

A comprehensive review of various security threats occurring in Cyber-physical systems and solutions with the help of machine learning are made transparent. The author has also discussed various challenges while deploying an ML solution for a Cyber-physical system. Various techniques are discussed to make the ML model resilient to attacks. One of them is Adaptive Adversarial training. The input data is modified to a specific limit, training the model to detect those deviations from the actual values. However, this technique has challenges. It makes the model complex and reduces accuracy to a great extent. Another method to make the ML model resilient is through a technique called Randomization. The technique is divided into two stages. Random resizing and random padding. In random resizing, the input image is randomly resized and in random padding, zero is padded around the input image. The technique requires few computations and introduces high robustness to the model (Chen et al., 2019).

An efficient Dropout-resilient Aggregation for a scalable privacy-preserving Machine learning scheme was proposed (Liu et al., 2022). It protects the system from various cyber-attacks and makes it more self-resilient. The author uses the secret sharing scheme, which is the main principle of the proposed solution. It divides the secret message into a group of small units that can be efficiently reconstructed. This proposed scheme divides the participants of a network into two classes. There is a central server and a group of clients. Each client has a locally trained model. The server will perform a summation of all the client models. Dropout-resilient scheme is introduced, and will not be affected even if any client withdraws from the network. The author chooses two types of threat models. In the first case, the participants are semi-honest. They will not try to attack the network directly but try to gain information from other honest participants. In the second case, the participants are malicious attackers who will try to tamper with the network with malicious code. Masking is performed on individual models stored in the client. This will enhance the security of the machine learning model (ML).

Attacks on Artificial Intelligence (AI) and ML-based systems are one of the topics which require attention in the current transforming digital world. As the adoption of technology is fast and widespread in many fields with state of an art spectrum, security of the same is often ignored. Security resilience and attack scenarios are to be thoroughly analyzed when a technology like AI is inherited into a domain. There are cyber-attack cases and contexts noted as the security of the system was ignored. One such example can be tampering with traffic signals to mislead a driverless car. Forensic investigations for such adversarial attacks play a key role in exploring digital pieces of evidence which act as expert testimony (Manasa et al., 2022).

Methodology

The need for self-healing and self-resilient servers is required because of cyber threats and cyber-attacks in local area networks. Methodology followed in this regard is shown in Fig. 1. Three contexts of a server associated with the cloud are depicted in which given an attack on any of these instances, the server would roll back to its regular functionality with the aid of machine learning techniques incorporated in them. From that specific point in time, any running application

over the cloud can resume processing the data and related activities.

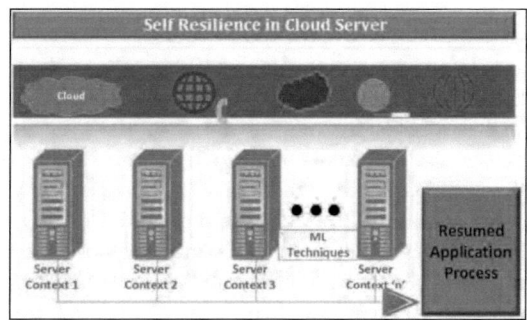

Results and Discussions

This section elicits three contexts of server with normal context, attack based, and rollback scenarios.

The network architecture has the following components:

- 4 End device (PC0, PC1, PC2, PC3).
- 2 Switches (2960-24TT).
- 3 Routers (Router1, Router2, Router3).
- 1 Cloud Server (Server0).

All three scenarios are simulated in Cisco Packet Tracer version 8.1.

Normal Context

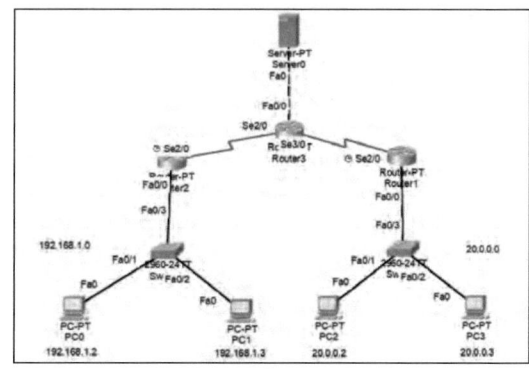

In Fig. 2, four end devices (PC0, PC1, PC2, PC3) belong to two different LAN networks with 192.168.1.0 and 20.0.0.0. These devices are eventually connected by two switches (Switch0 and Switch1). Switches eventually connect to two routers (Router2 and Router1) and the server (Server0). The connection between the router and server is indicated by the green arrow. This signifies there is no attack on the cloud server.

Attack Context

In Fig. 3, four end devices (PC0, PC1, PC2, PC3) belongs to two different LAN network with address 192.168.1.0 and 20.0.0.0. These devices are eventually connected by two switches (Switch0 and Switch1). Switches eventually connect to two routers (Router2 and Router1) and the server (Server0). The connection between the server and router is indicated by the red arrow. This signifies there is an attack on the cloud server.

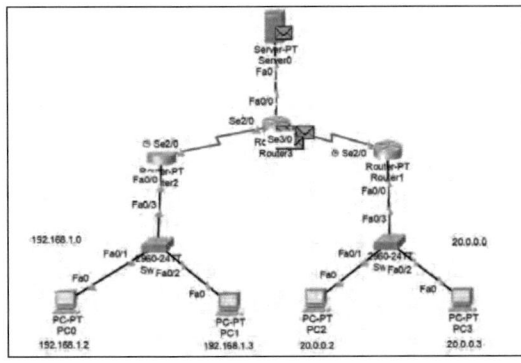

Self-Resilient Context

In Fig. 4, four end devices (PC0, PC1, PC2, PC3) belongs to two different LAN network with address 192.168.1.0 and 20.0.0.0. These devices are eventually connected by two switches (Switch0 and Switch1). Switches eventually connect to two routers (Router2 and Router1) and finally to the server (Server0). In this scenario, the server tries to come back to a normal state after the attack.

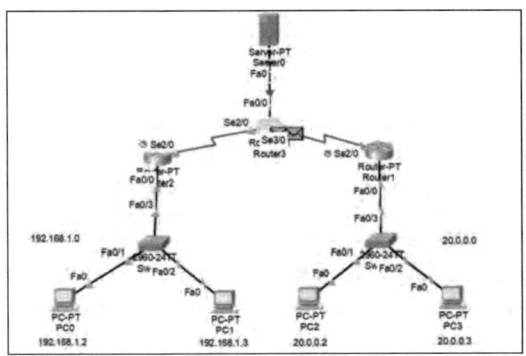

Machine Learning Perspective

Linear regression can be used to understand and predict the situations of malicious attacks in the local area network by the intruder/ cracker. Karl person correlation is calculated to analyze the relationship between attack scenarios for finalizing the outcome of self-resilience to the proposed structure. Machine learning techniques namely regression and correlation along with statistical observations can be carried out to understand the degree of relationship between the selected variables.

Such usage of technology can help resolve the issues faced by defensive networks, in this case, a cloud server. A combination of supervised and unsupervised learning methods can be applied to both qualitative and quantitative data sets for obtaining appropriate results.

Even if ML is a trendy choice for improving the system's security, it has various risks. ML algorithms are data-dependent; hence, a slight modification in the data can disrupt the entire model, leading to wrong outputs. This challenge needs to resolve to make ML-based security reliable. Measures must be taken to make data secure and cannot be tampered with by hackers.

Conclusion

An analysis is positioned in the area of self-resilient networks, especially with a specific server. Literature on the same is studied with regard to allied fields. Three situations of network architecture are depicted in correlation with end devices, switches, routers, and a server. The last scene shows that the server could able to recover to an as-usual state performing its regular operation and services to the clients. The resilience in this regard is obtained by the self-repair nature of network systems associated and linked with devices and transmission lines. This piece of work can throw a window for the readers and researchers to work on better servers that are built with strong and stubborn features to defend the attacks in any scenario.

References

Benkhelifa, E., Welsh, T., and Hamouda, W. (2018). A critical review of practices and challenges in intrusion detection systems for IoT: Toward universal and resilient systems. IEEE Communications Surveys & Tutorials, 20(4), 3496–3509.

Chen, M., Challita, U., Saad, W., Yin, C., and Debbah, M. (2019). Artificial neural networks-based machine learning for wireless networks: A tutorial. IEEE Communications Surveys & Tutorials, 21(4), 3039–3071.

Kumar, K. P., and Soumya, E. (2014). The Major Traits of Cyber Security: Case Study on Server Hardening. International Journal of Advanced Trends in Computer Science and Engineering, 3(1), 196–200.

Liu, Z., Guo, J., Lam, K. Y., and Zhao, J. (2022). Efficient dropout-resilient aggregation for privacy-preserving machine learning. IEEE Transactions on Information Forensics and Security.

Pradeep Kumar, K., Pillai, V. J., Sarath Chandra, K., and Chowdary, C. R. (2021). Disaster recovery and risk management over private networks using data provenance: Cyber security perspective. Indian Journal of Science and Technology, 14(8), 725–737.

Manasa, S., and Kumar, K. P. (2022). Digital Forensics Investigation for Attacks on Artificial Intelligence. ECS Transactions, 107(1), 19639.

Singh, V. K., Vaughan, E., and Rivera, J. (2020, February). Sharp-net: Platform for self-healing and attack resilient PMU networks. In 2020 IEEE Power & Energy Society Innovative Smart Grid Technologies Conference (ISGT) (pp. 1–5). IEEE.

Winarno, I., Ishida, Y., and Okamoto, T. (2019). A performance evaluation of resilient server with a self-repair network model. Mobile Networks and Applications, 24(3), 1095–1103.

Winarno, I., and Ishida, Y. (2015). Simulating resilient server using XEN virtualization. Procedia Computer Science, 60, 1745–1752.

Performance Study of Dual Rotor Induction Motor: Simulation Study

Chetan Bobade[a] and Dr. Swapnil B. Mohod[b]

[a]Research Scholar, S. G. B. University, Amravati, Badnera, Maharashtra (India)
[b]Department of Electrical Engineering, Prof. Ram Meghe College of Engineering and Management, Badnera, Maharashtra (India)
E-mail: [a]bobade.cm@gmail.com, [b]swapnilmohod@gmail.com

Abstract

Dual rotor Machines (DRMs) are an excellent solution to multi-motor drivetrains configurations for electric vehicle applications that require maximum power density and better vehicle driving performance. Using electromechanical energy conversion concepts, a steady-state model of a Dual rotor machine is created in this research. Equation of torque and Electrical equivalent circuit is incorporated into the developed model. The model aids in the knowledge of the motor operation and can be used to quickly assess performance Then, Results are compared, including torque-speed characteristics, speed, stator and rotor currents, real and reactive power, efficiency, and power factor. The effectiveness of the DRIM has also been examined under various load and speed circumstances. For validation purposes, the suggested model is created in MATLAB/Simulink.

Keywords: Dual-rotor induction motor (DRIM), Saturation and Unsaturation, mathematical model, simulation.

Introduction

Electric machines consume a huge portion of electrical energy in power sectors and Control systems, and drives even have a higher share than other motors. As a result, rising the effectiveness of induction motors might result in severe energy savings. Many researchers have already intended to refine and enhance the efficiency of induction motors in this respect. The efficiency of induction motors can be boosted by trying to better cool, using better materials, and developing electromagnetic stability. Pioneered the theory of an Induction Machine with a twin-rotor that also has inner and outer rotors (Machines, 2002). To enhance the efficiency of induction motors and automate power parameters. In the following subsections, the DRIM's structure and procedure framework are discussed. The literature has discussed various kinds of twice-rotor machines (TRMs). These machines are one-of-a-kind and are typically created for single use (Sun et al., 2015). For Ex. Hybrid electrical vehicle researchers and inventors are engaged to research novel techniques on the machine with a revolving double rotor; This machine can be assumed as the asynchronous squirrel-cage machine with winded rotor. Equipped with a new hybrid type motor stationary type of IM-PMM-DRM (induction motor-permanent magnet motor-DRM) (Dalal and Kumar, 2014). It includes a modified electrical equivalent circuit and the in terms of torque, even proposed an empirical model for combining IM-PMM- DRM, which is suitable for the rapid study of the specially designed motor (Dalal and Kumar,

2014; Choobdari Omran and Mosallanejad, 2018).

Introduces the concept of an analog circuit for a two-rotor, double-cage induction motor. Two inner rotors in this machine are axially and spatially separated from one another and are capable of rotating independently of one another at various speeds (Sinha et al., 2017; Choobdari Omran and Mosallanejad, 2018). The Dual rotor squirrel cage Induction motor is a unique style of an induction machine with a configuration that differs significantly from traditional induction machines. The DRIM is therefore more efficient and has a higher power factor. In contrast to the traditional induction machine, however, DRIM modeling has yet to be studied in the literature. This paper would analyze DRIM modeling and its efficiency under various conditions by the way. Electric motors typically have a power limit. High torque is expected at times in several electrical uses, such as locomotive motors. Motor torque increase can cause core saturation in these scenarios. The stator current magnitude, the magnetic absorptivity of the core, the motor's start-up time, the heat output of the motor, the insulation of the windings, and the pulsating torque are mainly affected by the saturation of the magnetic circuit (Choobdari Omran and Mosallanejad, 2018; Faiz and Seifi, 1995). Furthermore, magnetizing flux saturation has an impact on the drive control system's efficiency (Lorenz and Novotny, 1990). On the outside, since magnetizing flux saturation influences flux propagation, it may also influence the electromagnetic force's amplitude and direction (Levi et al., 2007).

The behavior of induction machines in saturated conditions differs greatly from that of linear magnetics models. The established air-gap length determines the magnetizing flux saturation (Choobdari Omran and Mosallanejad, 2018; Bottiglieri et al., 2007). The magnetizing current influences the primary flux saturation a great deal. Furthermore, the main flux has been observed to be highly dependent on the load or rotor present (Yahiaoui and Bouillault, 1995). Core saturation may also occur as a result of an abnormal rise in stator voltage.

In the study of electrical power systems, the complicated behavior of induction machines is key aspect. To determine the induction motor's transient function, a typical model is needed to estimate current, torque, rpm, and other parameters. Several scholars have used the conventional approach of induction system simulation. The concept of reorienting the q–d axis is used in several approaches to realize the effects of saturation on the main magnetization flux directions(Choobdari Omran and Mosallanejad, 2018). The consequences of saturation can be considered in a more sophisticated system by consuming the adjusted magnitude of the magnetizing flux and saturation parameters. In the simulation modeling of the induction machines with saturation magnetic field, certain approaches use stator and rotor currents as state parameters, and employ stator and rotor flux connections as state parameters, as well. The authors proposed a conventional approach for normal modeling of Alternating Current Machines(ACM) in the q-d axes, considering core saturation (Choobdari Omran and Mosallanejad, 2018).

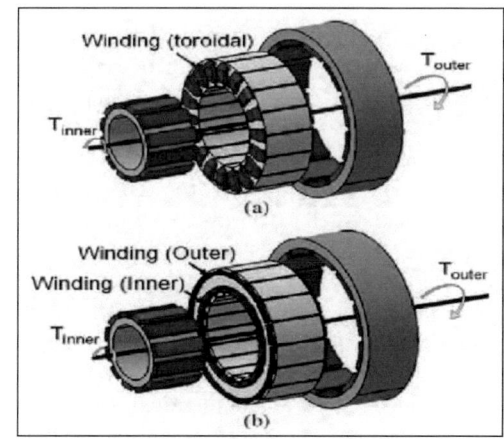

Fig. 1. Basic Topology of DRIM (Yeh et al., 2012)

Brief overview of the Induction Machine with dual rotor: The structure and ideals of the DRIM have undergone substantial examination. This section provides a concise summary of the DRIM's structure and concepts to Principles of design and structure: The dual rotor Induction machine is designed/ built so that two different machines are nested single machine. The flux of the outer rotor's squirrel cage type is directed outward at the outer surface of the external air gap, while the flux of the inner rotor's squirrel cage type is directed inward at the inner surface of the internal air gap (Machines, 2002). Accurate The double Stator coil sets are toroidally coiled back-to-back, sharing a group's back iron (Machines, 2002). The squirrel cage rotor in this design would connect the flux from the stator core across the two radial air gaps. The stator flux is moving radially inward and outward as it flows, the linking flux to both rotors would possible. This phenomenon increases the torque

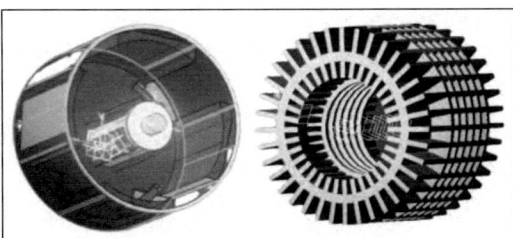

Fig. 2. Rotor and Stator structure of DRIM

density (Machines, 2002) to understand the modeling. As per the requirement of two fluxes, the construction of stator core would make change from the regular induction motor. This machine features two squirrel cage rotors and a stator layout. The outer rotor is shaped like a cylindrical shell (Choobdari Omran and Mosallanejad, 2018). The stator conductors are inserted into slots and teeth on the DRIM stator's inner and outer surfaces (Choobdari Omran and Mosallanejad, 2018). As a result of the mechanical connection between the inner and outer rotors, both rotors revolve at an identical speed. The cross-section view of the stator winding of DRIM is shown in Fig. 3. "The end winding conductors are much shorter in length in comparison to the traditional induction motor, as seen in Fig. 1." As a result, when opposed to traditional induction motors, the leak inductance and resistance are lower in the stator winding of asynchronous machines with dual rotors. The DRIM's performance, reliability, and power factor are all improved as a result of this function. The stator

winding's leakage inductances can be reduced to 71–81% of the IM using the 2 or 4 poles of DRIM. As a result, the DRIM's performance and power factor can be improved by over 3% and 0.08, respectively (Choobdari Omran and Mosallanejad, 2018). The DRIM functions in a similar manner to IM in terms of service. The stator coils generate a spinning magnetic field as symmetrical three-phase current flows through them. As seen in Fig. 2, the magnetic flux is broken into two directions that travel between air gaps and rotors. The revolving field generates currents in the squirrel cage dual rotors machine. The spinning stator magnetic field is produced currently in the rotor windings because, as per Lenz's law Theorem, It prevents the rotor's winding currents from shifting, which makes the rotor spin in the way of the revolving stator magnetizing field. Rotor rotation continues till the magnitude of generated rotor current and torque equals to the supplied burden.

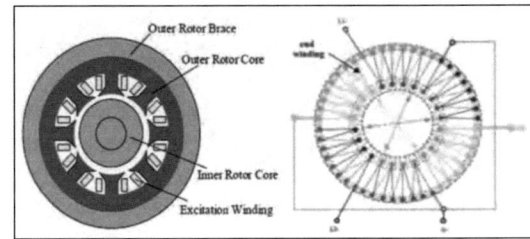

Fig. 3. End winding of machine with two stator and Cross-section of DRIM.

Analysis of Magnetic Fields: In order to make the magnetic behavior of the DRIM, Iron core failure is disregarded since it is assumed that the iron core's magnetic permeability is limitless. A condensed version of the DRIM's magnetic equivalent circuit is shown in Fig. 4 and is depends on Fig. 2. The path between the stator and the inner rotor has a magnetic reluctance of iR.; The outer rotor's magnetic path between the stator and it is either or oR. Windings of stator and both rotors, respectively, have turn numbers of Ns, Nro, and Nri. As a result, self and reciprocal inductances can be written as (1)–(4) (Choobdari Omran and Mosallanejad, 2018)

$$L_s = N^2 \left(P_{o\_} P_i \right) = L_{ss,o} + L_{ss,i} \quad \ldots \ldots (1)$$

$$L_{sro} = L_{ros} = N_s N_{ro} P_o \quad \ldots (2)$$

$$L_{sri} = L_{ris} = N_s N_{ri} P_i \quad \ldots \ldots (3)$$

Wherever P_o and P_i are same to $\frac{1}{R_o}$ and $\frac{1}{R_i}$, correspondingly and

$$L_{oi} = N_{ro}^{\ 2} P_o$$

$$L_{ii} = N_{ri}^{\ 2} P_i \quad \ldots (4)$$

L_s : stator self-inductance

L_{oi} and L_{ii} : self-inductances of rotors respectively

L_{sro} : min. value of mutual inductance between both rotors

L_{sri} : max. value of mutual inductance between both rotors

Rotor positioning affects the mutual inductances between the stator and rotors. To insure that the DRIM functions properly, To increase power density, the air gaps' lengths need to be adjusted. Because unit has the highest power density when the differences in the outside and inner air are equal, the magnetic flux distribution makes more sense (Choobdari Omran and Mosallanejad, 2018).Simulation Outcomes: The simulation is divided into 4 main sections in order to test the model shown in Fig. 4.

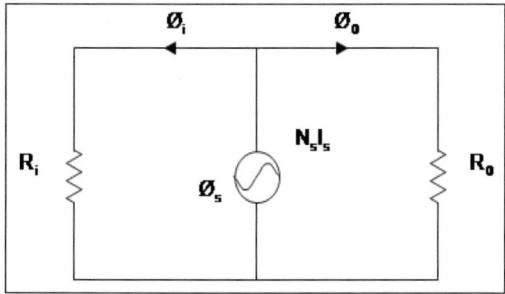

Fig. 4. Equivalent magnetic circuit of the DRIM

To assess the effects of saturation on DRIM results, the simulation of the machine includes and ignores the effects of saturation and skin effect. Furthermore, the DRIM's efficiency has been studied under various speed and load conditions. Torque, rpm, currents (Amp) in the stator winding and rotor winding, real and reactive power, performance, PF, and torque-RPM performance characteristic are all included in the simulation results. Table 1 summarizes the motor details.

Table 1. Title of the table

Symbol(S)	Parameter(P)	Value(V)
P	Pole number	4
Prated	Rated power	4 kw
Trated	Rated torque	20 N.m.
Vrated	Rated	310 V
Js.rated	Rated stator freq.	60Hz

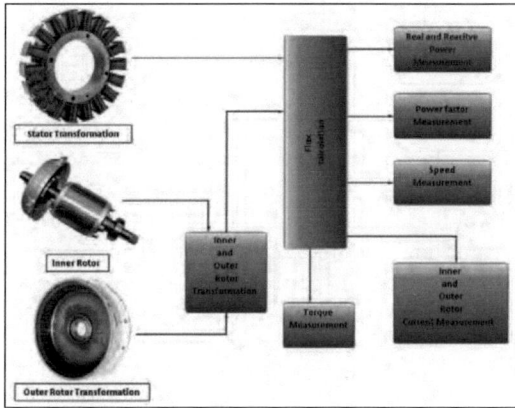

Fig. 5. Block Diagram of the DRIM

DRIM as a result of load: The findings of dual rotor induction motor simulation model are presented in the section. In no load condition, the torque of Twenty N-m is provided to the unit at time setting t=2. The load torque is decreased by fifty percent when the time setting at t=4. Figure 6. Depicts the electromotive Torque (T), rpm, and current parameters of both rotor and stator. It can be shown that as the load torque grows, the stator current rises and the motor speed falls, and the motor torque (Tm) matches the load torque (Tl). When the load torque is decreased, the stator current falls and the motor speed rises. Figure 6 indicates the effects of load variations on the current parameters of both rotor and stator (Choobdari Omran and Mosallanejad, 2018). The magnitude and freq. of the rotor currents

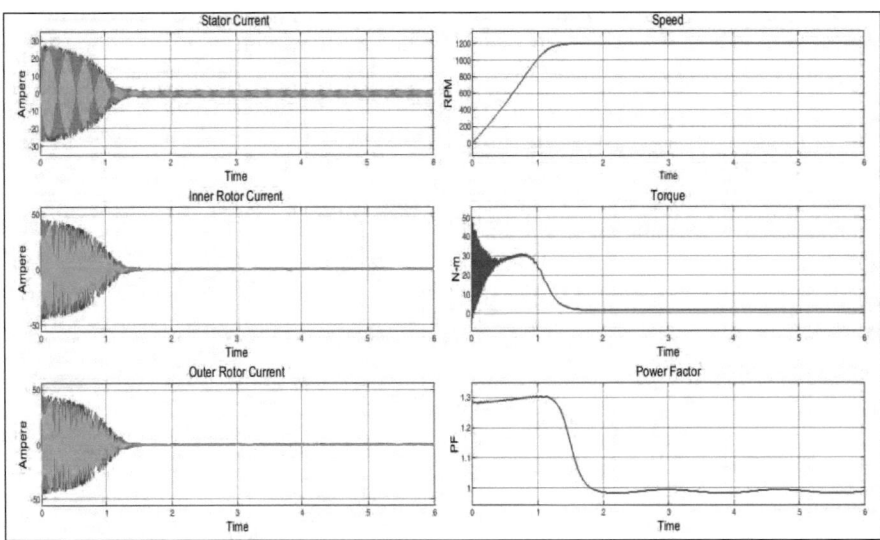

Fig. 6. (a) Stator Current (b) Inner Rotor Current (c) Inner Rotor Current (d) Speed (e) Torque (f) Power Factor

rise as the burden is added to the DRIM at a time setting of 2, approaching a steady state value. Since the load is instantly decreased current Magnitude and frequency drop. The real and reactive strength, performance, PF, and torque-rpm performance characteristics are shown in Fig. 6. At a time setting of 4, the rotor current (Ir) frequency slightly grows, and then the rotor.

Conclusion

In this paper, we propose a basic model of a double-rotor asynchronous machine to examine both Stable and transient responses with and without magnetic saturation of an Induction motor with two rotors. The DRIM model uses the discrepancy between the magnitude of the unsaturated and saturated magnetization fluxes to account for the saturation of the fluxes. This model also takes into account the effects of skin and temperature not compensating for rotor and stator resistance during operation.

References

Qu, R., and Lipo, T. A. (2002). Dual-rotor, radial-flux, toroidally-wound, permanent-magnet machines. Conference Record of the 2002 IEEE Industry Applications Conference. 37th IAS Annual Meeting (Cat. No.02CH37344), 2, 1281–1288, vol.2.

Bottiglieri, G., Consoli, A., and Lipo, T. A. (2007). Modeling of Saturated Induction Machines with Injected High-Frequency Signals. IEEE Transactions on Energy Conversion, 22, 819–828.

Omran, K. C., and Mosallanejad, A. (2018). Modeling and simulation of saturated double rotor induction machine. Compel-the International Journal for Computation and Mathematics in Electrical and Electronic Engineering, 37, 1139–1165.

Dalal, A., and Kumar, P. (2014). Analytical model of a permanent magnet brushless DC motor with non-linear ferromagnetic material. 2014 International Conference on Electrical Machines (ICEM), 2674–2680.

Levi, E., Bojoi, R., Profumo, F., Toliyat, H. A., and Williamson, S. (2007). Multiphase induction motor drives -: a technology status review. Iet Electric Power Applications, 1, 489–516.

Faiz, J., and Seifi, A. R. (1995). Dynamic analysis of induction motors with saturable inductances. Electric Power Systems Research, 34, 205–210.

Lorenz, R. D., and Novotny, D. W. (1990). Saturation effects in field-oriented induction machines. IEEE Transactions on Industry Applications, 26, 283–289.

Sinha, S., Deb, N. K., and Biswas, S. K. (2017). The Design and its Verification of the Double Rotor Double Cage Induction Motor. Journal of the Institution of Engineers (India): Series B, 98, 107–113.

Hausdorff Measures in Caratheodary Construction, Vital Covering Theorem, Steiner Symmetrization & Isodiametric Inequality

Sakina Rizvi[a], Chitra Singh[a]*

[a]Department of Mathematics, Rabindranath Tagore University, Raisen Bhopal - 464993, Madhya Pradesh, India
E-mail: chitra.singh@aisectuniversity.ac.in

Abstract

In this study we examine about Hausdorff measure and its dimension on a class of Borel regular measure and by the construction of Hausdorff measures, we depend on metric space X which may be used to generate a plenty of Borel measures on X with geometric extract (i.e., construction of Caratheodory). We describe some properties of Hausdorff measure and its dimension. Next, we study Vitali's covering theorem for a Lebesgue measure i.e. Let $A \subset \mathbb{R}^n$ and $\mathcal{F}$ a collection of non-degenerate closed balls in $\mathbb{R}^n$ which covers A finely. Then, for every $\in > 0$, there exists a disjoint subfamily $\mathcal{G} \subset \mathcal{F}$ such as

$$\mathcal{L}^n (\bigcup \mathcal{G} \leq \mathcal{L}^n (A) + \in \text{ and } \mathcal{L}^n (A \setminus \bigcup \mathcal{G}) = 0$$

After Study brief some properties of the operation called Steiner Symmetrization; this operation will be used to prove the Isodiametric inequality $\mathcal{L}^n = \mathcal{H}^n$).

Keywords: Hausdorff measure, Hausdorff dimension, Vitali's covering theorem and isodiametric inequality

Introduction

In this article we study about Hausdorff measure and its dimension with some of its properties (Hausdorff et al. 1918). It is a type of outer measure, named for Felix Hausdorff, that assigns a number in $[0, \infty]$ to each set in $\mathbb{R}^n$ or, in any metric space. It generalized the notion of area and volume to non-integer dimensions, specifically fractals and their dimensions. We focus on Caratheodory's construction which results in Hausdorff m-dimensional measure on X (C. Caratheodory et al. 1914). After define for a closed ball and fine cover we reached the prove of Vitali's covering theorem for a Lebesgue measure. The covering theorem is credited to the Italian mathematician Giuseppe Vitali and this states that it is possible to cover, up to a Lebesgue-negligible set a given subset A of $\mathbb{R}^n$ by a disjoint family extracted from a Vitali covering of A (Mattila et al., 1995). Another term Introduced by Jakob Steiner in 1836 known as Steiner symmetrization This operation will be used to prove the isodiametric inequality to show that $\mathcal{L}^n = \mathcal{H}^n$ in $\mathbb{R}^n$ (J.Steiner et. al. 1838). The Steiner Symmetrization is a process that takes a bounded Borel measurable set A and transforms it into a symmetric set $\bar{A}$ having the same measure and such that diam $\bar{A} \leq$ diam A. In this context, a subset A of $\mathbb{R}^n$ is symmetric iff A = −A; that is, x $\in$A if and only if −x $\in$A.

The Basic definition: Let X be a metric space, $\mathcal{F} \subset 2^X$ and $\zeta : \mathcal{F} \rightarrow [0, \infty]$. Roughly speaking, the idea is to "measure" the elements of $\mathcal{F}$ by means ofthe method or gauge ζ and use that to define a Borel measure on X, abstracting the geometric idea underlying the construction of the Lebesgue measure (Steven et al., 2008).

Caratheodory's construction

We define such measure in two steps:

1. For $0 < \delta \leq \infty$ we define $\forall \acute{A} \subset \times$,

$$\psi_\delta(A) := inf \left\{ \sum_{S \in \mathcal{G}} \zeta(S) \mid \begin{array}{c} G \subset F \cap \{S \mid diam S \leq \delta\}, \\ G \\ countable\,cover\,of\,A \end{array} \right\}$$

2. Define, for each $A \subset X$,

$$\psi(A) = \sup\{\psi_\delta(A) \mid 0 < \delta \leq \infty\} \in [0, \infty].$$

Note that, for fixed $A \subset X$, $\{\psi_\delta(A)\}_\delta$ is decreasing in δ, so that the sup in the definition above coincides with $\lim_{\delta \to 1} \psi_\delta(A)$.

With the notation above, we call ψ the result of Caratheodory's construction from the gauge ζ on $\mathcal{F}$, and we call ψ_δ the size δ approximating measure.

Result1

Let X be a metric space and ψ be the result of Caratheodory's construction from the gauge ζ on $\mathcal{F} \subset 2^X$. Then ψ is a Borel measure. Besides, if $\mathcal{F} \subset B_X$, ψ is Borel regular.

Definition 1

Let X be a metric space and m a non-negative real number.

Take $\mathcal{F} = 2^X$ and $\zeta : 2^X \to [0, \infty]$. Given by

$$\zeta(S) = a(m) \frac{(diam S)^m}{2^m}, \text{ were}$$

$$a(m) = \frac{\pi^{m/2}}{\tilde{A}\left(\dfrac{m}{2} + 1\right)}$$

(i.e., the Euclidean volume of $\mathbb{B}^m$ if m integer).

The result of Caratheodory's construction from the gauge ζ on 2^X is called Hausdorff m-dimensional on X, denoted by $\mathcal{H}^m$. We use the notation $\mathcal{H}^m_\delta$ for the size δ approximation of $\mathcal{H}^m$

Immediate properties of Hausdorff measure

Let X be a metric space and m a nonnegative real number. The following properties hold for

$$\mathcal{H}^m$$

1. The Hausdorff measure is compatible with the operation of taking traces. That is, if X is a metric space and $A \subset X$, the trace of $\mathcal{H}^m$ on A coincides with the m-dimensional Hausdorff measure on A (as a metric subspace of X).

2. The Hausdorff measure is invariant by isometries. That is, if Y is another metric space and f: $X \to Y$ is an isometry onto Y, then the push forward f# $\mathcal{H}^m$ coincides with the Hausdorff m-dimensional measure on Y.

3. If Y is another metric space and $f : X \to Y$ has Lipschitz constant Lip $f < \infty$, then $\forall A \subset X$, $\mathcal{H}^m(f(A)) \leq (Lip\,f)^m \mathcal{H}^m(A)$.

4. $\mathcal{H}^m$ also coincides with the result of Caratheodory's construction from ζ On $\mathcal{F}' = \{closed\,subsets\,of\,X\}$ Or $\mathcal{F}'' = \{Open\,subsets\,of\,X\}$. If X is a normed vector space, we also take $\mathcal{F}''' \{Closed\,convex\,subsets\,of\,X\}$.

5. $\mathcal{H}^m$ is a Borel regular measure on X.

6. $\mathcal{H}^0$ coincides with the counting measure on X.

Definition 2

Let X be a metric space and $A \subset X$. The extended real number $\inf\{m \in [0, \infty) \mid \mathcal{H}^m(A) = 0\} = \sup\{m \in [0, \infty) \mid \mathcal{H}^m(A) = \infty\} \in [0, \infty]$ is called Hausdorff dimension of A, denoted by $\mathcal{H} - \dim A$.

Properties of Hausdorff dimension

Let X be a metric space

1. If $Y \subset X$ is a metric subspace of X and If $Y \subset X$ is a metric subspace of X and $A \subset Y$, the Hausdorff dimension of A as a subset of the metric space Y is the same for A as a subset of the metric space X.

2. The Hausdorff dimension is invariant by isometries, i.e., if Y is a metric space, $f:X \to Y$ an isometry into Y and $A \subset X$, then $\mathcal{H}$-dim A = $\mathcal{H}$-dim f(A).

3. Let X, Y be metric spaces and $f:X \to Y$ be a Lipschitz map. For all $A \subset X$, $\mathcal{H}$-dim f(A) $\leq$ $\mathcal{H}$-dim A. In particular, if f is bi-Lipschitz onto its image (i.e., f is Lipschitz and has a Lipschitz inverse $f^{-1}: Im f \to X$, then for all $A \subset X$, $\mathcal{H}$-dim f(A) = $\mathcal{H}$-dim A.

4. If $A \subset B \subset X$, $\mathcal{H}$-dim $A \leq \mathcal{H}$-dimB. (Monotonicity)

5. If $A = \bigcup_{n \in \mathbb{N}} A_n \subset X$, then $\mathcal{H}$-dim A = sup { $\mathcal{H}$-dim An | n $\in \mathbb{N}$}

(Stability with respect to countable unions)

Vitali's covering theorem

For a closed ball $\hat{A} = \mathbb{B}(x,r)$ in $\mathbb{R}^n$ and

$$0 < t < \infty$$

Define $t\hat{A} = \mathbb{B}(x,tr)$. Let us say t = 5 and

$5B = \cup \{\hat{A}' \subset X \text{ closed ball} | \hat{A}' \cap B \neq \phi, diam \hat{A}' \leq 2 diam B\}|$

Fine cover

Let X be a metric space, $\mathcal{F}$ a collection of balls in X and $A \subset X$. Then $\mathcal{F}$ is a fine cover A, or that $\mathcal{F}$ covers A finely. if $\mathcal{F}$ is a cover of A such that, for all x$\in$A inf {diam B | x$\in$B$\in \mathcal{F}$} = 0

Vitali's Covering Theorem

(For Lebesgue measure)

We do not assume A to be $\mathcal{L}^n$ measurable.

Statement: Let $A \subset \mathbb{R}^n \mathbb{R}^n$ and $\mathcal{F}$ a collection of non-degenerate closed balls in $\mathbb{R}^n$ which covers A finely. Then, for every $\in$ > 0, there esists a disjoint subfamily $\mathcal{G} \subset \mathcal{F}$ such that

$\mathcal{L}^n (\bigcup \mathcal{G} \leq \mathcal{L}^n (A) + \in$ and $\mathcal{L}^n (A \setminus \bigcup \mathcal{G}) = 0$

(Mattila et al., 1995).

Proof

Assume first that $\mathcal{L}^n(A) < \infty$. We may assume that $\mathcal{L}^n(A) > 0$, otherwise the thesis is trivial.

Fix $0 < \delta < 5^{-n}$ (so that $1 - 5^{-n} + \delta < 1$) with $\delta \mathcal{L}^n(A) < \dot{o}$. Since $\mathcal{L}^n$ is a borel regular, by theorem

(Let μ be an open σ-finite Borel regular measure on a topological space (X, τ) for which each closed set is a G_δ the following property hold

$\forall A \subset X, \mu(A) = \inf \{\mu(U) | A \subset U \in \tau$ by open sets from outside)

$\exists$ an open set $\mathbb{U} \supset A$ such that

$$\mathcal{L}^n(\mathbb{U}) < (1 + \delta) \mathcal{L}^n(A).$$

Now to show that $\exists$ a disjoint subfamily $\mathcal{G} \subset \mathcal{F}$ whose balls are contained in $\mathbb{U}$ and $\mathcal{L}^n(A \setminus \bigcup \mathcal{G}) = 0$, $\mathcal{L}^n(\bigcup \mathcal{G}) \leq \mathcal{L}^n(\mathbb{U})(1 + \delta) \mathcal{L}^n(A) < L^n(A) + \in$.

Fix $\phi \in (1 - 5^{-n} + \delta, 1)$

1. Place $\mathcal{F}_\mathbb{U} = \{B \in \mathcal{F} | B \subset \mathbb{U}, diam B \leq 1\}$. Therefore $\mathcal{F}$ covers A finely, it is clear that $\mathcal{F}_\mathbb{U}$ is still a fine cover of A now we use this statement (Let X be a metric space and $\mathcal{F} \subset 2^X$ a family of nondegenerate closed balls in X such that sup {diam B | B$\in \mathcal{F}$}$<\infty$. Then there exists a disjoint subfamily $\mathcal{G} \subset \mathcal{F}$ such that $\bigcup_{B \in \mathcal{F}} B \subset \bigcup_{B \in \mathcal{G}} 5B$). Applying to $\mathcal{F}_\mathbb{U}$, there exists a disjoint subfamily $\mathcal{G}_\mathbb{U} \subset \mathcal{F}_\mathbb{U}$ such that $A \subset \cup \mathcal{F}_\mathbb{U} \subset \bigcup_{B \in \mathcal{G}_\mathbb{U}} 5B$.

Since $\mathcal{G}_\mathbb{U}$ is disjoint and it is countable. If X is separable, then G is countable (since any disjoint family of sets with nonempty interiors in X is countable) there exists a disjoint subfamily $\mathcal{G} \subset \mathcal{F}$ such that, for any $\hat{A} \in \mathcal{F}$, $\exists B' \in \mathcal{G}$ with B$\cap$B' $\neq \phi$ and diam B<2 diam B' (B$\subset$ 5B')

It follows that

$$\mathcal{L}^n(A) \leq \mathcal{L}^n \left(\bigcup_{B \in \mathcal{G}_\mathbb{U}} 5B \right) \leq \sum_{B \in \mathcal{G}_\mathbb{U}} \mathcal{L}^n(5B) =$$

$$5^n \sum_{B \in \mathcal{G}_\mathbb{U}} \mathcal{L}^n(B) = 5^n \mathcal{L}^n(\cup \mathcal{G}_\mathbb{U}).$$

Thus

$$\mathcal{L}^n(A \setminus \cup \mathcal{G}_\mathbb{U}) \leq \mathcal{L}^n(\mathbb{U} \setminus \cup \mathcal{G}_\mathbb{U}) =$$

$$\mathcal{L}^n(\mathbb{U}) - \mathcal{L}^n(\cup \mathcal{G}_\mathbb{U}) \leq (1 - 5^{-n} + \delta) \mathcal{L}^n(A) < \phi \mathcal{L}^n(A)$$

Moreover, since $\mathcal{L}^n(A) < \infty$

Now apply the continuity from the continuity property of measures.

(For a measure μ on X, the following property hold (Continuity from above)

If $(E_n)_{n\in\mathbb{N}}$ is a decreasing sequence in σ (μ), then

$$\mu\,(E_1) < \infty,\ \mu\,(\bigcap_{n=1}^{\infty} E_n) = \lim_{n\to\infty}\mu(E_n)).$$

By continuity from above for the Borel measure $\mathcal{L}^n\angle A'$ to obtain a finite subfamily $\mathcal{G}_1 \subset \mathcal{G}_\mathbb{U}$ such that $\mathcal{L}^n\left(A\setminus\cup\mathcal{G}_1\right) < \phi\mathcal{L}^n(A)$.

2. Given $2 \le j \in \mathbb{N}$, assume we have defined finite disjoint subfamilies $\mathcal{G}_1 \subset ... \mathcal{G}_{j-1}.\ \subset \mathcal{F}$ such that, for $1 \le i \le j-1$, the balls of $\mathcal{G}_i$ are contained in $\mathbb{U}$ and $\mathcal{L}^n\left(A\setminus\mathcal{G}_i\right) < \varnothing^i\mathcal{L}^n(A)$.

If $\mathcal{L}^n\left(A\setminus\mathcal{G}_{i-1}\right) = 0$ stop and it follows $\mathcal{G}_1 = \mathcal{G}_{i-1}$, other wise reapply the argument to the open set $\mathbb{U}' = \mathbb{U}\setminus\cup\mathcal{G}_{j-1}$ in place of $\mathbb{U}$ and to $A' = A\setminus\cup\mathcal{G}_{j-1}\subset\mathbb{U}'$ in place of A'

By using (5 times covering lemma), let X be a metric space and $\mathcal{F} \subset 2^x$ a family of nondegenerate closed balls in X such that sup {diam B | B $\in \mathcal{F}$} < ∞. Then there exists a disjoint subfamily $\mathcal{G} \subset \mathcal{F}$ such that

$$\bigcup_{B\in j} B \subset \bigcup_{B\in\mathcal{G}} 5B.$$

We extract a disjoint subfamily $\mathcal{G}_{\mathbb{U}'} \subset \mathcal{F}_{\mathbb{U}'}$ such that

$$A' \subset \cup\mathcal{F} \subset \bigcup_{B\in\mathbb{U}'} 5B$$

so that

$$\mathcal{L}^n\left(A'\setminus\cup\mathcal{G}_{\mathbb{U}'}\right) \le \left(1+\delta-5^{-n}\right)\mathcal{L}^n(A').$$

Apply continuity from above for the Borel measure $\mathcal{L}^n\angle A'$ there exists a finite set $\mathcal{G}'_j \subset \mathcal{G}_{\mathbb{U}'}$ Such that

$$\mathcal{L}^n\left(A'\setminus\cup\mathcal{G}'_j\right) < \phi\mathcal{L}^n(A') < \phi^j\mathcal{L}^n(A).$$

Put $\mathcal{G} = \mathcal{G}_{j-1}\cup\mathcal{G}'_j$. Then $\mathcal{G}_j \subset \mathcal{F}$ is a finite disjoint family whose balls are contained in $\mathbb{U}$, and $A\setminus\cup\mathcal{G}_j = A'\setminus\cup\mathcal{G}_j$ has Lebesgue measure $< \varnothing^j\mathcal{L}^n(A)$.

3. Define an increasing sequence such that $(\mathcal{G}_j)_{j\in\mathbb{N}}$ such that, for each $j\in\mathbb{N}$, $\mathcal{G}_j$ is a finite disjoint sub-family of $\mathcal{F}$ whose balls are contained in $\mathbb{U}$, with $\mathcal{L}^n\left(A\setminus\cup\mathcal{G}_j\right) < \varnothing^j\mathcal{L}^n(A)$.

Define $\mathcal{G} = \bigcup_{i\in\mathbb{N}}\mathcal{G}_i$; then $\mathcal{G}$ is a disjoint subfamily of $\mathcal{F}$ whose balls are contained in $\mathbb{U}$, with $\forall j\in\mathbb{N}$

$$\mathcal{L}^n\left(A\setminus\cup\mathcal{G}\right) \le \varnothing^j\mathcal{L}^n(A) \overset{j\to\infty}{\to} 0.$$

Hence $\mathcal{L}^n\left(A\setminus\cup\mathcal{G}\right) = 0$,

which concludes the proof in the case $\mathcal{L}^n(A) < \infty$

Steiner Symmetrization

Introduced by Jakob Steiner in 1836. This operation will be used to prove the isodiametric inequality to show that $\mathcal{L}^n = \mathcal{H}^n$ in $\mathbb{R}^n$ (J.Steiner et. al. 1838)

Definition3

Let $(e_1, ..., e_n)$ be the standard basis of $\mathbb{R}^n$ and identify $\mathbb{R}^{n-1} \equiv (e_1, ..., e_{n-1})$, $\mathbb{R} \equiv (e_n)$ so that $\mathbb{R}^n \equiv \mathbb{R}^{n-1}\times\mathbb{R}$. Define the Steiner symmetrization with respect to $\mathbb{R}^{n-1}$ to be the map $\mathcal{S}_{e_n}:2^{\mathbb{R}^n}\to 2^{\mathbb{R}^n}$ defined as

$$\mathcal{S}_{e_n}(A) = \bigcup_{x'\in\mathbb{R}^{n-1}|A_{x'}\neq\varnothing}\{(x',x_n\,||x_n|\le\frac{1}{2}\mathcal{L}^1(A_{x'})\}.$$

Properties of Steiner Symmetrization

Let $a\in\mathbb{S}^{n-1}$

1. $\forall A\subset\mathbb{R}^n$, $\mathrm{diam}\mathcal{S}_a(A)\le\mathrm{diam}\,A$.

2. if $A\subset\mathbb{R}^n$ is   – measurable, then so is $S_0(A)$ and $\mathcal{L}^n(A) = \mathcal{L}^n\left(\mathcal{S}_a(A)\right)$.

The Isodiametric Inequality

$(\mathcal{L}^n = \mathcal{H}^n)$

Statement

The Lebesgue measure of any subset of $\mathbb{R}^n$ is at most the measure of a Euclidean ball with the same diameter. That is, for all $A \subset \mathbb{R}^n$,

$$\mathcal{L}^n(A) \le a(n)\left(\frac{\mathrm{diam}\,A}{2}\right)^n$$

Proof

Let us assume that diam A<∞, Otherwise, it is trivial

1. Suppose $((a_1, ..., a_n)$ be the standard basis of $\mathbb{R}^n$. Define $\mathcal{S}_0 = \mathcal{S}_{a_n}....\mathcal{S}_{a_{n-1}}\ ...\ \mathcal{S}_{a_1}$.

By the properties of Steiner Symmetrization which holds for S_0 by applying n times in a row.

2. We contend that, for all $B \subset \mathbb{R}^n$, for $1 \leq j \leq n$, $S_0(B)$ is symmetric with respect to the hyperplane $(a_j)^\perp$ that is, denoting by the Rj: $\mathbb{R}^n \to \mathbb{R}^n$ reflection with respect to $(a_j)^\perp$,

$$R_j\left(S_0(B)\right) = S_0(B)$$

Indeed, for $1 \leq j \leq n$,

Let $B_j = S_{a_j} . o . .... S_{a_{j-1}} ...o. S_{a_1}(B)$.

i. By the definition of Steiner Symmetrization, it is clear that $B_1 = S_{a_1}(B)$ is invariant by R_1.

ii. Assume that, given $2 \leq j \leq n$, B_{j-1} is invariant by R_i for $2 \leq i \leq j-1$ to show that $B_j = S_{a_j}(B_{j-1})$ is by R_i for $1 \leq i \leq j$. It is clear for i=j, by the Steiner Symmetrization definition for i<j , since B_{j-1} is invariant by R_i, we have, denoting by P_j the orthogonal projection on $\mathbb{R}^{n-1} \equiv (a_j)^\perp$ and by $P_j^\perp$ the orthogonal projection on $\mathbb{R} \equiv (a_j)$, $\forall x \in \mathbb{R}^{n-1}$

$$B_{j-1} \cap P_j^{-1}(x) = R_i\left(B_{j-1}\right) \cap P_j^{-1}(x)$$

$$= R_i\left(B_{j-1} \cap P_j^{-1}\left(R_i^{-1}x\right)\right)$$

As $R_i^{-1} = R_i$ and $P_j^{-1} o R_i = P_j^\perp$, it then follows that, $\forall x \in \mathbb{R}^{n-1}$

$$(B_{j-1})_x = P_j^{-1}(B_{j-1} \cap P_j^{-1}(x))$$

$$= P_j^{-1} o R_i ((B_{j-1} \cap P_j^{-1}(R_i .x))$$

$$= P_j^{-1}(B_{j-1} \cap P_j^{-1}((R_i .x))$$

$$= (B_{j-1})_{R_i .x}$$

By the arbitrariness of $x \in \mathbb{R}^{n-1}$, the equality above implies $B_j = S_{a_j}(B_{j-1})$ is invariant by R_i as asserted

3. From the contention in the previous item, it follows that, given $B \subset \mathbb{R}^n$, $S_0(B)$ is invariant by R_0 o. . . . R_{n-1} . . .o R_1. i.e., $S_0(B)$ is symmetric with respect to the origin. Thus, $\forall x \in S_0(B)$, so $-x \in S_0(B)$, so that 2 $x \leq diam\, S_0(B) \leq diam B$, i. e

$$S_0(B) \subset \mathbb{B}\left(0, \frac{diam B}{2}\right).$$

4. It follows from the previous item applied to $B = \bar{A}$ That

$$\mathcal{L}^n(A) \leq \mathcal{L}^n(\bar{A})$$

$$= \mathcal{L}^n\left(S_0(\bar{A})\right) \leq \mathcal{L}^n\left(\mathbb{B}\left(0, \frac{diam \bar{A}}{2}\right)\right) = a(n)\left(\frac{diam \bar{A}}{2}\right)^n$$

$$= a(n)\left(\frac{diam A}{2}\right)^n.$$

This completes the proof

Conclusion:

To conclude the construction of Caratheodory allows several other outer measures in the Euclidean space, most of which coincide with the Hausdorff K-dimensional measures for C^1 submanifolds when k is an integer, but differ in general sets. One example is the Farvard measure also called integral geometric measure. In some common generalizations of the Hausdorff measures one restricts the class of admissible coverings in. For instance, one can use covering of balls. Our Goal is to show Hausdorff Measures in Caratheodary Construction, Vitali's Covering Theorem, Steiner Symmetrization & Isodiametric Inequality which are shown and results are clear and proved.

References

Caratheodory C. 1914, Uber das lineare maß von punkimengen-eine verallge meinerung des langenbe griffs, Nachrichten von der Gesells chaft der Wissenchaften zu Gottingen, Mathematisch-Physikalische Klasse 404 426.

Hausdorff F. 1918" Dimension and äusseres Mass" Math. Ann., 79:157 159 MR 1511917.

Mattila P. 1995, Geometry of sets and measures in Euclidean spaces, Cambridge Studies in Advanced Mathematics, 44, Cambridge University Press, Cambridge, Fractals and rectifiability, MR 1333890.

J. Steiner E. (1838) Beweise der isoperimetrischen Hauptsatze, J. Reine Angew. Math. 18, 281-296. MR 1578194

Steven G. Krantz and Harold R. Parks (2008), Geometric integration theory, Cornerstones, Birkhauser Boston, Inc., Boston, MA, 2008. MR 2427002.

Heuristic for Symmetric Euclidean TSP with 3/2-approximation Ratio

Alok Chauhan[*,a] *and Ramesh Ragala*[a]

[a]VIT University Chennai, India
E-mail: [*]alok.chauhan@vit.ac.in

Abstract

Travelling Salesman Problem (TSP) lacks efficient algorithms that provide optimal solutions to the problem (which is unrealistic also as TSP is NP-Hard) or give a guarantee of how far the solution will be from optimum. Christofides' algorithm possesses the best approximation ratio of 3/2 for symmetric TSP. A newly proposed heuristic called 2-repetitive nearest neighbor (RNN) is presented here. Experimental results show that 5/4 is the approximation ratio; however, based on upper bound investigation for approximation ratio a bound of 3/2 is conjectured. The average performance of 2-RNN is found to be better than that of Christofides' algorithm albeit at the cost of one order more of time complexity.

Keywords: Low degree spanning Tree, Open loop TSP, Conversion algorithm, Minimum spanning tree

Introduction

The *traveling salesman problem* (TSP) is widely studied problem in *algorithm engineering* and *combinatorial optimization*. Despite being very simple, no efficient algorithm exists for this problem till now. The problem belongs to the complexity class *NP-hard*. One has to find the shortest *Hamiltonian Cycle* in a given *complete graph* which is a permutation of vertices so that the length of the path is the minimum. *Symmetric* TSP is a TSP in an undirected complete graph. Hamiltonian cycle is also referred to as a tour in literature.

Being a very famous problem having several important applications, a large number of heuristics and approximation algorithms have been reported so far. Important ones are the class of local search algorithms such as *2-Opt* heuristic (Croes, 1958), the *3-Opt* algorithm (Lin, 1965), and *the k-opt* version by Lin and Kernigham (Lin et al., 1973) which applies a dynamic value of *k*. These algorithms have an upper bound of 2 (Rosenkrantz et al., 1997). A hierarchical hybrid algorithm in terms of density peaks clustering algorithm, ant colony optimization algorithm and *k*-Opt algorithm has been reported recently (Liao et al., 2018). For large-scale traveling salesman problems, H. D. Nguyen et al. proposed a hybrid genetic algorithm (Nguyen et al., 2007).

Motivation

Improvement in solution quality is an important aspect of approximation algorithms as well as motivation for the current work.

2-RNN Algorithm for Open Loop TSP

The input of TSP is a set of *n* vertices $\{v_1, v_2 \ldots v_n\}$ and the goal is to find the shortest path that travels through all vertices. *k*-RNN algorithm (Klug et al., 2019) considers all possible permutations of k vertices as the starting point, which is then completed to TSP_{path} by adding the remaining vertices one by one where TSP_{path} is defined as a path between two vertices of a graph that visits each vertex exactly once. Algorithm has steps as given below:

Step 1: For each permutation of the k vertices $v_1, v_2 .. v_k$ generate the partial tour $T = (v_1, v_2 ... v_k)$ and set the vertices $v_1, v_2 .... v_k$ as visited.

Step 2: Fix $i = k$. Till there are unvisited vertices left: identify v_{i+1} as the nearest unvisited neighbor of v_i and add v_{i+1} to T. If there are many nearest neighbors, choose anyone. Set v_{i+1} as visited and increase i by 1.

Step 3: Out of all $n! / (n-k)!$ paths generated, print the minimum path as the output.

Basically, k-RNN corresponds to NN algorithm repeated $n! / (n-k)!$ times. Special case $k=0$ corresponds to the standard nearest neighbor (NN), and $k=1$ to its repeated variant RNN. Here we are specifically interested in 2-RNN.

Association of 2-RNN Algorithm with Conversion Algorithm

Association results from a crucial observation that 2-RNN might be employed during certain computations in a conversion algorithm TREE-3(V, MST_5) (Khuller et al., 1996) which transforms a degree five minimum spanning tree (MST_5) corresponding to a set of points V in a Euclidean plane to a degree three spanning tree (ST_3) by rooting MST_5 at any of its leaf nodes, replacing all the edges to its children from each vertex v in V by the shortest path beginning from v and going through all its children and finally taking union of all such shortest paths. ST_3 is of interest to us as its cost is known to be less than or equal to 1.5 times the cost of MST_5 (Khuller et al., 1996). So, we start by first confirming that indeed 2-RNN can be used during the processing of TREE-3(V, MST_5). Second important observation is that TREE-3(V, MST_5) can be thought of as decomposing the graph G induced by V into a set of subgraphs each of which could be processed by 2-RNN. As 2-RNN can also be applied to G as a whole, it leads to the possibility of relating its output with that of TREE-3(V, MST_5).

Optimality of 2-RNN for n ≤ 4

Theorem 1: Given any 4-node undirected complete graph G ({V}, E), 2-RNN produces optimal TSP_{path}.

Proof: Let us consider $V = \{A, B, C, D\}$ and assume that $ABCD$ is the optimal path (Fig. 1). Since 2-RNN considers all pairs then it will consider also AB. Starting from AB there are only two possible continuations: BC and BD, of which 2-RNN will always choose the shorter (BC). However, the shorter of these edges must also belong to the optimal solution because if $BD > BC$ then |$ABDC$| > |$ABCD$|. Hence it is proven.

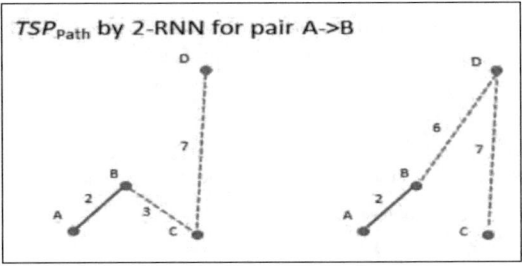

Fig. 1: TSP_{PATH} choice by 2-RNN for pair A->B: B->C->D. Other option is B->D->C.

Corollary 1: Given any 4-node undirected complete graph G (V, E), 2-RNN produces TSP_{path_min} (V) of G where TSP_{path_min} (V) is a minimum length TSP_{path} starting from vertex V.

Local Application of 2-RNN in Conversion Algorithm

The conversion algorithm selects any node as a root. It then processes the tree structure from root to leaf and every node that has degree higher than 3 is broken into parts. Consider subgraph containing vertex V and its children (Fig. 2). Since V is already connected to its parent, we keep it as it is but re-structure the children. We do this by creating TSP_{path_min} (V), starting from V and traveling via its all children. As a result, V will have only one child plus at most two neighbors in the chain created by solving its parent node. The process is then repeated for all the children.

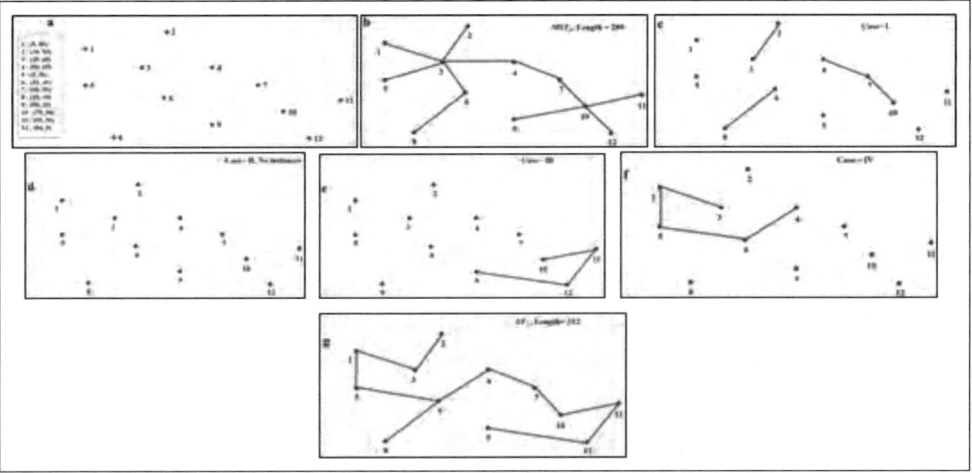

Fig. 2: Generation of ST₃ from MST₅, a: Graph Points; b: MST₅ (280 units); c: Case-I; d: Case-II; e: Case-III; f: Case-IV; g: ST₃ (312 units)

Due to the property of MST_5, every vertex V we process has at most four children. Four cases can happen during the restructuring.

Case I (0 or 1 children): The path is trivially the vertex V connected to its only child (if any).

Case II (V has 2 children v_1, v_2): $TSP_{path\_min}(V)$ is the shorter of the two paths: V-v_1-v_2 or V-v_2-v_1.

Case III (V has 3 children): apply 2-RNN to find $TSP_{path\_min}(V)$ (Corollary 1).

Case IV (V has 4 children, v_1, v_2, v_3, and v_4): apply 2-RNN to find $TSP_{path\_min}(V)$ by first finding $TSP_{path\_min}(V)$ (Corollary 1) in 4-node subgraph (v_1, v_2, v_3, v_4). Then minimum of $[TSP_{path\_min}(v_i) + V v_i]$, $i = 1,2,3,4$ is chosen.

Discussion

It has been shown in the conversion algorithm that for every vertex v in V, the shortest path beginning from v and passing through all its children weights most 1.5 times the weight of the edges it replaces, hence (Khuller et al., 1996):

$$|ST_3| \leq 1.5 * |MST|$$

This is proved by showing the upper bound of the shortest path beginning from v and passing through all its children for each of the four cases mentioned above. It makes use of some important properties of minimum spanning trees in a plane such as an angle formed at a vertex in MST by a pair of incident edges is largest angle in the triangle induced by the edges as well as the upper bound on the perimeter of a triangle (Khuller et al., 1996).

As $|MST| \leq |TSP|$

$|ST_3| \leq 1.5 * |TSP|$, where TSP is optimal TSP_{path}

The conversion algorithm induces graph decomposition where 2-RNN is applied to all the subgraphs to find $TSP(V)$ of each subgraph which is then joined together to produce ST_3. 2-RNN finds TSP_{path} of the original graph, which is also an ST_2. In both cases, 2-RNN is used in computing a spanning tree (ST_3 and ST_2 respectively) only (Fig. 3). Therefore, spanning tree computation can be done by decomposing the graph as well as without decomposition as illustrated by TREE-3(V, MST_5) and 2-RNN respectively. As 2-RNN is applied in both cases for a similar type of computation (spanning tree), the results in both cases are close to each other (Fig. 4).

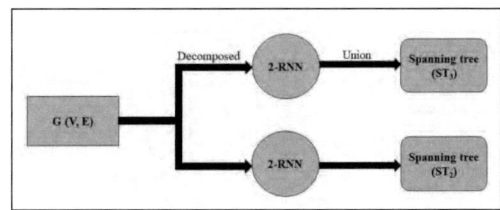

Fig. 3. 2-RNN application

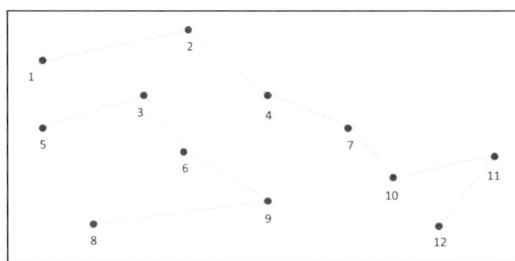

Fig. 4. 2-RNN output (ST_2, 303 units)

Experimental Results

Below we report some experimental outcomes of the 2- RNN algorithm. The experiments were planned for TSPLIB dataset (8 examples of the Symmetric TSP) (Reinelt, 1991).

Comparison was done with the following algorithms:

1. NN
2. 1-RNN
3. 2-opt
4. Christofides

1-RNN is basically well known Repetitive Nearest Neighbor algorithm. Due to exorbitant execution time, bigger k-values could not be considered for experimentation. When compared with Christofides, 2-RNN performs better than Christofides on average (Table 1).

Table 1. Output for 8 examples of the Symmetric TSP taken from TSPLIB (Reinelt, 1991). The gap represents the percentage by which the output goes beyond the optimum

Algorithm	Gap (%)	
	Average	Worst Case
NN	23.02	31.94
RNN (1-RNN)	19.58	26.47
2-RNN	18.75	23.01
Christofides	28.09	44.43
2-opt	7.25	9.87

Conclusion

Based on the above analysis, it is speculated that due to this relationship between 2-RNN and the conversion algorithm, an upper bound of 2-RNN could be 1.5.

References

Croes, G. A .(1958). A method for solving Traveling-Salesman problems. Oper Res, 6(6), 791–812.

Lin S (1965) Computer solutions of the traveling salesman problem. Bell Syst Tech J, 44(10), 2245–2269.

Lin, S., and Kernighan, B. W. (1973) An effective heuristic algorithm for the Traveling-Salesman problem. Oper Res 21(2), 498–516.

Rosenkrantz, Daniel J., Stearns, Richard E., and Lewis, Philip M. (1997). An Analysis of Several Heuristics for the Traveling Salesman Problem. SIAM Journal on Computing, 6(3), 563–581.

Liao, E., and Liu, C. (2018) A hierarchical algorithm based on density peaks clustering and ant colony optimization for traveling salesman problem. IEEE Access, 6, 38921–38933.

Nguyen H. D., Yoshihara I., Yamamori, K., Yasunaga, M. (2007). Implementation of an effective hybrid GA for large-scale traveling salesman problems. IEEE Trans SystMan Cybern B Cybern, 37(1), 92–99.

Klug, N., Chauhan, A., Vijayakumar, V., and Ragala, R. (2019). K-RNN. Extending NN-heuristics for the TSP. Mobile Networks and Applications, pp. 1–4, January 2019.

Khuller, S., Raghavachari, B., and Young, N. E. (1996). Low-Degree Spanning Trees of Small Weight. SIAM Journal on Computing, 25(2), 355–368.

Reinelt, G. (1991). TSPLIB - A traveling salesman problem library. ORSA J Comput 3(4), 376–384.

New Approach to Solve COVID-19 Via Fuzzy Mathematical Model

S. Sweatha[a] and S. Sindu Devi[b],*

[a]Research Scholar, SRM Institute of Science and Technology, Ramapuram, Chennai, India
[b]Faculty of Engineering and Technology, SRM Institute of Science and Technology, Ramapuram, Chennai, India
E-mail: *sindudes@srmist.edu.in

Abstract

We have developed a new fuzzy mathematical model which provides a framework to investigate the spread of the disease. The considered mathematical model consists of four compartments namely the susceptible class, the resistive class, the infected class, and the quarantined class(SRIQ). We have carried out the stability analysis for the possible equilibrium points along with the investigation of reproduction numbers and fuzzy reproduction numbers. Further, we have shown that backward bifurcation exists for this dynamic system.

Keywords: Stability, fuzzy reproduction number, backward bifurcation

Introduction

Currently, COVID-19 has captured the attention of scientists, researchers, governments, and all the people around the world, since it has a significant number of death rates. COVID-19 has been declared a communicable disease as it spreads from person to person through the droplets of infected persons during the act of coughing or sneezing. The total population is divided into four compartments where the transmission rate, the recovery rate, and the death rates are considered fuzzy parameters to make a unique investigation. The authors provided useful insight into the control of bacterial meningitis with a non-linear recovery rate further they discussed the conditions for forwarding and backward bifurcations (Asamoah et al., 2020). Roumen Anguelov et al. (2014) proposed a mathematical model which exhibits the phenomenon of bifurcation for a few values of parameters. The authors of the paper (Swetha Selvakumar et al., 2022) have proposed a model with fuzzy parameters and have analyzed the stability of the disease. Using Euler's method, graphically the authors have shown that there is decrease in malarial disease from time to time (Monisha Pandiyan et al., 2022). The organization of the paper is Preliminaries, Fuzzy Mathematical model, stability analysis, positivity and boundedness, Fuzzy reproduction number, numerical solution, and Conclusion.

Preliminaries

Fuzzy set: Let X be nonempty. A fuzzy subset S of X is denoled by S and is defined as

$$S = \{(x, \mu_{s(x)}): x \in X\}$$

Where $\mu_s: X \to [0,1]$ is a membership function associated with a fuzzy set S which deseribes the degree of belongingness of x with X. Here we use the membership function $\mu(x)$ to indicate the fuzzy subsets S. Also, $\mu(x)$ is called a fuzzy nember if X is the set of real numbers.

Fuzzy Measure

Let Ω be a non-empty set and $p(\Omega)$ denote the set of all subsets of Ω. Then $\mu:\Omega \rightarrow [0,1]$ is a fuzzy measure if

1. $\mu(\phi) = 0$ and $\mu(\Omega) = 1$
2. for A,B $P(\Omega)$, $\mu(A) \leq \mu(B)$ if $A \subset B$

Let $\mu:\Omega \rightarrow [0,1]$ be an uncertain variable, i.e.) is a fuzzy subset and a fuzzy measure on Ω. Then fuzzy expected value (FEV) of is the real number, defined by the sugeno measure.

$$FEV(\mu) = \int \mu d\mu = sup\{min(\alpha, k(\alpha))\}, \; 0 \leq \alpha \leq 1$$

Where $k(\alpha) = \mu\{\omega \epsilon \Omega : \mu(\omega) \geq \alpha\}$

Fuzzy Mathematical Model

As mentioned, the model is divided into four compartments, S as the susceptible population, R as the resistive population, I as the infected population, and Q as the quarantined population. The following are the governing fuzzy differential equations. All the parameters considered are non-negative constants.

$$\frac{dS}{dt} = \Lambda - \beta(\sigma)S(t)I(t) - (d + \delta(\sigma))S(t)$$

$$\frac{dR}{dt} = \alpha - \gamma H(t)I(t) + \omega(\sigma)I(t) - (d+\delta(\sigma))H(t)$$

$$\frac{dI}{dt} = \beta(\sigma)S(t)I(t) + \gamma H(t)I(t) + \varepsilon Q(t) - (d+\delta(\sigma) + \tau + \omega(\sigma))I(t)$$

$$\frac{dQ}{dt} = \tau I(t) - ((d + \delta(\sigma) + \varepsilon)Q(t)$$

Where Λ is the recruitment rate of susceptibles, β is the transferral rate of the disease, d is a natural death, δ is the death caused by the disease, α is the recruitment rate of a healthy human, γ is the transferral rate of a healthy human, ω is the recuperation rate of quarantined class, ε is the rate at which quarantined population gets infected, τ is the contact rate of the infected and healthy population. We have considered the transferral rate of disease $\beta(\sigma)$, disease-related deaths $\delta(\sigma,)$ and recuperation rate of quarantined class $\omega(\sigma)$ as fuzzy numbers with σ as virus load since there is an assortment in population.

Analysis of Fuzzy System

The transferral rate of the disease with the amount of virus load σ is given as $\beta(\sigma)$ (Barros et al., 2004). σm represents that the virus load is minimum and hence there is only negligible transferral of the infection taking place. For a transferral of infection to take place, there should be a certain amount of virus which is represented as $\sigma 0$ and σM is the limited amount of virus.

$$\beta(\sigma) = \begin{cases} 0, & if \;\; \sigma < \sigma_m, \\ \frac{\sigma - \sigma_m}{\sigma_0 - \sigma_m}, & if \;\; \sigma_m \leq \sigma \leq \sigma_0, \\ 1, & if \sigma_0 \leq \sigma \leq \sigma_M \end{cases}$$

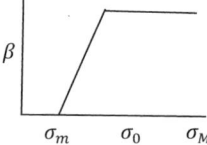

Fig. 1. Membership function of $\boldsymbol{\beta} = \boldsymbol{\beta(\sigma)}$

Let $\omega(\sigma)$ be the recuperation rate of the infected population who were quarantined and this is dependent on the amount of virus load (Verma et al., 2017). As long as the amount of virus load increases, the duration of the recuperation rate also increases. The following is the fuzzy membership function of $\omega(\sigma)$.

$$\omega(\sigma) = \frac{(\omega_0 - 1)}{\sigma_M} \sigma + 1 \; if \; 0 \leq \sigma \leq \sigma_M$$

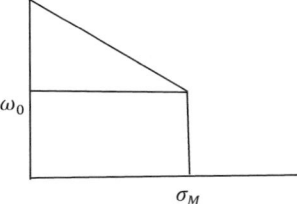

Fig. 2. Membership function of $\boldsymbol{\omega} = \boldsymbol{\omega(\sigma)}$

Where $\omega_0 > 0$ is the lowest recuperation rate.

We consider $\delta = \delta(\sigma)$ as the death due to the disease rate as a fuzzy number(Verma et al., 2017). Since the virus load is an increasing function, the death rate will be higher when the virus load is high. The maximum value of the death rate due to the disease is assumed as $(1-\eta)$, $\eta \geq 0$.

$$\delta(\sigma) = \begin{cases} \frac{(1-\eta)-\delta_0}{\sigma_0}\sigma + \delta_0, & if\ 0 \leq \sigma \leq \sigma_0 \\ (1-\eta) & if\ \sigma_0 < \sigma \end{cases}$$

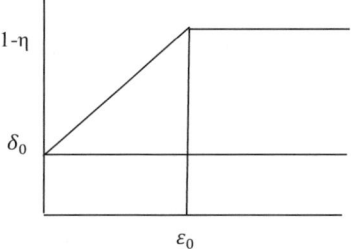

Fig. 3. Membership function of $\delta = \delta(\sigma)$

$$\Gamma(\chi) = \begin{cases} 0, & if\quad \sigma < \bar{\sigma} - \delta, \\ \frac{\sigma-\bar{\sigma}+\delta}{\delta}, & if\ \bar{\sigma}-\delta \leq \sigma \leq \bar{\sigma}, \\ \frac{-(\sigma-\bar{\sigma}-\delta)}{\delta}, & if\ \bar{\sigma} < \sigma \leq \bar{\sigma}+\delta, \\ 1, & if\quad \sigma > \bar{\sigma}+\delta. \end{cases}$$

Stability Analysis

The disease-free equilibrium is the state where there is no infection in the community. The disease-free equilibrium is represented as E_0 and this is evaluated by considering I and Q as zero.

$$E_0 = (\frac{\Lambda}{d+\delta}, \frac{\alpha}{d+\delta}, 0, 0)$$

Basic Reproduction Number

Basic reproduction number is defined as the average number of secondary infections caused by single infectious individual during entire infectious lifetime (Van den Driessche et.al, 2022)

$$R_0 = \frac{(\beta\Lambda+\gamma\alpha)\ (d+\delta+\varepsilon)}{(d+\delta)[(d+\delta+\tau+\omega)(d+\delta+\varepsilon)-\varepsilon\tau]}$$

Using Routh-Hurwitz criteria the disease-free equilibrium is asymptotically stable.

Fuzzy basic Reproduction

The basic reproduction number is R_0 increase with an increase in virus load and this cannot be a fuzzy set as it can be greater than 1.

Thus $0\leq \omega_0\ R0\ (\sigma) \leq 1$, where $\omega_0\ R0\ (\sigma)$ is a fuzzy set and hence $FEV\ [\omega_0 R0\ (\sigma)]$ is well-defined (Verma et al., 2017)

The fuzzy basic reproduction number

$$R_0^f = \frac{1}{\omega_0}\ FEV\ (\omega_0\ R_0(\sigma))$$

Where $FEV(\omega_0 R_0(\sigma))=\sup\{\inf(\alpha, k(\alpha)\}, 0\leq\alpha \leq 1$, $k(\alpha) = \delta\{\sigma: \omega0\ R0\ (\sigma) \geq \alpha\} = \mu(X)$, which is a fuzzy measure. To acquire $FEV(\omega_0\ R_0(\sigma))$ we define fuzzy measure, the possibility measure(Barros et.al,2003)

$$\mu(X) = sup\Gamma(\sigma)\ \sigma \in X, X \subset R$$

$R_0(\sigma)$ is not decreasing with $\sigma X = [\bar{\sigma}, \sigma_M]$, $\bar{\sigma}$, is the solution of the equation.

$$\alpha = \omega_0 \frac{(\beta\Lambda+\gamma\alpha)\ (d+\delta+\varepsilon)}{(d+\delta)[(d+\delta+\tau+\omega)(d+\delta+\varepsilon)-\varepsilon\tau]}$$

Here, $k(\alpha) = \mu\ [\sigma', \sigma_M] = \sup\ \Gamma(\sigma)$ with $\sigma' \leq \sigma \leq \sigma_M$, where $k(0) = 1$ and $k(1) = \Gamma\ (\sigma_M)$.

By linguistic meaning, the amount of virus has been classified into three categories.

<u>CASE i):</u> Weak virus load

$\bar{\sigma} + \delta \leq \sigma m$

$\beta(\sigma) = 0$, $R_0(\sigma) = 0$,

$FEV\ (\omega_0\ R_0(\sigma))= 0 < \omega_0 \Leftrightarrow R_0^f < 1$

Here, the disease will no longer exist.

<u>CASE ii):</u> Medium virus load

$$\beta(\sigma) = \frac{\sigma-\sigma_m}{\sigma_0-\sigma_m}, if\quad \sigma_m \leq \sigma \leq \sigma_0$$

$$R_0(\sigma) = \frac{(\beta\Lambda+\gamma\alpha)\ (d+\delta+\varepsilon)}{(d+\delta)[(d+\delta+\tau+\omega)(d+\delta+\varepsilon)-\varepsilon\tau]}$$

$$k(\alpha) = \begin{cases} 1, & if\ 0 < \alpha \leq \omega_0\ R_0(\bar{\sigma}), \\ \Gamma(\sigma'), & if\ \omega_0 R_0(\bar{\sigma}) < \alpha \leq \omega_0 R_0(\bar{\sigma}+\delta) \\ 0, & if\ \omega_0 R_0(\bar{\sigma}+\delta) < \alpha \leq 1 \end{cases}$$

so, if $\delta > 0$, $k(\alpha)$ is a continuous and decreasing function with $k(0) = 1$ and $k(1) = 0$. Hence, $FEV(\omega_0 R_0(\bar{\sigma}))$ is the fixed point of k and $\omega_0 R_0(\bar{\sigma}) \leq FEV(\omega_0 R_0(\sigma)) \leq \omega_0 R_0(\bar{\sigma}+\delta)$,

$$R_0(\bar{\sigma}) \leq R_0^f \leq R_0\ (\bar{\sigma}+\delta)$$

As the function, $R_0(\bar{\sigma})$ is an increasing and continuous function then by the intermediate value theorem there exists σ with $\bar{\sigma} < \sigma < \bar{\sigma} + \delta$ such that

$$R_0^f = R_0(\sigma) > R_0(\bar{\sigma}).$$

There exist virus load σ such that R_0^f and $R_0(\sigma)$ coincide. Additionally, the fuzzy average number of secondary cases R_0^f is highter than the number of secondary cases $R_0(\bar{\sigma})$ due to the medium virus amount.

<u>CASE iii)</u>: Strong virus load

$$\beta(\sigma) = 1, R_0(\sigma) = \frac{(\beta\Lambda + \gamma\alpha)\,(d+\delta+\varepsilon)}{(d+\delta)[(d+\delta+\tau+\omega)(d+\delta+\varepsilon)-\varepsilon\tau]}$$

$$k\,(\alpha) = \begin{cases} 1, & if \ \ 0 < \alpha \le \omega_0 R_0(\bar{\sigma}), \\ \Gamma(\sigma'), & if \ \ \omega_0 R_0(\bar{\sigma}) < \alpha \le \omega_0 R_0(\bar{\sigma} + \delta) \\ 0, & if \ \ \omega_0 R_0(\bar{\sigma} + \delta) < \alpha \le 1 \end{cases}$$

similar to case ii), we have

$$R_0(\bar{\sigma}) \le R_0^f \le R_0(\bar{\sigma} + \delta)$$

Thus, at $R_0^f > 1$; We can infer that the illness will be widespread.

Existence of Backward Bifurcation

By making use of the center manifold theory, (Castillo-Chavez and Song, 2004) have given the conditions to state the existence of backward bifurcation. Let

$$F_1 = \Lambda - \beta(\sigma)S(t)I(t) - (d + \delta(\sigma))S(t)$$

$$F_2 = \alpha - \gamma H(t)I(t) + \omega(\sigma)I(t) - (d + \delta(\sigma))H(t)$$

$$F_3 = \beta(\sigma)S(t)I(t) + \gamma H(t)I(t) + \varepsilon Q(t) - (d + \delta(\sigma) + \tau + \omega(\sigma))I(t)$$

$$F_4 = \tau I(t) - ((d + \delta(\sigma) + \varepsilon)Q(t)$$

Let $\beta = \phi$ be the bifurcation parameter. The value of the right and left eigenvectors is obtained by solving the above equations, $x = (x_1, x_2, x_3, x_4)$, $y = (y_1, y_2, y_3, y_4)$

$$x = \left(\frac{-\beta\Lambda}{(d+\delta)^2}, \frac{\alpha\gamma - \omega d - \omega\delta}{-(d+\delta)^2}, 1, \frac{\tau}{(d+\delta+\varepsilon)} \right),$$

$$y_1 = y_2 = -(d + \delta), \ y_3 = 1, y_4 = \frac{\varepsilon}{d+\delta+\varepsilon}$$

From the conditions of (Castillo et.al,2004),

$$\xi = \sum_{k,i,j=1}^{3} v_k w_i w_j \frac{\partial^2 F_k}{\partial x_i \partial x_j}(E_0, \phi)$$

$$\varphi = \sum_{k,i=1}^{3} v_k w_i \frac{\partial^2 F_k}{\partial x_i \partial \beta}(E_0, \phi)$$

Since $R_0 < 1$, we conclude that backward bifurcation exists.

Conclusion

In this paper, we have taken an assortment in population which leads to assuming a few of the parameters as fuzzy numbers. From the proposed mathematical model, we have done the stability analysis and computed basic and fuzzy basic reproduction numbers. We have shown that there exists backward bifurcation.

Reference

Asamoah, J. K. K., Nyabadza, F., Jin, Z., Bonyah, E., Khan, M. A., Li, M. Y., and Hayat, T. (2020). Backward bifurcation and sensitivity analysis for bacterial meningitis transmission dynamics with a nonlinear recovery rate. Chaos, Solitons & Fractals, 140, 110237.

Anguelov, R., Garba, S.M. and Usaini, S., 2014. Backward bifurcation analysis of epidemiological model with partial immunity. Computers & Mathematics with Applications, 68(9), 931–940.

Selvakumar, S. and Devi, S.S., 2022. Fuzzy Epidemic Model for the Transmission of Zika Virus. ECS Transactions, 107(1), 16851.

Pandiyan, M. and Devi, S.S., 2022. Fuzzy SIR-Epidemic Model for Transmission of Malaria. ECS Transactions, 107(1), 17691.

Massad, E., Ortega, N.R.S., de Barros, L.C. and Struchiner, C.J., 2009. Fuzzy logic in action:

Applications in epidemiology and beyond, Springer Science & Business Media, (232).

Barros, L.D., Leite, M.F. and Bassanezi, R.C., 2003. The SI epidemiological models with a fuzzy transmission parameter. Computers & Mathematics with Applications, 45(10-11), 1619-1628.

Verma, R., Tiwari, S. P., and Upadhyay, R. K. (2017). Dynamical behaviors of fuzzy SIR epidemic model. In Advances in Fuzzy Logic and Technology, 482–492.

Van den Driessche, P. and Watmough, J., 2002. Reproduction numbers and sub-threshold endemic equilibria for compartmental models of disease transmission. Mathematical biosciences, 180(1–2), 29–48.

Castillo-Chavez, C. and Song, B., 2004. Dynamical models of tuberculosis and their applications. Mathematical Biosciences & Engineering, 1(2), 361.

Machine Learning Strategies for Understanding Autistic Neuro Images

Meenakshi Malviya[a] and J. Chandra[b]

[a]Christ deemed to be University of Bangalore, India
[b]Christ deemed to be University of Bangalore, India
E-mail: meenakshimalviya@res.christuniversity.in

Abstract

The use of medical scans for identifying high-risk subjects is instigated but the diagnosis of autism is still a challenge for clinical experts due to the heterogeneous nature of symptoms. Extensive variability in phenotypic features of autism has comorbidity patterns including language impairment, intellectual disability, neuropsychiatric disorders, schizophrenia, epilepsy, and typical development with other conditions. The primary aim of the presented work is to review the diagnostic methodologies applied to a neurological condition known as autism using medical imaging like DTI, EEG, MRI, and fMRI with higher accuracy. The best-fitted imaging techniques for cost-effective faster and early diagnosis of the condition are suggested based on the achieved results and findings in various studies using single, fusion, and multimodality datasets.

Keywords: Autism, Multimodality, Fusion, Neurological, Imaging, MRI, DTI, EEG

Introduction

The Diagnosis of cognitive conditions in children is difficult due to the overlapping of symptoms and the heterogeneous nature of characteristics that affects at various levels on the spectrum differently for each person. It is an umbrella condition containing autistic disorder, Asperger's syndrome, Kanner's syndrome, pervasive developmental disorder not otherwise specified (PDD-NOS), and atypical autism. The global prevalence rate of the condition is 1 in 68 in the reports of the World Health Organization. According to the Centres for Disease Control and Prevention report, there is no medical test available for diagnosis of Autism but it can be identified at an early age of 18 months up to later years. Developmental monitoring and screening conducted by trained experts or clinical professionals are reliable but not accurate always. The global average age of diagnosis is 60 months and can start from 30 to 90 months but early diagnosis is still a challenge.

Some of the common symptoms are delay in communication (including speech, response to own names and gestures), repetitive behavior, learning difficulties, social interaction, and no or very less eye contact while communicating. The primary causes of the condition are- parental age, environmental effects, and family history/ background. Treatment is not available but therapies and training sessions are advised to support autistic people in leading a normal life.

Recent progress in autism identification is the use of multiple types of medical imaging. Medical imaging provides detailed and digital information on affected body parts with specific findings.

Diagnostic Approaches

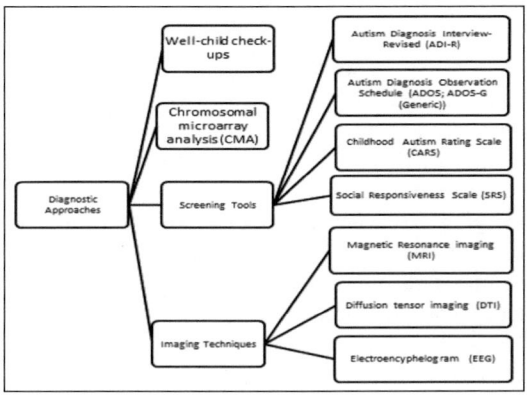

Fig. 1. Various screening tools and imaging modalities for diagnosis of autism

Autism has a genetic heritable condition with a higher rate of concordance at 70–90% in twins and genetic markers exhibit variations in DNA sequence whereas infants' and toddlers' samples represent a functional readout. CMA was used as a first-tier clinical diagnostic tool. SVM using microarray expression profiling produced 91% accuracy in autism diagnosis. An increased microglial activation is also observed in post-mortem studies. Some other important biomarkers are head circumference, Serotonin (blood biomarker), mitochondrial and Metabolic Markers, Oxidative Stress, etc.

Imaging Tools and the Findings

Diffusion Tensor imaging (DTI) is an analysis of white matter microstructure, that displays decreased FA, and reduced coherence in fiber tract directionality in the ventromedial prefrontal cortex, orbitofrontal cortex, and superior temporal gyrus. Due to the excellent contrast property with spatial and temporal resolutions, MRI has been proven to be an investigative tool for ASD. EEG signals are generated in various frequencies and measured to observe the functional connectivity in brain regions. EEG signals are helpful to identify autism as early as the age of 9 months.

Many functional brain fMRI showed overall brain volume where greater white matter volume and decreased grey matter volume in the left inferior frontal gyrus. Likewise, various observations are recorded with the use of different imaging modalities that are briefly mentioned in Table 1.

A model-level fusion was conducted by calculating functional network connectivity (FNC) from rs-fMRI and gray matter volume from sMRI to classify schizophrenia and autism. The structural brain was observed with increased cerebellum volume and Total Brain Volume (TBV). Autism fMRI studies found a decreased long-distance connectivity in brain regions at resting state. The default mode network (DMN) dysfunction in the brain is an important endophenotype for clinical behavioral deficits in ASD. The study of endophenotypes provides information belonging to ASD genetic expressions and the associated external phenotype.

Methodology

The study inclusion criteria for the review work were the use of a verified and valid dataset, various imaging approaches, and their findings for the successful identification of autistic subjects. Another criterion for the selection of articles was the use of multimodal imaging techniques for the successful identification of autism or to understand the functional connectivity in the autistic brain. The study selected the articles, reports, and book chapters published in clinical and non-clinical reputed publications. The included research work here belongs to highly precious publications like Frontiers neuroscience, Elsevier, NIMH, and IEEE.

Detailed Analysis Of Multiple Modalities For Early Prognosis

Data integration from multiple sources helps to better understand the biomarkers and their prevalence to increase diagnostic accuracy in the field of cognitive research.

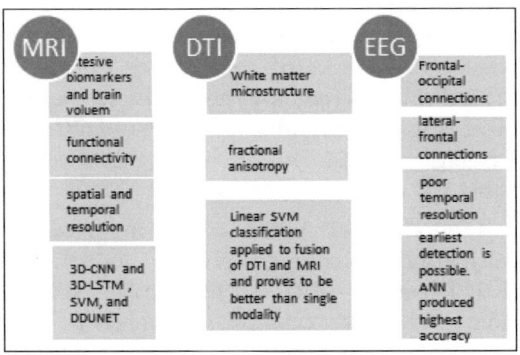

Fig. 2 Autism in multiple modalities

Autistic brain fMRI has two forms of image one is task-based which shows reduced functional connectivity between frontal and parietal regions of the brain, and another is a resting state study that shows decreased connectivity between the anterior and posterior insula, and reduced functional connectivity within and between resting state networks. Structural MRI yields information about brain anatomy, volumes of gray and white matter, and information about the gyrus and sulcus. Autistic brains have large head circumference and volume, and an increased frontal lobe but smaller cerebellar vermis, amygdala, and hippocampus. Superior temporal sulcus is implicated as a critical region to study social cognition in autistic control. The association of parts of the brain with social activities is briefly discussed, medial prefrontal cortex in making inferences, the posterior superior temporal sulcus in biological motion, and the inferior frontal gyrus in emotional judgment. Facial expressions are found associated with the interparietal sulcus in fetching special social attention, and amygdala in recognizing emotions. DTI is used to assess white matter structure in frontal lobes and corpus callosum using an index of white matter fiber density called fractional anisotropy but it shows an increased fractional anisotropy for 2-to-3-year-old subjects. The anterior Temporal Lobe displays a significant relationship with social and figurative communication (language). Lack of sleep spindles found in autistic patients can be identified as a biological marker that can be detected using EEG and/or fNIRS. The spatial resolution of EEG is limited and MRI has a poor temporal resolution,

both have common findings on the autistic brain - frontal-occipital connections recorded as lower long-range connectivity, and higher short-range connectivity in lateral frontal connections. Mostly used parameters of DTI imaging for autism detection are fractional anisotropy (FA) and mean diffusivity (MD) which is diffusion of water in the brain to explain the WM connectivity. Using DTI imaging 8 classification methods were applied to two various feature sets and produced best results with the use of feature set extracted using ML techniques. Reduced FA in WM, and reduced FA with increased MD in WM (Mostapha et al., 2015) were reported for autistic subjects. A study on language-impaired autistic classification using DTI produced 80% classification accuracy. Multimodal analysis using automated techniques helps to represent the high featured view of multisite data for building classification models (Calhoun and Sui, 2016).

Discussion

The findings incidents more than previous findings due to the continuous development in the field of neuroradiological datasets but the scale of biological changes affecting biomarkers requires large data size (Ambrosino et al., 2022), secondly, to automatize the classification, availability of larger datasets can be accomplished by using multisite and multimodal datasets (Shi et al., 2022). A small neuroimaging dataset is a hurdle for technicians to develop software or devices for predictive analysis. The Review work reported multiple neuroimaging datasets and their findings along with the ML algorithms applied for the classification of autism. The implementation of multimodality data introduced the use of fusion/adjoining techniques to convert the data into a single format. Multimodality classification is an emerging trend in the field of cognitive classification with neuroimaging with a few challenges to be considered-common representation, translation into a single cognitive classification with neuroimaging with a few challenges to be considered-common representation, translation into a single modality, feature alignment, and adjoining the information for successful classification.

Table 1. Automated models and modalities with classification accuracy

Ref. No.	Modalities	Findings	ML approaches	Classification accuracy
Grossi et al. (2021)	EEG (resting state EEG)	lower long-range frontal-occipital connections, and higher short-range connectivity in lateral–frontal connections	ANN	Avg 80% for ages under 18 months
Du et al. (2022)	sMRI and fMRI	sMRI-Increased GMV for the middle frontal gyrus	SVM	83.08%
		fMRI-weaker activation in the left middle frontal gyrus		
Hazlett et al. (2005) and Carper et al. 2002)	sMRI	Increased volume of Cerebellum, WM, GM, and TBV	NA	NA
Lu et al. (2022)	genes and imaging data	Successful classification	Multiple Kernel Learning	83.6%
Barttfeld et al. (2011)	DTI	decreased FA	NA	NA
Redcay (2008)	3D-T1-W-MRI	decreased superior temporal sulcus	Voxel-based morphometry	NA
Dekhil et al. (2019)	sMRI, fMRI fusion	Common feature set vector	Global random forest	81%
Crimi et al. (2017)	fMRI, DTI and T1-W fusion	Common feature set vector	Linear SVM	72%
Zhou et al. (2019)	Rs-fMRI	Spatio-temporal features	3D-CNN and 3D-LSTM model	Highest multisite accuracy
Li et al. (2019)	T1-weighted MR	Volume overgrowths of amygdala and CA1-3 of the hippocampus	Dilated-Dense U-Net (DDUNET)	insight information of infant's neurobiology
Chen et al. (2020)	Eye fixation, facial expression, cognitive features	Multimodality fusion helped to produce a more accurate and cost-effective model for autism detection	Random Forest	91%

References

Grossi, E., Valbusa, G., and Buscema, M. (2021). Detection of an autism EEG signature from only two EEG channels through features extraction and advanced machine learning analysis. Clinical EEG and Neuroscience, 52(5), 330–337.

Du, Y., He, X., Kochunov, P., Pearlson, G., Hong, L. E., van Erp, T. G., and Calhoun, V. D. (2022). A new multimodality fusion classification approach to explore the uniqueness of schizophrenia and autism spectrum disorder. Human Brain Mapping.

Hazlett, H. C., Poe, M., Gerig, G., Smith, R. G., Provenzale, J., Ross, A., et al. Magnetic resonance imaging and head circumference study of brain size in autism: birth through age 2 years. Arch Gen Psychiatry (2005) 62:1366–76

Carper, R. A., Moses, P., Tigue, Z. D., and Courchesne, E. (2002). Cerebral lobes in autism: early hyperplasia and abnormal age effects, NeuroImage, 16(4), pp. 1038–1051.

Lu, P., Li, X., Hu, L., et al. (2022). Integrating genomic and resting State fMRI for efficient autism spectrum disorder classification. Multimed Tools Appl 81, 19183–19194.

Barttfeld, P., Wicker, B., Cukier, S., Navarta, S., Lew, S., and Sigman, M. (2011). A big-world network in ASD: dynamical connectivity analysis reflects a deficit in long-range connections and an excess of short-range connections. Neuropsychologia. 2011, 49, 254–263.

Calhoun, Vince D. and Sui, Jing (2016). Multimodal fusion of brain imaging data: A key to finding the missing link(s) in complex mental illness, Biological Psychiatry: Cognitive Neuroscience and Neuroimaging.

Ambrosino, S., Elbendary, H., Lequin, M., Rijkelijkhuizen, D., Banaschewski, T., Baron-Cohen, S., and Durston, S. (2022). In-depth characterization of neuroradiological findings in a large sample of individuals with Autism Spectrum Disorder and controls. NeuroImage: Clinical, 103118.

Shi, C. L., Xin, X. W., and Zhang, J. C. (2022). Domain adaptation based on rough adjoint inconsistency and optimal transport for identifying autistic patients. Computer Methods and Programs in Biomedicine, 215, 106615.

Dekhil, O., Ali, M., El-Nakieb, Y., Shalaby, A., Soliman, A., Switala, A., .. and Barnes, G. (2021). A personalized autism diagnosis CAD system using a fusion of structural MRI and resting-state functional MRI data. Frontiers in psychiatry, 392.

Crimi, A., Dodero, L., Murino, V., and Sona, D. (2017). Case-control discrimination through effective brain connectivity, 2017 IEEE 14th International Symposium on Biomedical Imaging (ISBI 2017), pp. 970–973.

El-Gazzar, A., Quaak, M., Cerliani, L., Bloem, P., van Wingen, G., Mani Thomas, R. (2019). A Hybrid 3DCNN and 3DC-LSTM Based Model for 4D Spatio-Temporal fMRI Data: An ABIDE Autism Classification Study. In:, et al. OR 2.0 Context-Aware Operating Theaters and Machine Learning in Clinical Neuroimaging. OR 2.0 MLCN 2019 2019. Lecture Notes in Computer Science, vol 11796. Springer, Cham.

Li, G., Chen, M. H., Li, G., Wu, D., Sun, Q., Shen, D., and Wang, L. (2019, April). A preliminary volumetric MRI study of amygdala and hippocampal subfields in autism during infancy. In 2019 IEEE 16th International Symposium on Biomedical Imaging (ISBI 2019) (pp. 1052-1056). IEEE

Chen, J., Liao, M., Wang, G., and Chen, C. (2020). An intelligent multimodal framework for identifying children with autism spectrum disorder. International Journal of Applied Mathematics and Computer Science, 30(3).

Solution Synthesis, Characterization, Photoluminescence and Afterglow Studies of Y_2O_3:Eu^{3+}, Ho^{3+} Phosphor

Monika Khale,[a] Anusha Upadhya,[a] Sapna Raghuwashi,[a] Prachi Tadge,[,a] and Sudeshna Ray*[,a]*

[a]Centre of Advanced Materials for Research, Faculty of Sciences,
Rabindranath Tagore University, Bhopal M.P. India

Abstract

The synthesis, structural characterization, photoluminescence, and afterglow study of Y_2O_3:Ho^{3+} (1%), Eu^{3+} (2%) phosphors synthesized by a complex-based precursor solution method followed by annealing at 600 °C and 1000 °C, are reported. XRD patterns confirm the formation of a pure cubic phase of Y_2O_3 in the samples whereas the size of the phosphor samples was measured by TEM. Under UV 254 nm excitation, both samples show intense red emission at 612 nm, which is attributed to the forced-electric dipole transition of Eu^{3+} ion. Importantly, Y_2O_3:Eu^{3+}, Ho^{3+} sample, annealed at 1000 °C exhibits a red afterglow as can be seen in by the naked eye and it can last for 2 min, after extinguishing the UV light and Eu^{3+} ion the luminescent center during the decay process. Ho^{3+} is responsible for the presence of traps and in turn, is responsible for the generated afterglow.

Keywords: Long-afterglow, Luminescent center, Decay, Red-emitting

Introduction

Long afterglow phosphor is a kind of eco-friendly material, which could absorb energy from sunlight, fluorescent lamp or UV light, store the absorbed energy, and then release the energy gradually in form of afterglow (Yu et al, 2009). Because of the long afterglow properties, these materials have been used in many fields, such as illumination in darkness, emergency sign, artwork, luminescent paints, detection of radiation, display, and photocatalytic degradation. Zinc sulfide (ZnS) doped with copper (and later co-doped with cobalt) was the most famous and widely used persistent phosphor. ZnS-based phosphors, such as ZnS:Cu⁺, were found to have persistent time of as long as 40 minutes (Wang et al., 2002) The effects of co-doping were also investigated, for example, co-doping of ZnS:Cu⁺ with Co^{2+} doubled the persistent time of the phosphor (Lehmann, et al., 1972). However, still the brightness and the lifetime of the materials were rather low for practical application. To resolve this problem, traces of radioactive elements such as promethium or tritium were often introduced in the powders to stimulate the brightness and lifetime of the light emission. But even then, a commercial glow-in-the-dark object had to contain a large amount of luminescent material to yield an acceptable afterglow (Matsuzawa et al., 1996). The next generation of long persistent phosphors comprised the alkaline earth sulfides, such as CaS, and SrS. These phosphors were known as Lenard's phosphors, activated with various dopants such as Bi^{3+}, Eu^{2+}, Ce^{3+} etc., and exhibited longer persistence over hours as compared to ZnS:Cu⁺. These phosphors could be excited by sunlight showing versatility in their applications. Among the afterglow phosphors,

$SrAl_2O_4$:Eu^{2+}, Dy^{3+}, and $CaA_{12}O_4$:Eu^{2+}, Nd^{3+} have received significant attention due to their intense afterglow.

In 2014, Xu and co-worker have reported a new NIR long-persistent $SrGa_{12}O_{19}$:Cr^{3+} phosphor with an emission duration of more than 2 h with broadband phosphorescence from 650 to 950 nm (Xu et al. 2014). In 2019, Zheng et al. reported a new cyan-emitting long afterglow phosphor with composition $CaSnO_3$:Lu^{3+} and in the same year, Jiang et al. disclosed a new green emitting phosphor with the composition $Ca_{14}Zn_6Ga_{10}O_{35}$:Mn^{2+},Ge^{4+}.

Recently, in 2022, rare earth doped long afterglow strontium aluminate phosphor has been reported for application in in-vitro optical imaging (Calatayud et al., 2022). In 2020, Poelman et al., in the review article, clearly mentioned the significance of the existence of metastable trap levels that can momentarily store the excitation energy, for the occurrence of persistent luminescence. While native defects like oxygen vacancies can sometimes cause these trap levels, usually co-dopants like Nd^{3+} or Dy^{3+} are frequently employed to introduce these trap levels in order to get persistent luminescence.

Although different host-based afterglow phosphors have been reported, for industrial applications, large quantity of phosphor is required. In this aspect, we have selected a host Y_2O_3, which although not a new host, can be synthesized by an easily scalable synthesis method. Y_2O_3 has been a conventional host for rare-earth doping. Y_2O_3:Eu^{3+} was widely used as the phosphor in color display. In this paper, a red-emitting long afterglow material Y_2O_3:Eu^{3+}, Ho^{3+} was synthesized and the structure, photoluminescence, and decay characteristics of the phosphor material were investigated. Ho^{3+} has been added to create a metastable defect, which is responsible for the afterglow of this phosphor.

Experimental Section

In our synthesis route, Y_2O_3 (Sigma-Aldrich), and Ho_2O_3 (Sigma-Aldrich) were dissolved in minimum volume of concentrated HNO_3 and boiled in water bath to get clear solutions of $Y(NO_3)_3$, $Eu(NO_3)_3$ and $Ho(NO_3)_3$. The nitrate solutions were taken in a beaker according to the required stoichiometry. After that, TEA (triethanolamine) was added to that nitrate solution, and the molecular ratio of metal salt:TEA was maintained 1:4. Then a certain amount of concentrated HNO_3 solution was added to maintain the pH 3–4 to avoid any kind of metal hydroxide precipitation. TEA was used as it is an efficient chelating ligand and possesses good coordination properties with metal ions. The clear solution of TEA complexed metal nitrate is evaporated on a hot plate at 180 °C with constant stirring. Continuous heating of the solution causes foaming and puffing. During evaporation of the precursor solution, the TEA present in the system probably led to the formation of vinyl functional groups that cause polymerization. The black fluffy has been calcined and then annealed at 600 °C and 1000 °C to get the required phosphors.

Characterization

The phase purity of the as-synthesized sample was identified by using powder X-ray XRD) XRD analysis was analyzed with a Bruker AXS D8 advanced automatic diffractometer with Cu-$K\alpha$ radiation (λ = 1.5418 Å), over the angular range $10° \leq 2\theta \leq 80°$, operating at 40 kV and 40 mA. Samples for TEM were deposited onto 300 mesh copper TEM grids coated with 50 nm carbon films. Nanocrystalline samples dispersed in acetone were placed on the grid drop-wise. The excess liquid was allowed to evaporate in air. The grids were examined in JEOL 2010 microscope with Ultra-High Resolution (UHR) pole piece using a LaB_6 filament operated at 200 kV. The photoluminescence (PL) excitation and emission spectra as well as the afterglow of the sample were analyzed by using an RF-5301PC (Shimadzu, Japan) equipped with a 150 W Xenon light source and the slit width was kept 1.5 nm.

Result and discussion

The phase purity of the synthesized samples has been checked by powder XRD. Figure 1 presents the XRD patterns of Y_2O_3: Eu^{3+} (2%), Ho^{3+} (1%), phosphor annealed at 600 °C and 1000 °C. In Fig. 1, the XRD patterns of both phosphors show the positions and relative intensity of the diffraction peaks, which can be indexed to the Ia-3 (206) space group with the standard card of cubic Y_2O_3 (JCPDS#88-1040). No impurity phase is observed in XRD patterns, demonstrating Eu^{3+}/ Ho^{3+} are well incorporated into Y_2O_3 and formed a solution structure. This confirms the effective doping of Eu^{3+} and Ho^{3+} ions on Y^{3+} site.

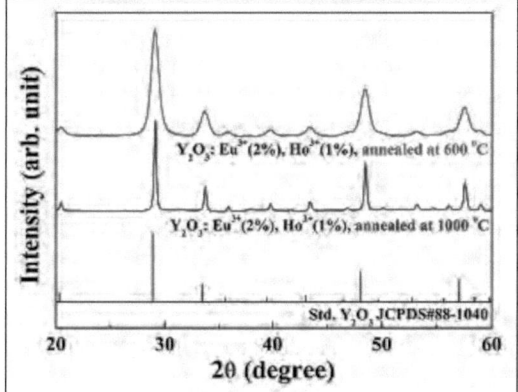

Fig. 1. XRD of Y_2O_3 doped with Eu^{3+} and Ho^{3+} phosphors annealed at 600 °C and 1000 °C

From Fig. 1, it can also be seen that the XRD of the sample annealed at 600 °C is comprised of sharp peaks reflecting the considerably good crystallinity of the phosphor. Both the samples exhibit good crystallinity which owe to the synthesis process. Phosphors samples of Y_2O_3:Eu^{3+} of similar crystallinity and with fine size tuning have been achieved by our group and reported in 2014, which clearly justifies the improved crystalline nature of as-synthesized samples annealed at 600 °C and 1000 °C (Ray et al. 2014).

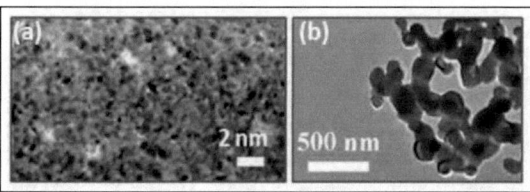

Fig. 2. TEM image of Y_2O_3 doped with Eu^{3+} and Ho^{3+} phosphor at (a) 600 °C and (b) 1000 °C

Figs. 2 (a) and (b) show the TEM images of Y_2O_3: Eu^{3+} (2%), Ho^{3+} (1%), phosphors annealed at 600 °C and 1000 °C. The particle average size

of the particle annealed at 600 °C and 1000 °C is 12 nm and 150 nm respectively.

Photoluminescence excitation and emission spectra of Y_2O_3:Eu^{2+}, Ho^{3+} phosphors have been presented in Figs. 3 and 4. The excitation spectra of Y_2O_3:Eu^{2+} (2%) Ho^{3+} (1%), samples monitored upon 612 nm emission at room temperature are shown in Fig. 3. Excitation spectra are comprised of broad brand absorption peaks located at 254 nm respectively. Figure 4 displays the photoluminescence spectra of the phosphors excited by UV light at the wavelength of 254 nm.

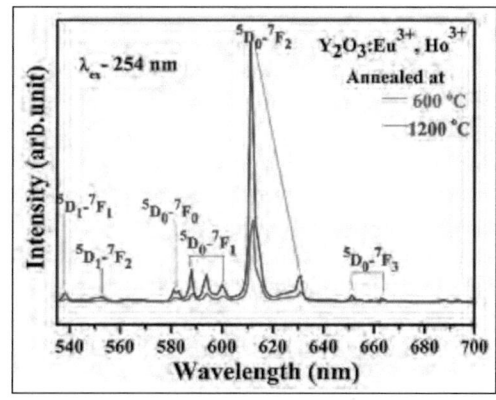

Fig. 3. Photoluminescence excitation of Y_2O_3:Eu^{3+}, Ho^{3+} samples at λem – 612 nm.

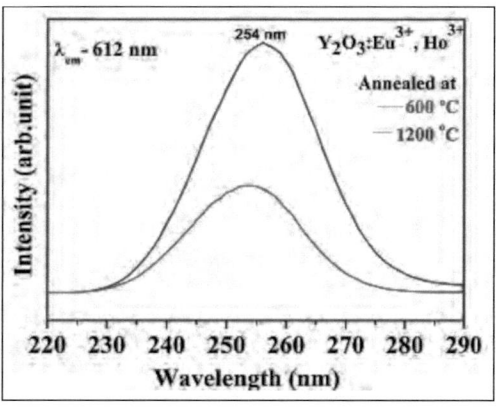

Fig. 4. Photoluminescence emission spectra of Y_2O_3:Eu^{3+}, Ho^{3+} samples at λem – 254 nm

Y_2O_3:Eu^{3+}, Ho^{3+} phosphors exhibited a strong red emission. Sharp emissions peaking at 582, 588, 594, 600 and 612 nm can be assigned to the $^5D_0 \rightarrow ^7F_J$ (where J is 0, 1, and 2) transitions of Eu^{3+} ion (marked in Fig. 3). No emission can be observed in the Y_2O_3:Eu^{3+}, Ho^{3+} sample, indicating that the doped Ho^{3+} ion do not show any significant emission under 254 nm excitation. The strongest red emission at 612 nm is due to the low symmetry position of Eu^{3+} ion with an inversion center and this forced-electric dipole transition is often dominant in the emission spectrum. So the strongest emission at 612 nm is ascribed to the $^5D_0$-$^7F_2$ transition of Eu^{3+} ion, which occupied on the C_2 site and these ions are dominant in the host.

Moreover, Y_2O_3:Eu^{3+}, Ho^{3+} samples were excited under UV 254 nm for 5 min, and then the excitation source was removed to examine whether the materials show long afterglow emission in the dark. Y_2O_3:Eu^{3+}, Ho^{3+} sample exhibits an obvious red afterglow which can be observed by naked eye for about 60 and 120 seconds as shown in the left inset of Fig. 5. Due to the higher photoluminescence emission intensity of the phosphor annealed at 1000 °C, was used for checking afterglow emission.

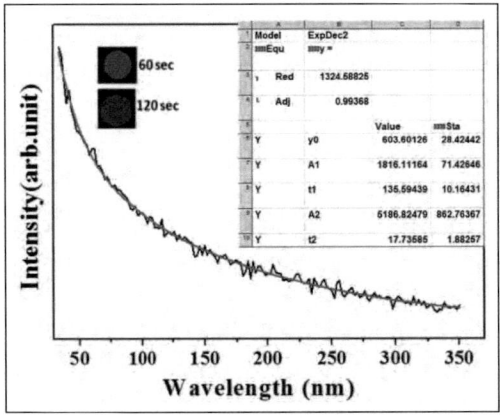

Fig. 5. The decay curve of the Y_2O_3:Eu^{3+}, Ho^{3+} sample, the left inset is the afterglow photograph of Y_2O_3:Eu^{3+}, Ho^{3+} phosphor which was examined at 60 second and 120 second after UV 254 nm light excitation for 5 min.

The decay curve of the Y_2O_3:Eu^{3+}, Ho^{3+} sample was recorded and shown in Fig. 5. It contains a rapid-decaying process and a slow-decaying process, the decay curve is similar to that of typical long afterglow materials, and it can be evaluated by fitting using a double exponential equation which reflects the trend of the decay.

The equation is as follows:

$$I = I_1 \exp \frac{(-t)}{\tau_2} + I_2 \exp \frac{(-t)}{\tau_2} \quad ......(1)$$

Where I represent the phosphorescent intensity; I_1 and I_2 are two constants; τ represents the time; τ_1 and τ_2 are decay constants that decide the decay rate for the rapid and the slow exponentially decay components. The fitting results of the parameters of τ_1 and τ_2 and the fitting curves are shown in Fig. 5.

It can be observed that the decay curve of an as-synthesized sample annealed at 1000 °C, is similar to that of typical long afterglow materials. The decay characteristic of Y_2O_3:Eu^{3+}, Ho^{3+} phosphor is similar to the rare-earth doped alkaline-earth aluminates long afterglow materials, which have drawn much attention from 1990s. One of the typical aluminates is $SrAl_2O_4$:Eu^{2+}, RE^{3+} (RE = Dy, Nd and Ho, etc.), which was successfully synthesized by Matsuzawa in 1996. Generally, the Eu^{2+} ion is the only luminescent center in $SrAl_2O_4$:Eu^{2+}, RE^{3+} phosphors, and the doped RE^{3+} ions often generate the trap level in the host. In order to exhibit long afterglow phenomena, the phosphor should have suitable trap level and trap density and the trapped carriers do recombine in the luminescent centers occurred with the afterglow. In this case, Ho^{3+} is responsible for the creation of trap responsible for the observed afterglow, as Y_2O_3:Eu^{3+} does not exhibit afterglow.

Conclusion

Y_2O_3 doped with Eu^{3+} and Ho^{3+} phosphors were synthesized by a complex-based precursor solution method followed by annealing at 600 °C and 1000 °C. The phase identification by XRD shows that both samples are comprised of pure cubic phases of Y_2O_3 with considerably good crystallinity. Both the samples display photoluminescence excitation and emission spectra, the emission spectra were dominated by the red emission corresponding to the forced electric-dipole allowed transition of Eu^{3+}. Y_2O_3:Eu^{3+}, Ho^{3+} phosphor, annealed at 1000 °C exhibits a red long afterglow phenomenon which is attributed to the codoping of Ho^{3+} into the matrix, as Y_2O_3:Eu^{3+} phosphor does not exhibit afterglow.

References

Duan, B. C., Zhang, J. C., Liu, X., Yuan, Q., and Wang, Y. H. (2014). Tunable blue-white-red photoluminescence and long-lasting phosphorescence properties of a novel zirconate phosphor, J. Alloys Comp., 587, 318–325.

Calatayud, D. G., Jardiel, T., Erica Cordero-Oyonarte, F., Amador, C., Caballero, A. C., Villegas, M., Noguera, A., Adalia, A., and Peiteado, M. (2022). Biocompatible Probes Based on Rare-Earth Doped Strontium Aluminates with Long-Lasting Phosphorescent Properties for In Vitro Optical IMAGING, Int. J. Mol. Sci., 23, 3410.

Jiang, B., Ch., F., Zhao, L., Wei, X., Chen, Y., and Yin, M. (2019). Luminescence properties of a new green emitting long afterglow phosphor Ca14Zn6Ga10O35:Mn2+, Ge4+, J. Lumin., 206, 224–239.

Lehmann, W. (1972). Activators and co-activators in calcium sulfide phosphors, J. Lumin. 5:87 107.

Matsuzawa, T., Aoki, Y., Takeuchi, N., and Murayama, Y. (1996). New long phosphorescent phosphor with high brightness. $SrAl_2O_4$:Eu^{2+}, Dy^{3+}, J. Electrochem. Soc., 143:2670 2673.

Poelman, D., Heggen, D., Du, J., Cosaert, E, and Smet, P. F. (2020). Persistent phosphors for the future: Fit for the right application, J. Appl. Phys. 128, 240903.

Ray, S., Luis, S., Manjon, F., Mollar, M. A., Gomis, O., Rodriguez, U., Agouram. M., Munoz, A., and Lavin, V. (2014). Broadband, site selective and time resolved photoluminescence spectroscopic studies of finely size-modulated Y_2O_3:Eu^{3+} **phosphors synthesized by a complex based precursor solution method, Current Appl. Phys. 14**, 72–81.

Xu, J., Chen, D. Q., Yu, Y. L., Zhu, W. J., Zhou, J. C., and Wang, Y. S. (2014). Cr^{3+}:$SrGa_{12}O_{19}$: A broadband near-infrared long persistent phosphor, Chem. Asian J., 9, 1020–1025.

Yu, N. Y., Liu, F., Li, X. F., and Pan, Z. W. (2009). Near infrared long-persistent phosphorescence in $SrAl_2O_4$:Eu^{2+}, Dy^{3+}, Er^{3+} phosphors based on persistent energy transfer. Appl. Phys. Lett. 95, 231110.

Zheng, C., and Liu, Q. (2019). Luminescent properties of a new cyan long afterglow phosphor $CaSnO_3$:Lu^{3+}, RSC Adv., 9, 33596–33601.

A Comprehensive Review of the Forensic Significance Toxic Medicinal Plants

Pratibha Thakre,[a] *Monika Khale,*[b] *Madhu Mishra,*[a] *Keyur Acharya,*[b]
Suchi Modi,[*,a] *Prachi Tadge,*[*,b] *and Sudeshna Ray*[*,b]

[a]Department of Life Science, Faculty of Science, Rabindranath Tagore University, Bhopal, M.P. India
[b]Centre of Advanced Materials Research, Faculty of Science, Rabindranath Tagore University, Bhopal, M.P. India
E-mail: sudeshna.ray@aisectuniversity.ac.in; prachi.tadge@aisectuniversity.ac.in; suchi.modi@aisectuniversity.ac.in

Abstract

Medicinal plants are a two-edged sword that might be used as a medicine for several diseases and as deadly poisonous substances to commit murder or suicide when administered in high doses. Plants have toxic properties, which they create as their self-defense in the form of secondary metabolites these toxic plants may be used as biological weapons and have significant application in "Forensic Botany, which is composed of the scientific use of plant materials to solve crimes and to gain information regarding possible crimes." Forensic Botany is similar to DNA Fingerprinting. Various plants and plant materials like pollen at the scene of crime or a rare plant type present near a murdered victim can be helpful in connection to a suspect or a victim. The present paper summarizes the application of two poisonous plants such as "Conium maculatum (Hemlock)" and "Datura Stramonium" (Datura) for medicinal purposes and in criminal activities.

Keywords: Medicinal Plant, Forensic Botany, Secondary Metabolites

Introduction

Numerous plant species are toxic to people. To protect themselves from danger, plants create toxins as metabolites. Organic substances such as alkaloids, diterpenes, flavonoids, tannins, cardiac and cyanogenic glycosides, pro-anthocyanidins, phenylpropanoids, lignans, nitrogen-containing compounds, resins, oxalates, and certain proteins or amino acids are the main dangerous elements found in plants. These harmful secondary metabolites have different properties depending on their origin and environment. In addition to being hazardous to humans, poisonous plants have also been linked to cattle deaths, health problems, decreased production, birth defects, and shorter life-spans, all of which have been linked to significant financial losses. Importantly, in homicide, burglary, and other criminal offenses, toxic plants can be used as evidence. Some poisonous plants have medicinal and nutraceutical potential and can treat some serious ailments despite their harmful effects on people and other animals. It is noteworthy to mention that medicinal herbs can be used to treat a variety of illnesses, but when used in excessive dosages, they can also be lethal poisonous chemicals that can be used to commit homicide or suicide. Toxic plants could be turned into biological weapons (Khajja et al., 2011).

At crime scenes, forensic professionals can gather traces and leftover elements from these poisonous medicinal plants to solve the crime. To further comprehend the importance of medicinal herbs in forensic investigations more research must be done. To give a thorough understanding of chemical substances that can have a beneficial or bad impact on human existence at various dosages as well as to find the ideal or overdose concentrations for either therapies or poisonous effects using current biotechnological technologies, are of great importance. The purpose of this overview is to highlight the application of medicinal plants in the area.

It has been reported that Nerium Oleander is a medicinal plant that is grown in two attractive colors, pink and white. It is grown in gardens, schools, and all over India, and has been used as medicine to treat skin problems. However, it also appears to have some toxic effects attributable to its roots, leaves, seeds, and stems. Phytochemicals are a class of biological substances found in plants. They are categorized as bioactive substrates, and studies have shown that they are advantageous to human health. Terpenes, polyphenols, and alkaloids are just a few examples of the biological substances that make up these phytochemicals or substrates. Alkaloids have the majority of pharmacological effects, including anti-cancer and anti-asthma effects. Additionally, several harmful substances could be brought into alkaloids like atropine and tubocurarine. The healthcare system has relied on herbal therapy for a long time, starting from 2600 BCE (World Health Organization (WHO)). Nowadays, especially in poorer nations, medicinal plants are thought to be a common component of pharmaceuticals used to cure patients. For a very long time, people have used medicinal herbs to treat ailments like parasite infections, inflammation, and coughing.

Therapeutic plants are frequently referred to as alternative medicine because it is known that they can be used to achieve several undesirable goals, such as inducing death, severe injury, and coma by changing the concentration of deadly substances. The oleander plant's cardiac glycosides primarily cause cardiotoxicity, though they can occasionally be fatal. Even while certain plants don't contain cardiac glycosides, they can nonetheless be hazardous to the human heart and oesophagus. The well-known plant Peganum harmala (P. harmala), which grows in semi-arid rangeland regions like the Middle East and North Africa, serves as another example. It is a plant that is poisonous and can hurt both people and animals and causes disorders of the neurological and digestive systems as symptoms. The P. harmala plant was once employed to promote abortion. A methanolic extract from P. harmala was given to female rats as part of an experiment in small doses ranging from 2 to 3.5 g/day for 30 days. While the rats' physical and dietary conditions did not change, it was seen that there was a stimulatory impact on the uterus. In the neurological system, opium, an opiate sedative derived from the poppy plant, binds to specific opioid receptors to reduce pain.

It is acknowledged that medicinal plants are a double-edged sword that can be used not only for medical treatments, where they are referred to as alternative medicine (AM) (Dar et al., 2017), but also for several harmful purposes, such as causing death, serious injury, and stuporing by manipulating the concentration of poisonous substances. The cardiac glycosides found in oleander plants primarily induce cardiotoxicity and, in rare circumstances, can be fatal.

Therefore, narcotics and hazardous plants offer substantial evidence for forensic investigations into crimes like suicide or burglary. Botanical weapons made of toxic compounds that can be employed for either self-defense or malicious intent have been used for a variety of reasons. Because they may be found for free, hazardous plant materials are preferred by professional criminals. As a result, the poison's concentration is determined by the criminal's motive. There are many techniques to identify crimes, such as using biotechnology and molecular tools to identify certain toxins or new technology that employs a probe to bind to a target molecule to identify specific poisons. This paper attempts to evaluate and advance knowledge on the effective forensic use of medicinal herbs to identify criminal offenses.

Table 1 (Dubey et al., 2018) shows the list of eight poisonous plants along with their "Family," "Active Compounds" and major symptoms of the impact of these poisonous plants on humans.

Table 1. Common Toxic plants with their Family, Active Compounds and major symptoms

Plant Species	Family	Active Compounds	Major Symptoms in Humans
Abrus precatorius (Dickers et al., 2003)	Fabaceae	Abrin abric acid	Gastrointestinal disease
Argemone Mexicana (Dalvi et al., 1985)	Papaveraceae	Toxic alkaloids sanguinarine, berberine, protopine	Generally causes edema, fruits are contaminated
Gutierrezia sarothrae (Ralphast et al., 2011)	Asteraceae	Saponin in leaves	Severely toxic if eaten in quantity and also causes abortion
Nerium indicum (Langford et al., 1996)	Apocynaceae	Cardioactive glycosides	Diarrhea, drowsiness, respiratory paralysis
Nicotiana tabacum (Benowitz et al., 1998)	Solanaceae	Nicotine	Vomiting, diarrhea, slow pulse, collapse, respiratory failure
Strychnos nuxvomica (Brown et al., 1975)	Loganiaceae	Strychnine, brucine, vomicine, protostrychnine, chlorogenic acid, n-oxystrychnine	Painful seizures, spasms, difficulty breathing, dizziness, confusion
Calotropis gigantean (Seiber et al., 1982)	Asclepiadaceae	Laurane, saccharose, B-amyrin, calotroposide; calactin, calotoxin; calotropins DI and DII, gigantin	Diarrhea, vomiting, effects in lactation
Aconitum napellus (Ameri et al., 1998)	Ranunculaceae	Aconitine, mesaconitine, hypaconitine, other Aconitum, alkaloids	Weakness or inability to move, sweating, breathing problems, heart problems, ventricular, arrhythmias

Toxic Plants of Forensic Significance

Plants are sometimes the main witness in criminal cases such as burglaries, suicides, and other heinous crimes. Plant fragments can be found in the clothing or possessions of a criminal and later serve as evidence in a court of law when they are interpreted by experts in taxonomy, molecular taxonomy, anatomy, and ecology. The first instance of this happened in January 1935 when Arthur Koehler, a wood technician, used his understanding of wood anatomy to identify the origin of one of the parts of a wooden ladder used in the crime to reach the baby. Plants that produce poisonous compounds as secondary metabolites can be utilized as evidence in forensic investigations. Criminals commonly use plant-derived or botanical weapons in rape, burglary, and murder cases. Conium, Cicuta, Nerium, Aconitum, Datura, and Ricinu are some poisonous plant species that are specifically utilized for forensic significance to arrest criminals. These plants are highly toxic and used for homicidal and suicide objectives. Datura is the only plant with such a criminal past, and the seeds are attractive poisons.

Poisonous Impact and Medicinal Use of Hemlock and Datura

Conium Maculatum (Hemlock), the lethal poison that the Greek philosopher Socrates was ordered to swallow, is claimed to have been given to offenders. Hemlock is harmful at all stages of growth, although it is most lethal in the early spring growth stages (Reynolds et al., 2005). Hemlock is a highly toxic plant. The plant is poisonous in every aspect. Hemlock seeds contain poisons that are so lethal that some individuals have passed away after eating game birds that have consumed hemlock seeds. Hemlock was brought to North America as an ornamental plant from its native regions in Europe and western Asia. Southern Canada and the US both commonly have it. Hemlock frequently grows close to barriers such as fences, roadsides, ditches, vacant lots, pastures, crops, and fields, where it can be mistaken for non-toxic plants and causes accidental death due to hemlock poisoning.

Hemlock leaves, roots, and seeds are used to create medicine despite major safety concerns. It is used for painful disorders like children's teething, swollen and painful joints, and cramps, as well as for respiratory difficulties like bronchitis, whooping cough, and asthma. Additionally used for mania and anxiety, hemlock. It is important to mention that Poison hemlock (*Conium maculatum*) is a plant that is poisonous for humans and animals and causes conditions like Central nervous system such as (1) Depression, (2) Respiratory failure, (3) Acute Rhabdomyolysis (4) Acute renal failure and even death (Vetter et al., 2004).

Datura stramonium has long been known for its hallucinogenic and euphoric effects. It was dried and smoked for hallucination and total relaxation. Accidental Datura poisoning is fairly common, and a variety of cases have been documented, including Datura poisoning from eating a hamburger, atropine poisoning after eating porridge tainted with D. stramonium, atropine poisoning after drinking tainted comfrey tea, and after consuming sorghum flour tainted with Datura seeds, many individuals in Botswana were ill. Several fatalities have happened, especially in kids who are drawn to the seeds and capsules. A fatal D. stramonium poisoning, the smoking and ingesting of D. stramonium for its psychedelic effects, and the eating and chewing of Datura seeds during a suicide attempt are all instances of intentional D. stramonium poisoning. The identification of two active compounds of Datura such as "Atropine" and "Scopolamine" by HPLC technique has been reported in the literature (Gaillard et al., 1999; Groeneveld et al. 1990).

Summary and Outlook

Different poisonous secondary metabolites that are deadly to human health are found in plants. The vast majority of these hazardous concepts have forensic importance. There haven't been many studies on this subject, but there is a lot more that needs to be done because there are many dangerous plants in nature. Forensic evidence in criminal investigations requires the right integration of botanical knowledge of these plants from allied fields, such as plant systems, plant anatomy, and plant ecology. To raise awareness while planting hazardous trees in social forestry and agro-forestry initiatives, more information is desired on the impacts of various environmental and geographic variables on the formation of toxic principles in plants. The examination of the toxin derived from the toxic plants must be carried out to create the database for solving any crime in order to solve the crime committed by toxic plants. Numerous therapeutic herbs can be fatal when taken in big doses, so further research is urgently needed with a focus on phytochemical analyses of Dhatura (Datura stramonium) and Hemlock (Conium maculatum), two plants that have medicinal benefits but can be lethal when used in large doses.

References

Ameri, A. (1998). The effects of Aconitum alkaloids on the central nervous system, Prog. Neurobiol.56(2): 211 235.

Benowitz N. L. (1998). Nicotine Safety and Toxicity, Oxford University Press, USA.

Dar, R. A., Shahnawaz, M., and Qazi, P. H. (2017). General overview of medicinal plants: A review, J. Phytopharmacol. 6, 349–351.

Brown, M., Vale (1975) W., Central nervous system effects of hypothalamic peptides, Endocrinology, 96(5), 1333–1336.

Dalvi, R. R. (1985). Sanguinarine: its potential, as a liver toxic alkaloid present in the seeds of Argemone mexicana, Experientia, 41(1), 77–78.

Dickers, K. J., Bradberry, S. M., Rice, P., Griffiths, G. D., and Vale, J. A. (2003), Abrin poisoning, Toxicol. Rev., 22(3), 137–142.

Dubey, N. K., Dwivedy, A., Chaudhari, A. and Das, A., (2018). Common Toxic Plants and Their Forensic Significance, Natural Products and Drug Discovery (349 374)

Gaillard, Y. and Pepin, G. (1999). Poisoning by plant material: review of human cases and analytical determination of main toxins by high-performance liquid chromatographye (tandem) mass spectrometry, J. Chromatogr. B, 733, 181–229.

Gaire, B. P. and Subedi, L., (2013). A review on the pharmacological and toxicological aspects of Datura stramonium L. Journal of Integrative Medicine, 11(2), 73–79.

Groeneveld, H. W., Steijl, H., Berg, B. and Elings, J. C. (1990). Rapid, quantitative HPLC analysis of Asclepias fruticosa L. and Danaus plexippus L. cardenolides, J. Chem. Ecol., 16, 3373–3382.

Khajja, B., Sharma, M. and Singh, R. (2011). Forensic Study of Indian Toxicological Plants as Botanical Weapon (BW) A Review, J. Environment Analytic. Toxicol., 01, 1–5.

Langford, S. D., and Boor, P. J. (1996). Oleander toxicity: an examination of human and animal toxic exposures, Toxicology, 109(1), 1–13.

Ralphs, M. H., McDaniel, K. C., (2011). Broom snakeweed (Gutierrezia sarothrae): toxicology, ecology, control, and management, Invasive Plant Sci. Manage. 4(1), 125–132.

Seiber, J. N., Nelson, C. J., and Lee, S. M. (1982), Cardenolides in the latex and leaves of seven Asclepias species and Calotropis procera, Phytochemistry, 21(9), 2343–2348.

Vetter, J. (2004). Poison hemlock (Conium maculatum L.), Food Chem. Toxicol., 42, 1373–1382.

Investigation on SIDO Quadratic Buck Converter for Battery Charging and LED Driver Applications

Chandrasekhar Azad Narlapati,ᵃ Nilanjan Tewari,,ᵃ J. Meenakshi,ᵃ and Sankar Narayan Mahatoᵇ*

ᵃVellore Institute of Technology, Chennai Campus, Chennai, India
ᵇNational Institute of Technology, Durgapur, India
E-mail: *nilanjantewari.ee@gmail.com

Abstract

In recent days, the application of power electronics has increased enormously in battery charging and LED driver applications. DC-DC buck converters are popular choice for both of these applications. This paper presents a single input dual output (SIDO) based DC-DC quadratic buck converter for both battery charging as well as LED driving simultaneously. The typical cascading feature of the quadratic buck converter is utilized here to obtain multiple output from the converter. A detailed design of the SIDO quadratic buck converter is presented in this article. Finally a prototype converter of 150W is fabricated for the purpose of investigation under various loading conditions. A 12V, 80Ah lead acid battery, 12V, 6W LED strip and different combinations of resistive loads are considered during this experimental work. The position of battery and LED are interchanged from first buck stage to second buck stage to obtain different parameters for analysis.

Keywords: Quadratic Buck, Battery charger, LED driver

Introduction

DC-DC buck converters are popularly used for battery charging and LED driving applications (Lucio dos Reis Barbosa et al., 1999; Vinicius Miranda Pacheco et al., (2000)). Different topologies of DC-DC buck converters have evolved for these applications (Malay Ranjan Khuntia et al., 2020; Shrikant Misal et al., (2020)). Interleaved buck converters, synchronous buck converters, and quadratic buck converters are very common among them (Ravindranath Tagore Yadlapalli et al., 2017; M. Veerachary et al., (2020)).

Interleaved buck converters are used to mainly reduce the ripple in output current which is desired especially for battery charging applications. On the other hand, in a synchronous buck converter, the diode is replaced with a switch for performance enhancement. The dynamic performance and controller design of a coupled inductor-based SIDO buck converter is presented by (Gayatri Nayak et al., 2020).

A single switch quadratic buck converter is suitable for a wide conversion ratio (Veerachary, 2017). The relation between output and input voltage is in quadratic ratio for this converter. This converter can also be used as a dual output converter for the two buck stages. This paper presents a SIDO quadratic buck converter for battery charging as well as LED driving applications.

Quadratic Buck Converter

The power circuit of the SIDO quadratic buck converter is demonstrated in Fig. 1. It consists of single switch, three diodes and two inductors.

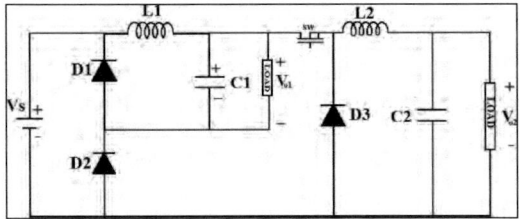

Fig. 1. SIDO Quadratic Buck Converter

Operation of this converter is explained through two modes.

Mode 1: Switch "SW" is turned on during this mode. Diodes D1 and D3 are reversed-biased during this mode of operation. The equivalent circuit for Mode 1 is shown in Fig. 2.

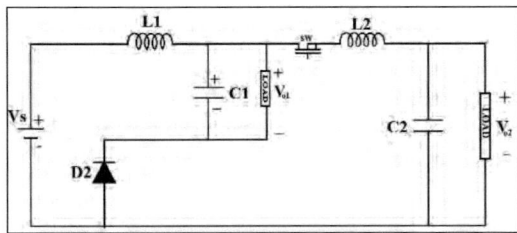

Fig. 2 Equivalent Circuit during Mode 1

Mode 2: Switch is turned off during Mode 2. Diode D1 and D3 are forward-biased during this time interval and they are providing a freewheeling path for the inductors. Equivalent circuit for this mode is shown in Fig. 3.

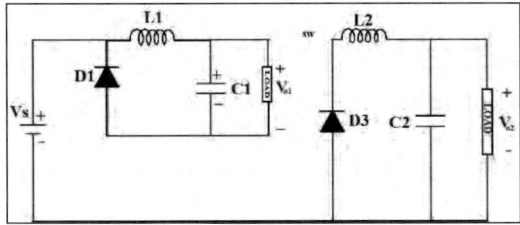

Fig. 3. Equivalent Circuit during Mode 2

The first stage buck output can be taken across the capacitor C1. The voltage across this capacitor is given by,

$$V_{C1} = DV_S \dots\dots\dots\dots\dots\dots\dots\dots(1)$$

The second stage buck output can be taken across the capacitor C2. The voltage across this capacitor is given by,

$$V_{C2} = V_0 = D^2V_S \dots\dots\dots\dots\dots\dots(2)$$

It is clear from the expression that the second stage output voltage is in quadratic relation with the duty cycle.

Design of Passive Elements

Inductors and capacitors of the SIDO quadratic buck converter are chosen by using the following equations,

$$L_1 = \frac{V_{L1}(1 - D)T}{\Delta I_{L1}} \dots\dots\dots\dots\dots\dots\dots(3)$$

$$L_2 = \frac{V_0(1 - D)}{\Delta I_0 . f_s} \dots\dots\dots\dots\dots\dots\dots(4)$$

$$C_1 = \frac{I_{L1} . D(1 - D)T}{\Delta V_{C1}} \dots\dots\dots\dots\dots\dots(5)$$

$$C_2 = \frac{\Delta I_{C1}}{8\Delta V_{C2}f_s} \dots\dots\dots\dots\dots\dots\dots(6)$$

Where, D is duty cycle and VS is the input voltage to the converter.

Hardware Prototype

A hardware prototype of quadratic buck converter of 250 W as shown in Fig. 4 is developed for analysis. The input voltage to the converter is selected as 35 V. The switching frequency of the converter is 50 KHz. The value of the L1, L2, C1 and C2 are chosen as 100µH-8A, 65µH-10A, 100µF-50V and 47µF-100V respectively. The part number of the switch and diodes are IRFP460 and MUR1610CT respectively.

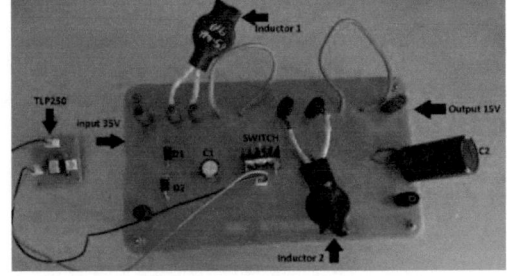

Fig. 4. Prototype of the Converter

Experimental Results

The dual output quadratic buck converter is tested with different combinations of load which includes LED, battery, and resistive load.

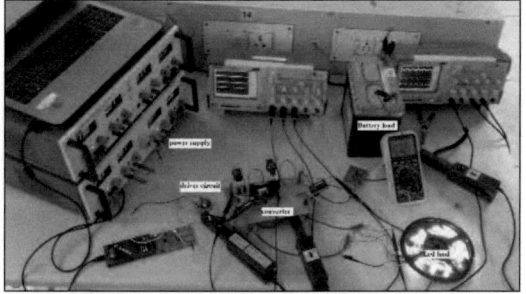

Fig. 5. Prototype Converter with LED and battery

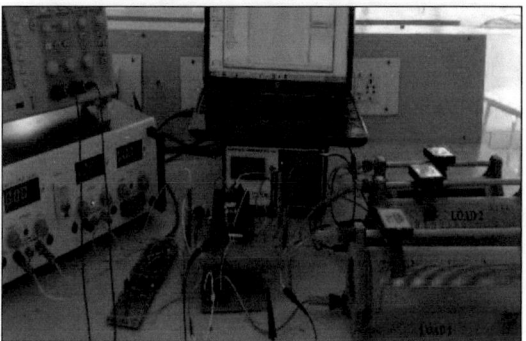

Fig. 6. Prototype Converter with two resistive loads

A. Battery load at first buck stage and variable resistive load at second buck stage

In this case, testing has been carried out by keeping the battery load at the first buck stage and a variable resistive load at the second buck stage. Table I shows the voltage and current values at different buck stages. Necessary waveforms are captured and shown in Fig. 7 and 8.

Table 1. Case number 1: Battery & Resistive Load

V_s	I_s	V_{01}	I_{01}	V_{02}	I_{02}
32V	267 mA	14.1V	180 mA	7.02 V	528 mA
32V	208 mA	14.1	198 mA	11.7 V	256 mA

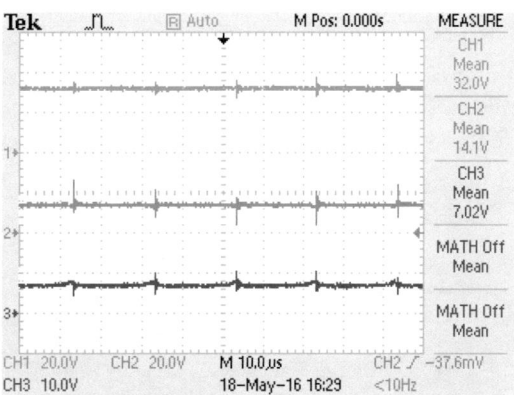

Fig. 7. CH1: Input Voltage; CH2: 1st Buck Stage Voltage with Battery; CH3: 2nd Buck Stage Voltage with resistive load

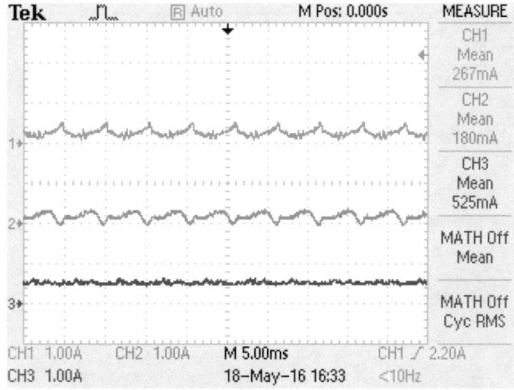

Fig. 8. CH1: Input Current; CH2: 1st Buck Stage Current; CH3: 2nd Buck Stage Current

B. Variable resistive load at first buck stage and battery at second buck stage

In case number 2, testing has been done with variable load at first buck stage and battery load at second buck stage. By varying the load at first buck stage, voltage and current waveforms have been captured and values have been noted.

Table 2. Case number 2: Resistive Load & Battery

V_s	I_s	V_{01}	I_{01}	V_{02}	I_{02}
32V	672 mA	10.6V	590 mA	14.3 V	664 mA
31.9V	377 mA	11.5V	206 mA	14.0 3V	355 mA

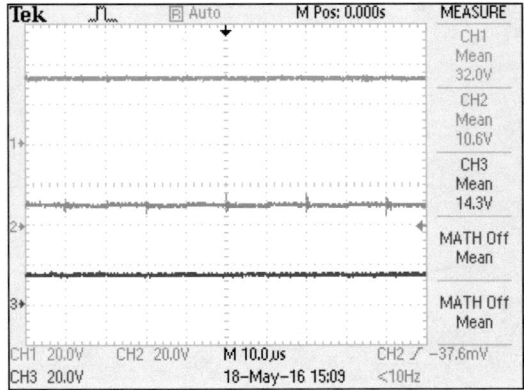

Fig. 9. CH1: Input Voltage; CH2: 1st Buck Stage Voltage with resistive load; CH3: 2nd Buck Stage Voltage with Battery

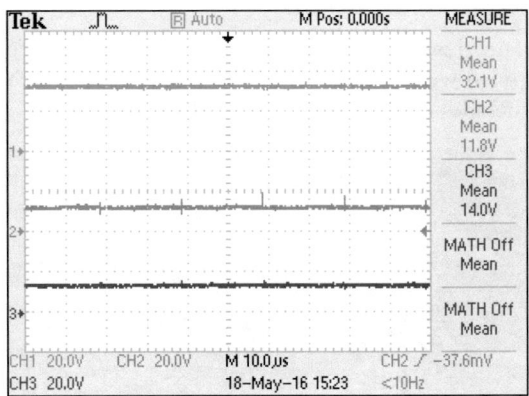

Fig. 11. CH1: Input Voltage; CH2: 1st Buck Stage Voltage with LED; CH3: 2nd Buck Stage Voltage with Battery

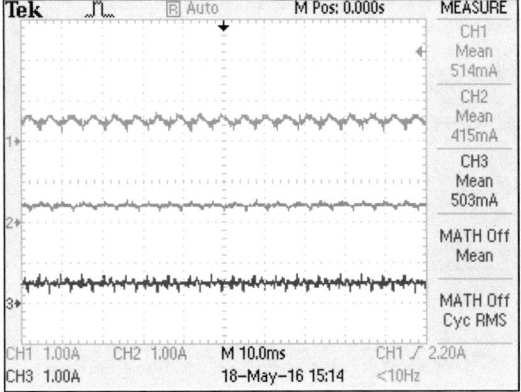

Fig. 10. CH1: Input Current; CH2: 1st Buck Stage Current; CH3: 2nd Buck Stage Current

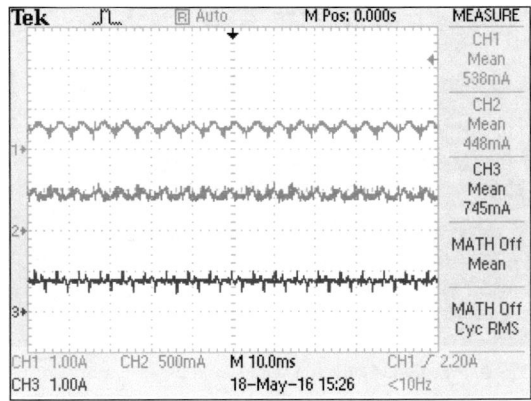

Fig. 12. CH1: Input Current; CH2: 1st Buck Stage Current; CH3: 2nd Buck Stage Current

C. LED load at first buck stage and *battery load at second buck stage*

Table 3. Case number 3: LED & Battery

V_s	I_s	V_{01}	I_{01}	V_{02}	I_{02}
32.1V	538 mA	11.8V	498 mA	14.0 V	745 mA
26.04 V	274 mA	9.84V	130 mA	12.8 V	466 mA

In this case, the circuit has tested with LED load at first buck stage and battery load at the second buck stage. By varying the input voltage, voltage at the loads has been noted and required voltage and current waveforms have been captured and shown in Figs. 11 and 12.

D. *Battery load at first buck stage and LED load at second buck stage*

In the last case, battery is connected to the first buck stage and a LED strip is connected to the second buck stage for experimentation. Table IV shows the different parameters for first and second buck stage output. The corresponding waveforms are shown in Figs. 13 and 14.

Table 4. Case number 4: Battery and Load

V_s	I_s	V_{01}	I_{01}	V_{02}	I_{02}
32V	661 mA	14.5V	614 mA	11.3 V	705 mA
26V	316 mA	14.1V	203 mA	9.48 V	328 mA

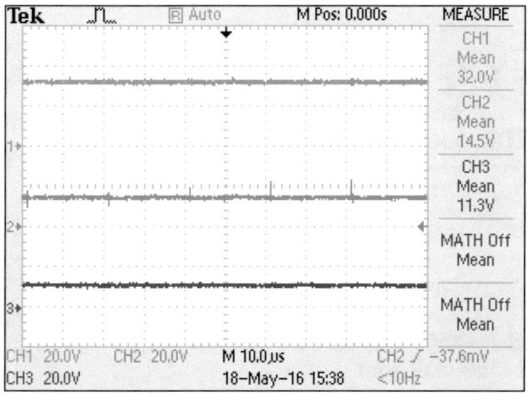

Fig. 13. CH1: Input Voltage; CH2: 1st Buck Stage Voltage with Battery; CH3: 2nd Buck Stage Voltage with LED

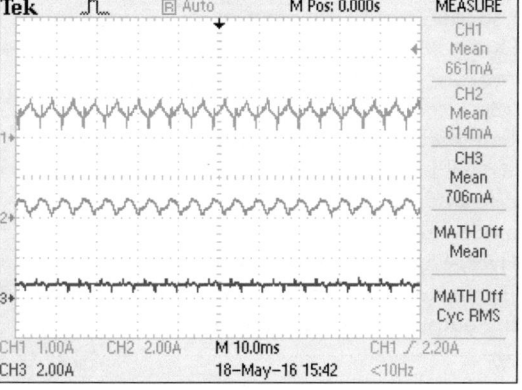

Fig. 14. CH1: Input Current; CH2: 1st Buck Stage Current; CH3: 2nd Buck Stage Current

Conclusion

The performance of the prototype SIDO quadratic buck converter is investigated for different cases to check its suitability as a dual output converter. A lead acid battery, LED strip, and variable resistance are connected to the converter by interchanging their position from the first buck stage to the second stage or vice versa. Figs. 8, 10, 12, and 14 confirm that the converter is capable of charging the battery along with the secondary load (either LED or variable resistor). As shown in the waveforms, in all cases the converter is capable of charging the battery and driving the LED. All these results confirm its ability for battery charging as well as LED driving applications.

References

Reis Barbosa, Lucio dos, Vieira, Joao Batista, Jr., de Freitas, Luiz Carlo, Silva Vilela, Marcio da, and Farias, Valdeir Jose. (1999). A Buck Quadratic PWM Soft-Switching Converter Using a Single Active Switch. IEEE Trans. on Power Electronics, 14(3), 445–453.

Pacheco, Vincius Miranda, do Nascimento, Acrisio Jose Jr., Farias, Valdeir Jose, Vieira, Joao Batista, Jr., and de Freitas, Luiz Carlos (2000). A Quadratic Buck Converter with Lossless Commutation. IEEE Transactions on Industrial Electronics, 47(2), 264–272.

Khuntia, Malay Ranjan and Veerachary, Mummadi (2020). Analysis and Control of Pseudo Quadratic Buck-Boost Converter. 3rd International Conference on Energy, Power and Environment: Towards Clean Energy Technologies.

Misal, Shrikant, Veerachary, Mummadi (2020). Analysis of a Fourth-Order Step-Down Converter. IEEE Trans. on Industry Applications, 56(3), 2773–2787.

Yadlapalli, Ravindranath Tagore, Kotapati, Anuradha (2017). Dynamic Performance Comparison of Quadratic Buck Converter with Analog and Digital Average Current-Mode Controllers. 3rd International Conference on Advances in Electrical, Electronics, Information, Communication and Bio-Informatics (AEEICB17).

Veerachary, Mummadi, Misal, Shrikant (2020). Single-switch Semi-Quadratic Buck Converter. 2020 IEEE International Conference on Power Electronics, Samrt Grid and Renewable Energy (PESGRE2020).

Veerachary, Mummadi (2017). Two-Switch Semiquadratic Buck Converter. IEEE Trans. on Industrial Electronics, 64(2), 1185–1194.

Nayak, Gayatri, Nath, Shabari (2020). Decoupled average current control of coupled inductor Single-Input Dual-Output Buck Converter. IEEE Journal of emerging and Selected topics in Industrial Electronics, 1(2), 152–161.

Secure RSA based Text File Encryption and Decryption using LabVIEW

T. Pavan Sai,[a] S. Rohan,[b] M. Yasmeen,[c] and S. Shafi[d]*

[a, b, c, d]Department of ECE, B V Raju Institute of Technology, Narsapur, Telangana, India
E-mail: *shaikshafi245@gmail.com

Abstract

In the recent past, the usage of digital platforms like, online banking, e-shopping and reservations have gained incredible popularity. This is due to the rapid development of information technology through the widespread of internet. However, security became a demanding issue during data dissemination. To overcome this problem, improved cryptographic encryption and decryption algorithms are proposed using plain text. Both encryption and decryption algorithms are implemented using LabVIEW for better security with ease of use. The proposed method makes use of a series of keys known as public, and private keys to encrypt and decrypt the data. The encryption key is employed to offer security to confidential data which thereby denies any unauthorized access. In this model, an efficient design has been proposed to encrypt plain text and decrypt it using RSA (Rivest-Shamir-Adelman) algorithm. The basic operations include key generation in RSA key generation module, encryption in encryptor module, and the retrieval of the original text at the output of the decryptor module.

Keywords: Encryption, Decryption, RSA, LabVIEW

Introduction

In the era of digital communication, the fear of personal data breaches and data altering has been increasing progressively. Therefore, safeguarding the data has become a major concern. To overcome these problems cryptography techniques have been introduced. This technique provides highly secured data communication and therefore plays a major role in the network of communication. Cryptography comes with amazing algorithms to supply the required protection against the encroachers of data. In recent years, there have been numerous reports of the exposer of personal information by hackers. The communication system of the multi-computer network security has been considerably curtailed, posing major hazards to network users' legitimate rights and interests, as well as the protection of citizens'

personal information. For the aim of security, all possible approaches are employed for personal messages (Narmatha et al., 2019).

In such cases, the encryption process comes into the picture to protect the data at both rest and transient sides. Encryption has also been used by civilians and armed forces to have secret communication. Many algorithms have been developed to satisfy security goals, but challenges were faced in implementing a highly secured process. There are issues that couldn't be solved by the existing research. To examine the role of cutting-edge encryption technology in network security protection. Liu, a computer specialist provided an in-depth introduction to computer data encryption technology and technique principles, analyzed challenges in the computer sector within present scientific and technological constraints, and developed related research techniques and technology (Cheon et al., 2015).

The proposed work overcome the drawbacks and fulfilled the gap by handling a secure encryption process. In a nutshell, data plaintext is converted into simple digital plaintext information and then changes into digital cipher text that ordinary people cannot easily comprehend by utilizing certain encryption technologies. Asymmetric cryptography is chosen over symmetric cryptography as it ensures highly secured data transmission. It is also known as public key cryptography as two keys are used.

The proposed work uses RSA, with asymmetric encryption and decryption keys, making it one of the finest. Because of the factorization problem, the initial layer of security in this study will be RSA. Here, the text is translated to a Unicode value and subsequently to the appropriate color code for secure transmission (Dong et al., 2018).

Literature Survey

While society, economy, and infrastructure have become very dependent on technology and computer networks a single negligence can result in disastrous cyber-attacks. Therefore, protecting data against unintentional, unlawful, or unauthorized access has become a challenge. Encryption of data and decrypting the output helps preserve the data and manage data confidentiality (Agrawal et al., 2018). RSA cryptosystem is an efficient method for computing the factorization of large numbers. This algorithm uses two prime numbers to generate public and private keys (Sangeeta et at., 2017). RSA algorithm works on the principle of digital signatures which is based on public key cryptography, (Hong et al., 2000) also cracking the longest key in RSA is difficult (Dorothy et al., 1984). The RSA algorithm provides a high level of security to encrypt the 8-bit plaintext data. This algorithm is the most popular among all the asymmetric key encryption algorithm (Satyaki Roy et al., 2012).

Proposed work

In this section, we elucidate the asymmetric encryption of a text using LabVIEW. The main aim of the proposed model is to improve the security of the data transmitted across the internet using the RSA algorithm.

RSA (Rivest–Shamir–Adleman) Algorithm

RSA Algorithm comes under an asymmetric cryptography process. This cryptography process consists of a public key and a private key. The main principle behind the RSA Algorithm is Factorization. As we know the fact that the factorization of a large Integer is a difficult task even for a modern computer. In RSA Public key consists of two integers where one is the multiplication of two large prime numbers. Similarly, the Private Key is also derived from the same two large prime numbers. But if the length of the Integer is small the possibility of a private key getting stolen is high. So, the larger the Integer value is the more secure data transmission. RSA Algorithm is divided into three many parts.

1. Key generation.
2. Encryption.
3. Decryption.

Key generation

In this section, we try to explain how to generate the keys using RSA Algorithm.

Key generation can be classified into.

- Public Key generation.
- Private Key generation.

In RSA we choose two distinct prime numbers. Let those two prime numbers be p and q. hear the numbers p and q are kept secret. Let us consider another number n which is equal to the multiplication of p and q. i.e., n = p

X q. n is used as the modulus for both public and private keys. Let there be another number $\phi(n)$. $\phi(n)$ is called the Euler totient function this is used in finding Public and Private keys. As we have discussed in the above section that to generate a public (or) private key we first need to have two prime numbers. So, to generate prime numbers we created a sub vi named "prime_gen.vi" this vi takes a range as input and generates a prime number bounded to the given range.

Prime Generation Algorithm.

STEP 1: Initialize

STEP 2: Create two variables

STEP 3: Read the values of the variables

STEP 4: Generate a random number between the variables.

STEP 5: Check whether the generated number is prime

STEP 6: IF it is prime end the loop and store the number in a Numeric

STEP 7: ELSE repeat the process from step 5 till you get a prime number

STEP 8: END

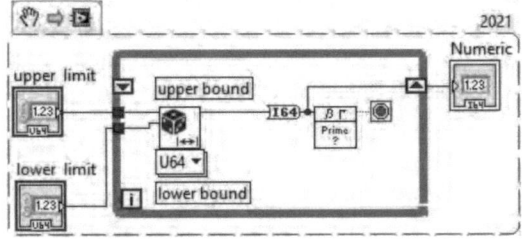

Fig. 1. Schematic of prime generation

Figure 1. The above figure shows how the prime generation algorithm has been implemented using LabVIEW. Before going into the public key generation we need to know about the Euler totient function $\phi(n)$.

$$\phi(n) = (p{-}1) \times (q{-}1)$$

Where P and Q Prime numbers are generated.

Public key Generation Algorithm

STEP 1: Initialize

STEP 2: Generate three variables p, q, e and fi_of_n (Euler totient function).

STEP 3: Read the values p, q (primes numbers generated).

STEP 4: Calculate fi_of_n = (p − 1) x (q − 1).

STEP 5: FOR i going 1 to fi_of_n

IF (i is prime) AND (GCD(i, fi_of_n)) e = i

STEP 6: RETURN e (The public key)

STEP 7: END

Private key Generation Algorithm

STEP 1: START

STEP 2: Create a variable d, private_key. STEP 3: d is multiplicative inverse of e(Public key). STEP 4: IF (e x d = 1 mod $\phi(n)$)

    Then private_key = d

STEP 5: RETURN private_key

STEP 6: END

Encryption

The encryption module facilitates the user by encrypting the message provided by him/her using the public key he/she has. This encryption converts a human-readable text message into a non-readable numeric format. In this project we have created a vi named "encryption.vi" this vi consists of two inputs that should be provided by the user they are the public key and message.

Encryption Algorithm

STEP 1: Initialize the variables

STEP 2: Generate three variables e, n, msg.

STEP 3: Read e, n, and msg from the user.

STEP 4: LOOP through the entire msg.

STEP 5: In each iteration take a character from msg and convert it to ASCII code.

STEP 6: Now offset these codes between the range 0 − 25.

STEP 7: Now create a variable cy (Cypher).

STEP 8: cy = (m^e mod n).

STEP 9: store these values into an array this the Encrypted data.

STEP 10: END.

Decryption

The decryption process facilitates the user by decryption the message that he/she had received using the **private key** that only they have access to. We have created a separate vi for decryption in this process the user will have access to the Encrypted data that he had received and this data will be decrypted using the **private key** to a human-readable form (to text form).

Decryption Algorithm.

STEP 1: START.

STEP 2: Take the Encrypted data from the user.

STEP 3: Read the **private key** from the user.

STEP 4: ITERATE through the entire Encrypted data.

STEP 5: In each iteration take the cypher code and do calculate the message from it.

STEP 6: m = (c^d **mod** n).

STEP 7: Convert this m to its respective ASCII codes.

STEP 8: Display the output as a string.

STEP 9: END.

Using LabVIEW, to implement of Decryption process, we used a python script for calculating the value of m as LabVIEW can't handle numbers greater than 64-bit length.

Results

Figure 2 shows *unveil* the real-time working of the RSA keygen vi. Hear the user has provided in the limits of (100, 10) as input and the vi generated *Public key, Private key, Modulo value.* Hear the user should feed in there *Public key, modulo value* to encrypt the text data provided.

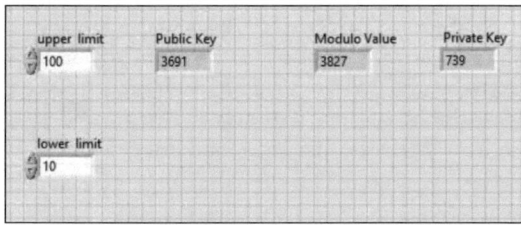

Fig. 2. Working of RSA using LabVIEW

Similarly, in the real-time working of Decryption.vi, the user should feed in their *Private key, modulo value* to decrypt the data sent by the Sender.

References

Narmatha, W. D., Jenifa, W., Moses, M. M., and Kumar, J. P. (2019). Text File Encryption and Decryption by FFT and IFFT Algorithm Using Lab view. Journal of Communication Engineering and Its Innovations, 5(3), 8–15.

Cheon, J. H., and Kim, J. (2015). A hybrid scheme of public-key encryption and somewhat homomorphic encryption. IEEE transactions on information forensics and security, 10(5), 1052–1063.

Dong, T., Wang, Y., and Lei, L. (2018, December). A File Encryption Algorithm Based on Dynamic Block Out of order Matrix Mapping. In 2018 International Conference on Security, Pattern Analysis, and Cybernetics (SPAC) (pp. 246–248). IEEE.

Agrawal, M., and Mishra, P. (2012). A comparative survey on symmetric key encryption techniques. International Journal on Computer Science and Engineering, 4(5), 877.

Kaur, A. (2017). A Review on Symmetric Key Cryptography Algorithms. International Journal of Advanced Research in Computer Science, 8(4).

Hong, J. H., and Wu, C. W. (2000). RSA public key crypto-processor core design and hierarchical system test using IEEE 1149 family. National Tsing-Hua University, Taiwan, Doctoral dissertation.

Denning, D. E. (1984). Digital signatures with RSA and other public-key cryptosystems. Communications of the ACM, 27(4), 388–392.

Roy, S., Maitra, N., Agarwal, S., Nath, J., and Nath, A. (2012). Ultra encryption algorithm (UEA): Bit level symmetric key cryptosystem with randomized bits and feedback mechanism. International Journal of Computer Applications, 975, 8887.

OPAL-RT-Based Implementation of a Power Controller for a PV-Fed Grid-Connected Inverter

Meenakshi Jayaraman,*,a Rajkiran Balakrishnan,b and Nilanjan Tewaric

a,b,cSchool of Electricial Engineering, Vellore Institute of Technology, Chennai, India
E-mail: *meenakshi.j@vit.ac.in

Abstract

This article presents a power flow controller for a photovoltaic (PV) fed grid-connected inverter connected to an LCL filter. The power flow controller is used to regulate the efficient exchange of power between the inverter and grid. The control method uses direct-quadrature transformation to control the inverter output current according to variations in the load. With this transformation, since the steady-state current components have direct current values, proportional integral controllers can be employed easily in the method. The system is simulated using MATLAB-Simulink and results are presented. Further, an experimental prototype model of the inverter-filter system is developed with the direct-quadrature controller implemented using OPAL-RT real-time simulator. Results obtained from simulation and experimentation validate the design of the controller for the PV-fed inverter system connected to the LCL filter. With the power flow controller, efficient power exchange is achieved between the inverter and the grid.

Keywords: Photovoltaic (PV), inverter, passive filter, Direct-Quadrature (DQ), Pulse Width Modulation (PWM), Proportional Integral (PI)

Introduction

With the exhaustion of fossil fuels and the need to cater to environmental constraints, the use of sustainable energy sources to meet energy demands has drawn more attention. With policy changes and promotional programs from utilities and countries worldwide towards the green planet, solar-based grid-connected and standalone systems have become an integral part in distributed energy generation (Gomes et al., 2018; Jayaraman et al., 2020). With a standalone photovoltaic (PV) system, energy can be supplied to a consumer without any grid connection. Grid-connected PV systems supply the surplus power available back to the grid. When PV systems are used as grid-connected PV systems, many things should be taken into the account. Main among them is that a good controller should be there for regulating the power flow between the PV-fed inverter and the grid. This is required for reactive and active power flow control between the grid and inverter. In this article, a digital current controller is used for regulating the power flow. A current controller is used because the current is the quantity that is easily dependent on the load variation. Direct-Quadrature (DQ) transformations are carried out and proportional-integral (PI) controllers are used to regulate the power exchange. While traditional pulse width modulation (PWM) is employed for generating inverter switching pulses, LCL filters are employed to connect the inverter to the grid.

Single Phase DQ Transform

Single-phase DQ transform is a synchronous reference frame transformation in which a pair of orthogonal quantities in stationary frame is transformed into a rotating frame (Crowhurst et al., 2010; Chatterjee et al., 2018). In this paper, a current controller is used for power flow regulation. For this controller, DQ transform is used, and then using a discrete PI controller proper regulation is obtained. The resultant D and Q vectors in this rotating reference frame will be a constant. This indicates that two orthogonal components are required for the transformation of the synchronous reference frame. To generate one missing orthogonal component transport delay method is used for generating fictitious phases and this is used for achieving $\pi/2$ delay. Now for transforming a current signal to DQ frame using the transport delay method, the actual current is denoted as Ir and $\pi/2$ delayed signal as imaginary component Ii. The matrix representation of the transformation of real and imaginary components to DQ frame is given by Eq. (1).

$$I_{dq} = \begin{bmatrix} I_d \\ I_q \end{bmatrix} = T\,I_{ri} = \begin{bmatrix} \sin\omega t & -\cos\omega t \\ \cos\omega t & \sin\omega t \end{bmatrix} \begin{bmatrix} I_r \\ I_i \end{bmatrix} \quad \ldots\ldots\ldots(1)$$

Here the original signal's DC component is mapped to the AC signals which are rotating at fundamental frequency. Thus the presence of DC component will make the steady state error as zero. The matrix representation for inverse transformation is presented in Eq. (2).

$$I_{ri} = \begin{bmatrix} I_r \\ I_i \end{bmatrix} = T^{-1} I_{dq} = \begin{bmatrix} \sin\omega t & \cos\omega t \\ \cos\omega t & -\sin\omega t \end{bmatrix} \begin{bmatrix} I_d \\ I_q \end{bmatrix} \ldots\ldots\ldots(2)$$

System Configuration and Simulation Results

The simulation circuit of the DQ controller for power flow regulation of the PV fed grid-connected inverter is done in MATLAB/Simulink environment. PWM inverter is employed. Further LCL filter is designed using Jayaraman et al. (2017) and Jayaraman et al. (2020). for connecting the inverter to grid/load. The Simulink system model is presented in Fig. 1. The simulation parameters are listed in Table 1. The main parts of the model are a grid-connected inverter, single-phase grid, DQ controller, and the load. It can be seen that the load is connected

in between the inverter and grid. The main aim of this DQ controller is to ensure that as long as the PV power is available it should be taken for supplying to the load. Whenever the PV power is not available then only the load should take power from the utility grid. Here R Load of 220 Ω and RL Load of 220 Ω and 24 mH are assumed as the two scenarios. During the simulation for the first 0.2 secs R Load operates and for next 0.2 secs both R Load and RL-Load will operate in parallel. Thus it can be seen that during these load variations, this DQ controller will generate a suitable reference waveform for the inverter sinusoidal PWM technique and thus ensure that the power is fed from the inverter to the load and not from the grid to the load. The DQ control circuit includes phase locked loop (PLL), DQ transform blocks, and PI controllers. PLL is used for the sine and cosine reference generation which are used in both DQ transformation and inverse DQ transformation. Here inverter output current is transformed into DQ. Then the load current is obtained by subtracting grid current from inverter current and then it is also transformed do DQ rotating reference frame. After these DQ transformations, load current is used as the command signal and then inverter output current is subtracted from it. This will generate error signals and then they are passed to PI controllers. The main function of this PI controller is to eliminate the error between the load current and the inverter output current. Figure 2a shows the inverter output current. It can be seen that in the output current that the current value is increased when the load is switched from R to both R & RL Load in parallel at 0.2 sec. With the DQ controller it is possible for regulating the power flow between PV powered inverter and single phase grid both connected to a common load. In the simulation, the load variation is incorporated by switching R-Load for first 0.2 secs and then R and RL Load in parallel for next 0.2 secs. Figure 2b shows the real power and reactive power output from the inverter. It can be seen that for the first 0.2 secs the power drawn from the inverter by the load is about 240W and for the next 0.2 to 0.4 secs power drawn by the load is 480 W and 8 VAR.

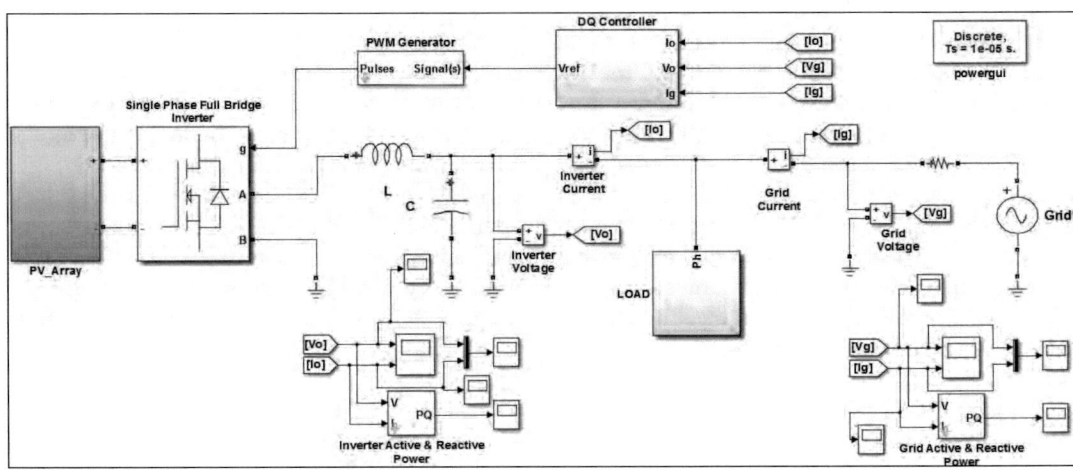

Fig. 1. System model

Table 1. Circuit parameters used for simulation

Parameters	Values
Output voltage from PV array	325 V
Switching Frequency	10KHz
Filter	L= 2 mH and 2mH; C= 10μF
Grid	230V,50Hz

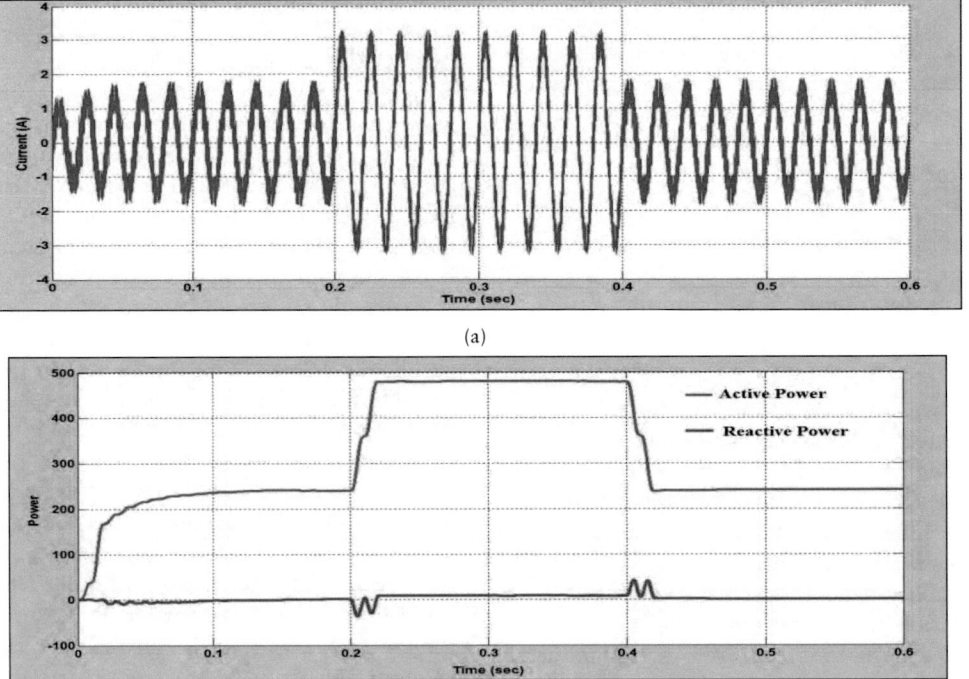

(a)

(b)

Fig. 2. (a) output current from inverter (b) real/reactive power output of grid-connected inverter

Power required for the load can be supplied from the PV fed grid-connected inverter. The grid voltage/currents are displayed in Fig. 3a. The grid voltage is maintained at 230 V. It is also seen that the grid current is almost zero throughout the operation. Thus it can also be inferred that the net power transfer from the grid is also almost zero. Hence, it can be noticed that the power drawn by the load from the single-phase grid is almost zero. Fig. 3b shows the active and reactive power flow between the load and the grid. The active and reactive power drawn from the grid by the loads is almost in the range of zero.

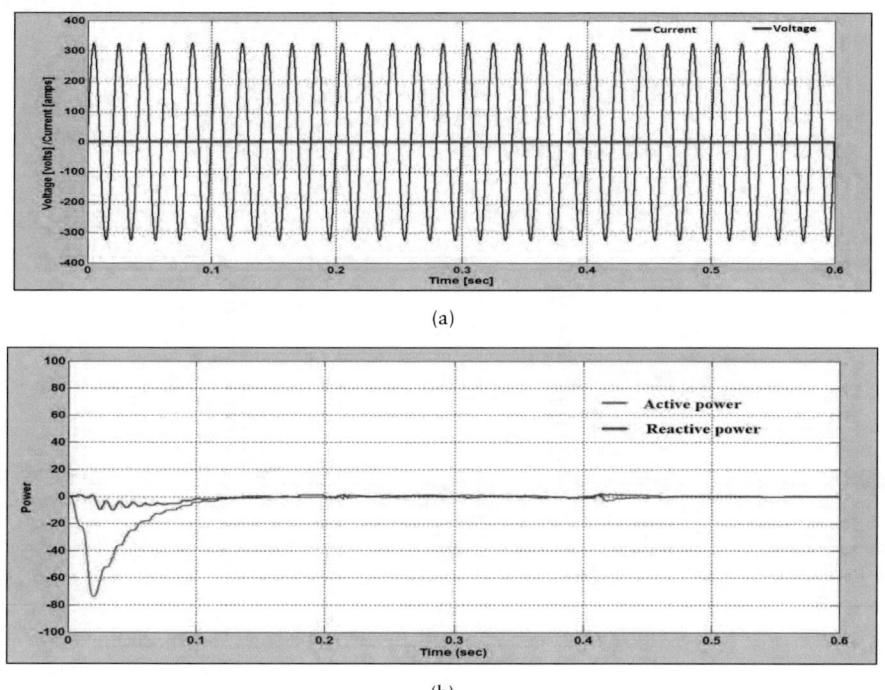

(a)

(b)

Fig. 3. (a) output voltage/current from grid (b) real/reactive power output from grid

Table 2: Circuit parameters used for experimentation

Parameters	Values
Input voltage to the inverter	17V
Switching Frequency	10KHz
Gate Driver and Power MOSFET	TLP250 and IRFP460
Load	R= 50Ω

Experimental Results

Hardware implementation of the power flow controller for the single-phase grid-connected inverter is carried out. The controller circuit is made to run in OPAL-RT simulator and the power circuit is configured on hardware layout. Table 2 shows the parameters and components used for the power flow controller for the grid connected inverter with filter. From Fig. 4a and Fig. 4b, it can be noted that the inverter output voltage is about 12 V RMS and inverter output current is about 200 mA. It is clear from Fig.4c that due to the DQ controller, the current drawn from grid is approximately equal to zero. Figures 4d and 4e show the output voltage and current from the inverter and grid taken in Power Quality (PQ) Analyzer. From the experimental results have shown, it is clear that with the DQ power flow controller, an efficient power flow control is achieved. Figures 4e and 4f show the output power drawn from the grid-connected inverter and the grid respectively. It can be noted that with the DQ power flow controller no power is drawn from the grid side by the load. Power required for the load is given by the grid-connected inverter. It can be seen that active and reactive power drawn from the grid by R load and both R and RL loads together are almost in the range of zero.

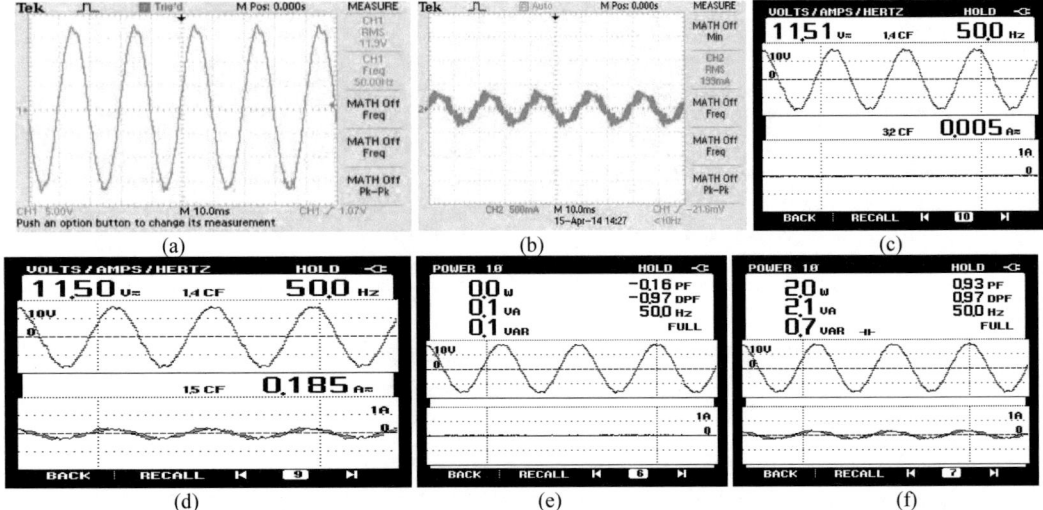

Fig. 4. (a) output voltage of inverter (b) output current of inverter (c) output voltage and current from inverter (d) output voltage and current from grid (e) power drawn from inverter (f) power drawn from grid

Conclusion

PV fed grid-connected inverters are gaining more importance nowadays. The main need for the controller in a grid-tied inverter is to regulate the power flow from the inverter to the load and from the grid to the load. The simulation of a power flow controller for a full bridge inverter with LCL filter connected to the grid is done in MATLAB/Simulink. A hardware model for the above controller is done using OPAL-RT. With the power flow controller, efficient power exchange is achieved between the inverter and grid.

References

Chatterjee, A. and Mohanty, K. B. (2018). Current control strategies for single phase grid integrated inverters for photovoltaic applications - a review. Renew. Sustain, Energy Rev. 92, 554–569.

Crowhurst, B., El-Saadany, E. F., El Chaar, L. and Lamont, L. (2010). A. Single-phase grid-tie inverter control using DQ transform for active and reactive load power compensation. Proc. 2010 IEEE International Conference on Power and Energy, 489–494.

Gomes, C. C., Cupertino, A. F., and Pereira, H. A. (2018). Damping techniques for grid connected voltage source converters based on LCL filter: An overview. Renew. Sustain, Energy Rev. 81, 116–135

Jayaraman, M., and Sreedevi, V.T. (2017). Implementation of LC and LCL passive filters for harmonic reduction in PV based renewable energy systems. Proc. 2017 National Power Electronics Conference (NPEC), 363–369.

Jayaraman, M., and Vellithiruthy Thazhathu, S. (2020). Analysis of a novel (LCR)trap-LC-RC filter with improved performance for standalone inverters. Int. J. Electron. 107(2), 310–330.

Extraction of Essential Oils From Medicinal and Aromatic Plants in Oman Using Solar Energy

Parimal S. Bhambare,[a,*] C. V. Sudhir,[a] Dinesh K. Kaithari,[a] Varghese Joy,[a] and Amira Al Gharibi[a]

[a]Department of Mechanical and Industrial Engineering, College of Engineering, National University of Science and Technology, Muscat, Oman
E-mail: *parimalbhambare@nu.edu.com

Abstract

The use of solar energy for the extraction of essential oils and Hydrosols is one of the promising applications of renewable energies. Oman has several herbal and medicinal plants, that are used as Hydrosols or essential oils in Omani traditional ethnic medicine practices. Traditional methods used to produce Hydrosol or essential oil make use of fossil fuels such as wood or coal. This paper presents a solar integrated unit used for producing essential oils from Frankincense material with the associated economic and solar energy availability analysis. This unit is based on the hydro and steam distillation methods that are integrated with the fixed-focus Scheffler concentrators. This novel approach will lessen the dependency on conventional energy sources while reducing the overall production cost. In addition, this system is expected to generate employment opportunities for unskilled youths and families in the rural settings of the country.

Keywords: Solar energy, Essential oil extraction, Hydro distillation, Steam distillation, Scheffler Concentrator, Frankincense

Introduction

Omani traditional ethnomedicine practice roots lie in the traditional Islamic medicine style, *Unani Tibb*. This method is based on humoral system that uses products such as syrups, essential oils, creams, infusions, decoctions, powders, inhalers, etc. mainly derived from medicinal and aromatic plants for the treatment of simple to complex diseases. These essential oils and aromatic solutions or Hydrosols are derived from different parts/products of medicinal and aromatic plants namely fruits, stems, roots, gums, barks, seeds, peels, or whole plants. There are different medicinal and aromatic plants available in Oman, such as, Myrtle or Myrtus communis called yas in Arabic, which usually grows on the banks of the wadis. Another plants, Frankincense (known as Boswellia carteri), Euphorbia larica, Rhazya stricta, Citrulus colocynth are found largely in the Dhofar region (Madhu Divakar, et al., 2016). In addition to the medicinal use, essential oils and hydrosols find their applications in different food, beverage, pharmaceutical, and FMCG industries as essence for giving certain flavors to their products or as fragrance in perfume industries. Moreover, certain essential oils are also used in animal farming and crop protection (Serdar Oztekin, et al., 2007).

Sultanate of Oman, being a tropical country in the northern hemisphere, has one of the highest clear sky index and solar densities in the world. A study on renewable energy in Oman has evaluated that solar insolation varies from 4.5 to 6.1 kWh/m² per day. This corresponds to 1,640 to 2,200 kWh per year (Authority for Electricity Regulation Oman, 2008). Solar thermal energy has been utilized for the extraction of essential oils and aromatic solutions or hydrosols by several researchers for different plant materials such as rosemary, orange, lemon, rose, lemongrass, clove, cumin, Fennel, Sambong, Oregano, Cymbopogon citrus, eucalyptus, peppermint, lavender, etc (Manyako, et al., 2022, Mona, et al. (2020), Munir (2010)). Distillation methods namely wet steam, dry steam, and hydro or water distillation are used for the extraction of almost 90% of essential oils (Serder Oztekin, 2007). Concentrating solar collectors such as Parabolic Trough collectors (PTC), Scheffler concentrators, and Parabolic dish reflectors have been used by almost all the researchers for the distillation process (Manyako, et al., 2022; Mona et al., 2020; Munir, 2010). Steam distillation method has been rarely integrated with the Scheffler concentrators for essential oil extraction. Scheffler concentrators offer several advantages over other solar concentrators in the rural settings of Oman (Parimal Bhambare, et al., 2018). In the present work standing-type Scheffler concentrators are used for the development of a batch-type standalone hydro cum steam distillation unit. This unit can be used for both hydro and steam distillation. Frankincense, a commercially important herbal product in Oman and the entire gulf region has been used for testing in the present setup. During experimentation, steam distillation has shown better yield over hydro distillation using Frankincense.

Experimental Setup and Methodology

Experimental setup consists of 16 m² Scheffler concentrators integrated with essential oil extraction units based on steam and/or hydro distillation method. Figure 1 shows the layout diagram of the complete set-up. Distilled water at 30 °C is pumped into the steam header and it is further naturally circulated through the dome-type receivers of the concentrators. The wet steam separated in the steam header vessel is directly supplied to the jacketed distiller vessel. For hydro distillation, the mixture of the plant material and the water is stored inside the vessel and valve 2 is

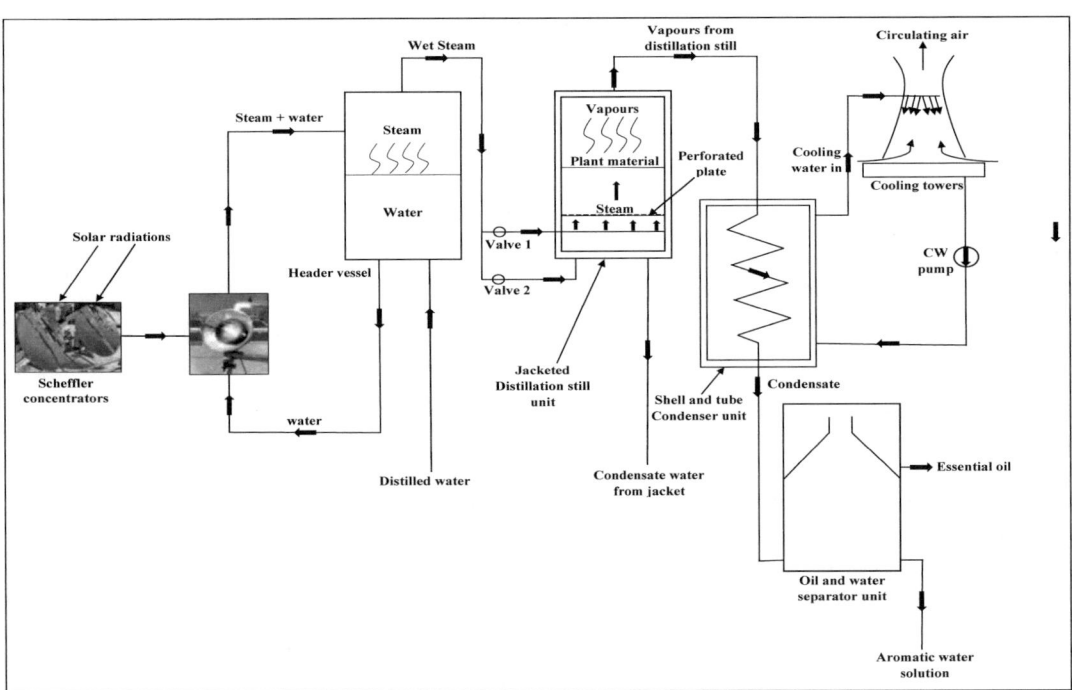

Fig. 1. Layout plan for the solar steam cum hydro distillation unit

regulated to circulate the steam inside the jacket of the vessel for heating. For steam distillation, the plant material is spread evenly over the perforated plate and valve 1 is operated to introduce the steam from the bottom of the vessel through the steam nozzle. The steam vapors from the distillation vessel, containing the aroma and the essential oil extract of the plant material, are led to the tube in tube type heat exchanger. The cooling water from the cooling tower is circulated through the heat exchanger using the cooling water (CW) pump. The condensate with essential oil enters the oil and water separator vessel. The hydrosol can be drained from the bottom of the separating vessel while the essential oil can be tapped out from the top. For the distillation process the pressure is maintained atmospheric inside the distillation vessel and the system is operated between 9 am to 3 pm, for 6 hours for the day.

Omani Frankincense or Luban is available in four types or grades commercially, Hoojri, Najdi, Shathari, and Shaabi as shown in "Fig. 2" (Salim Al-Saidi, et al. 2012). From these four types, Hoojri grade Frankincense with particle size of 3 mm is used for the analysis in the present study due to its best essential oil yield (Salim Al-Saidi, et al. 2012).

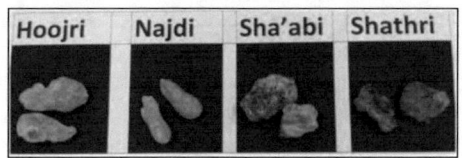

Fig. 2. Omani Frankincense commercially available grades (Salim Al-Saidi et al., 2012)

The percentage yield has been calculated using the following formula

$$
\begin{aligned}
& Percentage\ yield\ (\eta_{EO}) \\
&= \frac{Vol.\ of\ essential\ oil\ (ml)}{Mass\ of\ Frankincense\ (gm)} \\
&\times 100
\end{aligned} \quad (1)
$$

The essential oil yield per kg of the Frankincense used is calculated as below

$$
\begin{aligned}
& Essential\ oil\ yield\ (ml/gm) \\
&= \frac{Vol.\ of\ essential\ oil\ collected\ (ml)}{Mass\ of\ Frankincense\ (gm)}
\end{aligned} \quad (2)
$$

Solar thermal energy available per day ($E_{day,\ thermal}$) considering the losses due to optical losses, tracking errors, mirror specular errors, receiver misalignments, etc., is calculated using the following Equation 3.

$$
E_{day,thermal} = 0.5777 \times A_s \times time \times I_{bn} \quad (3)
$$

I_{bn} is the direct beam radiation. The average essential oil yield per day for each month is calculated using the following equation,

$$
\begin{aligned}
& Essential\ oil\ yield\ per\ day, ml/day \\
&= V_{essential\ per\ kWh} \\
&\times E_{day,thermal} \\
&\times \eta_{thermal}
\end{aligned} \quad (4)
$$

$V_{essential\ per\ kWh}$ is the volume of essential oil produced per kWh of energy input. This value for Frankincense oil is experimentally estimated as 4.5 ml/kWh.

Results and discussion

Solar energy and thermal energy availability analysis

Due to the elliptical shape of the Scheffler concentrator dish, aperture area (As) is the actual area that is visible to the solar radiations. The aperture area varies throughout the year. As shown in "Fig. 3," the average aperture area is 72.4% of the total Scheffler dish area. From "Fig. 4" it can be noted that the maximum average hourly solar insolation of 743 W/m² is measured in April month while the minimum of 443 W/m² is measured in January month. The maximum thermal energy of 37 kWh is estimated for April and a minimum of 28 kWh is estimated for January. This could be noted in Fig. 5.

Essential oil yield analysis

Steam and hydrodistillation methods are used in the process. Table 1 below shows the average percentage yield obtained from both these methods during experimentation. Hydro distillation has been carried out using two methods. Both these methods have given similar results. From Table 1, it can be observed that steam distillation gives better results over the hydro distillation. The maximum yield of 8.2% is obtained using the steam distillation process. The estimated yearly essential oil with thermal efficiency of 72% for a 16 m² Scheffler concentrator is shown in Fig. 6.

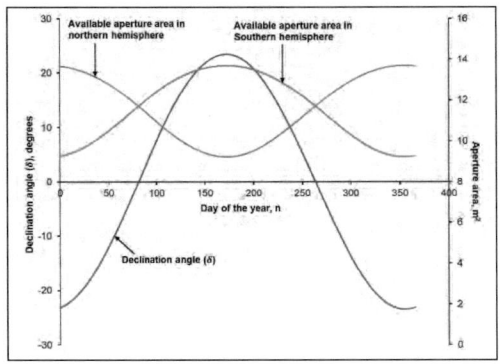

Fig. 3. Aperture area and declination angle (Parimal Bhambare et al., 2022)

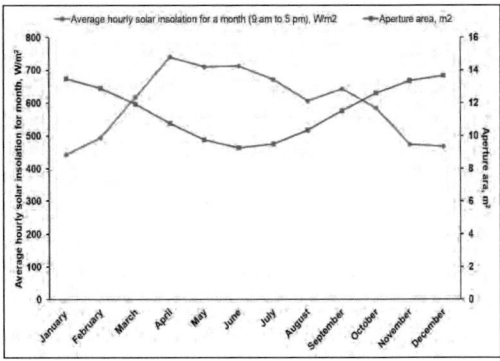

Fig. 4. Average hourly solar insolation at Muscat

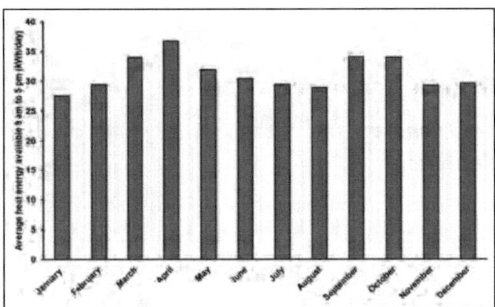

Fig. 5. Average daily and monthly available heat energy at Muscat for 16 m² Scheffler dish

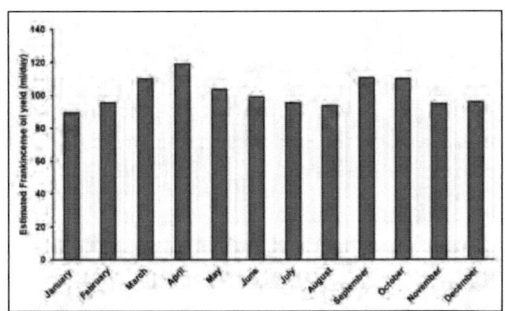

Fig. 6. Estimated Frankincense oil yield per day for each month (9 am to 5 pm daily)

Table 1. Comparison of percentage yield for steam and hydro distillation

Method	Average Percentage yield (%)
Hydro distillation	5.8
Steam distillation	6.7

The maximum yield of 120 ml/day is expected in April with 37 kWh of solar thermal energy available while lowest yield of 89 ml/day is expected in January month.

An average yield of 101 ml/day is expected for the entire year with average hourly solar insolation of 597 W/m² for each day of the year. If the system is operated for average of 9 months a year, total essential oil yield of 28 to 30 liters is expected for the year.

Economic analysis

For the present study, Hojari grade 3 type Frankincense is used.

The complete set-up cost with the distillation unit and one 16 m² Scheffler concentrator is 12000 OMR. From Table 2, for the expected annual yield of 28 to 30 liters of essential oil and about 2300 liters of Hydrosol, with the overhead cost of 200 OMR/month for maintenance and unskilled labor, the simple payback period could be estimated as 3.05 years for the unit.

Table 2. The expected payback period calculation

Cost details	OMR/year	Earning details	OMR/year
Raw material	2100	Essential oil	4800
Distilled water	1073	Hydrosol	4600
Overhead cost	2400		
Total cost	5473	Total earnings	9400

Expected profit per year = 9400 – 5473 = 3927 OMR/year

The expected simple payback period = 12000/3927 = 3.05 years

Conclusion

This paper has presented a solar integrated unit used for producing essential oils and hydrosols from medicinal and aromatic plants in Oman. The experimental analysis has shown that steam distillation provides better yield over the hydro distillation process for Frankincense material. An overall yield of 6.7 to 8.1% is obtained during the experimentation. The maximum essential oil and hydrosol yield is expected for April and the minimum for January. If the system operates 9 months annually, an overall essential oil yield of 28-30 liters with 2300 liters of hydrosol is expected annually, with a single 16 m² Scheffler concentrator for Frankincense. A simple payback period of 3.05 years is expected for the system. This system is expected to provide employment opportunities to unskilled youths and families in the rural settings of the country itself.

Acknowledgment

The research leading to these results has received Project Funding from The Research Council of the Sultanate of Oman under Research Grant No. BFP/RGP/EI/20/187. Authors would acknowledge the support and encouragement received from The Research Council as well as College of Engineering, National University of Science and Technology, Muscat Sultanate of Oman.

References

Authority for Electricity Regulation Oman. (2008). Study on Renewable Energy Resources, Oman. Final Report.

Divakar, Madhu, Al-Siyabi, Amani, Varghese, Shirley and Al-Rubaie, Mohammed. (2016). The Practice of Ethnomedicine in the Northern and Southern Provinces of Oman. Oman Medical Journal, 31(4), 245–252.

Manyako, K. E., Chiyanzu, Idan, Mulopo, Jean, and Abdulsalam, Jibril. (2022). Pilot-Scale Evaluation of Concentrating Solar Thermal Technology for Essential Oil Extraction and Comparison with Conventional Heating Sources for Use in Agro-Based Industrial Applications. ACS Omega. 7 (24), 20477–20485.

Radwan, Mona N., Morad, M. M., Ali, M. M., Wasfy, Kamal I. (2020). A solar steam distillation system for extracting lavender volatile oil. Energy reports, 6, 3080–3087.

Munir, A. (2010). Design, development and modeling of a solar distillation system for the processing of medicinal and aromatic plants. Institute of Agricultural Engineering, Universität Kassel. Universität Kassel.

Bhambare, Parimal, Majumder, M., Sudhir, C. V. (2018). Solar Thermal Desalination: A Sustainable Alternative for Sultanate of Oman. International Journal of Renewable Energy Research, 8(2), 733–751.

Bhambare, Parimal Sharad, Chitrapady Vishweshwara, Sudhir. (2022). Design aspects of a fixed focus type Scheffler concentrator and its receiver for its utilization in thermal processing units. Energy Nexus, 7(100103), 1–13.

Al-Saidi, Salim, Rameshkumar, K. B., Hisham, Abdulkhader, Sivakumar, Nallusamy, and Al-Kindy, Salma (2012). Composition and Antibacterial Activity of the Essential Oils of Four commercially available Frankincense. Chemistry and Biodiversity. 9, 615–624.

Serdar Oztekin, M. M. (2007). Medicinal and Aromatic Crops: Harvesting, Drying, and Processing. 1st ed. New York: CRC Press.

Oman Luban. Frankincense (Boswellia sacra) Essential oil & Hydrosol prices. Available from: Available from: https://omanluban.com/

A Review on Multi-Modal Classification for Emotional Intelligence

P. Y. Preema,[*,a] *J. Chandra,*[2] *and Alwin Joseph*[3]

[1,2,3] Christ University Bangalore, India
E-mail: [*]preema.py@res.christuniversity.in

Abstract

Emotions are unavoidable in human beings and play an essential role in perceiving things. This paper reviews various methods used to capture human emotions effectively. The work analyses the effectiveness of unimodal, bimodal, and multimodal machine learning models created for the processing of emotions. These models effectively handle various signals and input modalities, including physiological and audio-visual signals, from which emotions are captured and analyzed. The study concludes that machine learning algorithms, including Support Vector Machine, K-Nearest Neighbour, Gaussian Mixture Model, and Convolutional Neural Network, are considered popular methods for processing the various inputs for classifying emotions from multiple data sources. The research also identifies that the combination of EEG and eye movements could give better accuracy compared with other data forms.

Keywords: Physiological signals, Audio-visual signals, K-Nearest Neighbour (KNN), Convolutional Neural Network (CNN), Support Vector Machine (SVM), Gaussian Mixture Model (GMM)

Introduction

Human-Computer Interaction is a challenging area in the real world. In the scientific world, scientists investigate a possibility in the future where the computer can identify the real emotions of a human. This could be revolutionary because feeling is essential for human life. Automated analysis of human emotions brings effective behavior and increasing attention from researchers in psychology, computer science, etc. However, some of the feelings are not identified accurately. Identifying accurate tracking of emotions can bring changes in the life of oneself and the work area. The mode of accessing emotions can be categorized into physical and physiological. Physical signals can be known as external signals (S. Chen et al., 2020), such as speech, face, and postures, and physiological signals are internal, including EEG, EMG ECG, galvanic skin response, etc. Among the emotion recognition methods, unimodal uses only single data and bimodal with any two data. In contrast, combining more than two data modules helps recognize humans' emotions. However, the actual appearance of the face is expressed by the context of the social situation, in which the person may not express their real emotions. In addition to physiological data, emotions are recognized from voice and facial expressions to improve the accuracy rate of emotion recognition. The research also reviews various data used for unimodal, bimodal, and multimodal approaches.

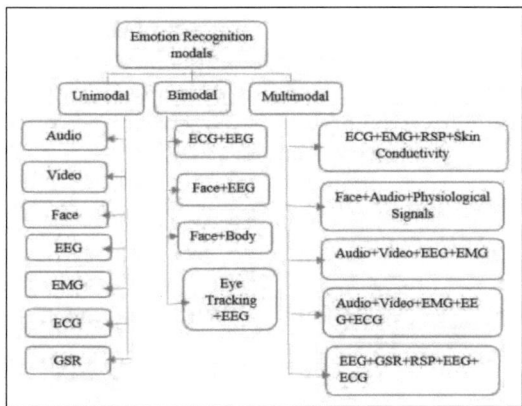

Fig. 1. Emotion Recognition Modals

Related Works

The emotional intelligence review aims to understand different modalities for uni, bi, and multi-modal classification approaches. Unimodal specifically shows emotion recognition based on one data. It can be either from the physiological or physical module. Fig1 depicts various approaches to identifying the emotion, and various data are used for different modules. It also shows the different combinations used in multimodal.

The audio-based emotion recognition system is used to identify the primary emotions by extracting different features from audio. MFCC method (Uddin and Nilsson, 2020) extracts signals from speech. Speech emotion recognition (SER) requires time to make the HCI more exciting and active. The performance of emotion recognition is identified by comparing two different datasets. The face is the main focusing area where expressions of emotion are tackled (Tarnows et al., 2017). Identification of emotions is being made on the movement and changes of muscles. Emotion assessment is based on the user's emotional expression and physiological signals. Physiological signals are collected by placing sensors on the different parts of the body. Among the physiological signals (Xu et al., 2019), EEG plays a significant role in identifying emotion. The review (Suhaimi et al., 2020) on EEG signals helps capture accurate brain signals for emotion recognition based on databases and classifiers. EEG signals' emotional parameters are ranked based on valence, arousal, liking, and dominance. With the help of NB, SVM, and CNN, the machine learning algorithm is used as classifier to get the accuracy. The DGCNN method (Song et al., 2020) is used to understand the fundamental relationship between different EEG signals using other datasets.

The combination of lyrics and audio modules identifies the music's emotional recognition. The other variety is physiological signals and facial expressions. The paper (Hassouneh et al., 2020) aims to help disabled children like autistic children, whose emotional identification can be gained through their facial expressions and EEG signals. The CNN and LSTM classifiers are used to identify real-time emotions and the LSTM model for the EEG signals and CNN for facial expression and acquired nearly 87.25% and 99.81% for both modules, respectively, EEG with Eye Tracking(ET)(Zhao et al., 2018) has attained. 91% accuracy using the SVM classifier, respectively. The comparison study concluded that the bimodal technique performed better than the unimodal. The paper (Ranganathan et al., 2016) focuses on four physiological signals: EEG, RSP, ECG, and GSR. The classifier SVM performed well while classification was done using single and multiple data. Multiple physiological signals with different classifiers (Song et al., 2019) like SVM, KNN, LSTM, and ALSTM identified seven emotions in which ALSTM was performed among all the classifiers.

Result Analysis

Unimodal models work with a single modality of data, whereas bimodal have various modalities, including EEG with ECG, facial expression, and Eye Tracking. Multimodal covers different combination types of physical or physiological modalities of data. The review shows that many classifiers (Singla et al., 2020) are being used to compare the study indicating that many reach the accuracy level of audio data. The highest value achieved in audio data is 93% with the help of NSL, and emotions are identified based on valence and arousal. The 3NN classifier could achieve an accuracy rate of 95.5%. The data EEG and video were performed with MLP, and CNN classifiers, respectively, and the accuracy level and the identified emotions are well explained in Table 1. It also depicts the four highest accuracies gained by bimodal and multimodal signals. The accuracy level attained by three different models is represented in Fig. 2 the multimodal performance is considered better than the other two models.

Table 1. Different Types of data, Classifiers, Accuracy, Emotions of Unimodal, Bimodal and Multimodal Emotional Recognition System.

Reference	Data	Classifier	Accuracy	Emotions
Uddin and Nilsson (2020)	Audio	NSL	93%	Happy, sad, anger, neutral
Schofield (2018)	Video	MLP	81.84%	Anger, disgust, fear, happiness, sadness, surprise
Xu et al. (2019)	EEG	CNN	81.14%	Valence
Tarnowski et al. (2017)	Face	3NN	95.5%	Neutral, joy, fear, surprise, anger, sadness, disgust,
Shin et al. (2017)	ECG + EEG	MLP SVM BN	83.97% 63.97% 99.32%	Comic, Fear, Sad, Joy, Angry, Disgust
Hassouneh et al. (2020)	Facial Expressions +EEG	CNN LSTM	Face: 99.81% EEG: 87.25%	Happiness, sadness, anger, fear, disgust, and surprise
Zhao et al. (2018)	EEG+ Eye movement	SVM	91.01 %	Positive, Negative, Neutral
Chen et al. (2020)	ECG+EMG, RSP+Skin conductivity	C4.5 Decision Tree	Joy and anger: -100% Pleasure: -92% Sadness: -88%	Joy, anger, sadness and pleasure.
Ranganathan et al. (2016)	Face+Audio+ Physiological Signals	SVM	Face: 95.4%, Audio: -90.62%, Physical: 78.6% Multimodal: -52.4%	Angry, Happy, Sad, Disgust, Fear, Surprise and Neutral
Chen et al. (2022)	Audio+ Video+ EEG+ EMG	KNN LSTM	EEG: - 39.70% EMG: -37.09% Audio: -1.32% Image: -7.20%	Happy, sad, anger, fear, disgust, surprise, neutral
(Abtahi et al. (2018)	Audio+Video+ EMG+EEG+ECG	DBN LSTM	EEG: - 51.7% EMG: - 62.8% Images: - 71% Voice: - 64.5% EEG: - 37.6% EMG: - 58.4% Images: - 81.1% Voice: - 70.3%	Neutral, Sad, Happy, Disgust, Fear, Surprise, Anger

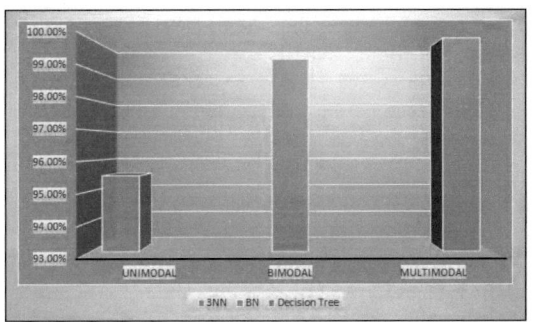

Fig. 2. Performance Evaluation on Various Emotional Intelligence modals

Conclusion

The paper reviewed various popular methods for unimodal, bimodal, and multimodal multiple human emotions. So, the review finds that the KNN classifier gives 95.5% accuracy as the best classification for unimodal and bimodal Bayesian Networks with 99.3% accuracy. Finally, the decision tree classification has secured 100% accuracy with different emotions for the multimodal. The study concludes that the SVM, KNN, and BN classifiers are efficient methods for handling data. The review finds that the influence of more than two classifiers gives better accuracy for handling the multiple data types concerning various emotions. So, the integrated approach is the best method for handling multiple heterogeneous data to identify different emotions.

References

Abtahi, F., Ro, T., Li, W., and Zhu, Z. (2018). Emotion Analysis Using Audio/Video, EMG and EEG: A Dataset and Comparison Study. Proceedings - 2018 IEEE Winter Conference on Applications of Computer Vision, WACV 2018, 2018-January (3), 10–19. https://doi.org/10.1109/WACV.2018.00008

Chen, J., Ro, T., and Zhu, Z. (2022). Emotion Recognition with Audio, Video, EEG, and EMG: A Dataset and Baseline Approaches. IEEE Access, 10, 13229–13242. https://doi.org/10.1109/ACCESS.2022.3146729

Chen, S., Zhang, L., Jiang, F., Chen, W., Miao, J., and Chen, H. (2020). Emotion Recognition Based on Multiple Physiological Signals. Zhongguo Yi Liao Qi Xie Za Zhi = Chinese Journal of Medical Instrumentation, 44(4),283–287. https://doi.org/10.3969/j.issn.1671-7104.2020.04.001

Hassouneh, A., Mutawa, A. M., and Murugappan, M. (2020). Development of a Real-Time Emotion Recognition System Using Facial Expressions and EEG based on machine learning and deep neural network methods. Informatics in Medicine Unlocked, 20, 100372. https://doi.org/10.1016/j.imu.2020.100372

Ranganathan, H., Chakraborty, S., and Panchanathan, S. (2016). Multimodal emotion recognition using deep learning architectures. 2016 IEEE Winter Conference on Applications of Computer Vision, WACV 2016. https://doi.org/10.1109/WACV.2016.7477679

Schofield, D. (2018). Human-Computer Interaction. Theories, Methods, and Human Issues. Human-Computer Interaction. Theories, Methods, and Human Issues, 10901(January), 325–344. https://doi.org/10.1007/978-3-319-91238-7

Shin, D., Shin, D., and Shin, D. (2017). Development of emotion recognition interface using complex EEG/ECG bio-signal for interactive contents. Multimedia Tools and Applications, 76(9), 11449–11470. https://doi.org/10.1007/s11042-016-4203-7

Singla, C., Singh, S., and Pathak, M. (2020). Automatic Audio Based Emotion Recognition System: Scope and Challenges. SSRN Electronic Journal, 1–6. https://doi.org/10.2139/ssrn.3565861

Song, T., Zheng, W., Lu, C., Zong, Y., Zhang, X., and Cui, Z. (2019). MPED: A multi-modal physiological emotion database for discrete emotion recognition. IEEE Access, 7(c), 12177–12191. https://doi.org/10.1109/ACCESS.2019.2891579

Song, T., Zheng, W., Song, P., and Cui, Z. (2020). EEG Emotion Recognition Using Dynamical Graph Convolutional Neural Networks. IEEE Transactions on Affective Computing, 11(3), 532–541. https://doi.org/10.1109/TAFFC.2018.2817622

Suhaimi, N. S., Mountstephens, J., and Teo, J. (2020). EEG-Based Emotion Recognition: A State-of-the-Art Review of Current Trends and Opportunities. In Computational Intelligence and Neuroscience (Vol. 2020). Hindawi Limited. https://doi.org/10.1155/2020/8875426

Tarnowski, P., Kołodziej, M., Majkowski, A., and Rak, R. J. (2017). Emotion recognition using facial expressions. Procedia Computer Science, 108, 1175–1184. https://doi.org/10.1016/j.procs.2017.05.025

Uddin, M. Z., and Nilsson, E. G. (2020). Emotion recognition uses speech and neural structured learning to facilitate edge intelligence. Engineering Applications of Artificial Intelligence, 94(January), 103775. https://doi.org/10.1016/j.engappai.2020.103775

Xu, J., Ren, F., and Bao, Y. (2019). EEG Emotion Classification Based on Baseline Strategy. Proceedings of 2018 5th IEEE International Conference on Cloud Computing and Intelligence Systems, CCIS 2018, 43–46. https://doi.org/10.1109/CCIS.2018.8691174

Zhao, J., Chen, S., Wang, S., and Jin, Q. (2018). Emotion Recognition using Multimodal Features. 2018 1st Asian Conference on Affective Computing and Intelligent Interaction, ACII Asia 2018, 3, 521–529. https://doi.org/10.1109/ACIIAsia.2018.8470385.

Factors Regulating the Human Microbiome

M. P. Namratha[*,a]

aCHRIST(Deemed to be University)
E-mail: *namratha.mpmp@gmail.com

Abstract

A healthy diet support both physical and mental well-being, whereas a poor diet can disrupt bodily processes and potentially increases the risk of developing many diseases. Various resources have been spent examining every conceivable aspect of the daily diet that could have an impact on health, including food ingredients, additives, and cooking methods. High-throughput sequencing technology is assisting in the slow accumulation of data that most of these factors primarily interact with the gut microbiome to cause downstream effects. In reaction to a person's daily diet, the gut flora may be able to function as an extremely sensitive mirror. The relationship between nutrition, gut microbiome, and health has been steadily shown to be complex, however, this relationship is rarely covered in greater detail. To assist people in making more informed recipe selections, this review shows the most recent developments in the interactions connecting diet and the gut microbiome. It also analyzed the major factors that have been implicated in this process and showed the potential for maintaining health or treating diseases through diet intervention.

Keywords: Microbiota, Health, Gastro-intestine, Bacteria

Introduction

While "microbiota" denotes all the microorganisms in the environment, "microbiome" mentions the collection of genomes since all the microorganisms are in a specific habitat. These phrases are usually used interchangeably in real life. To stress the variety of the microbiota and, in specific, the human intestinal microbiome typically contains not only microorganisms but also archaea, viruses, fungi, and parasites, the term "microbiota" has replaced "flora." We now understand that it constitutes a highly complex and developed environment that is crucial to the emergence and upkeep of homeostasis. Technologies that identify microorganisms and explain the genetic makeup and metabolism have dramatically advanced and are constantly developing, which has allowed us to understand the composition and activities of the gut microbiome. We can now annotate a given microbiome's composition, the abilities of its members (via an examination of their genomes), and the outcome products they create (using metabolomics and other methods) in ever-increasing depths of information. We now understand the function of gut microbiota in the growth and continued function of the central CNS, once more through the use of animal models.

The adult microbiome may be influenced by a various range of variables, although age, location, nutrition, and drugs have emerged as the major sources of inter-individual variance. There has recently been a lot of interest in methods for altering the gut microbiota and also in the microbiota's potential as a source of novel physiologically active chemicals and as a gauge for how effectively different interventions would work. There has recently been a lot of interest in methods for altering the gut microbiota and also in the microbiota's potential to produce novel physiologically active chemicals and serve as a prediction of how effectively certain therapies will work. Although such studies allow for a great deal of flexibility in terms of phenotypic and genotypic manipulation, allow for a lot of potential interventions, and make it easier to collect a variety of biological trials, that are not without flaws, and the extrapolations of human being condition must be made with caution (Quigley and Gajula, 2020).

• Modifying the microbiome

When evaluating all interference that aims to effectively and, we hope, constructively modulate the microbiota, it's important to keep in mind the strong and complex environment that awaits. Simple ideas about how a supplement or drug can affect the microbiota–host interface have sparked a lot of excitement and also disappointment. An understanding of a variety of likely interactions between involvement and the hosts' nutrition, immune system, genomes, and resident commensals should raise awareness of the challenges ahead. One should be aware of the obstacles ahead by understanding the variety of potential exchanges amongst an intervention and host food, immune system, as well as with local commensals (Singh et al., 2017).

Variety of interferences that modulate the microbiome.

1. On Lifestyle modifications
   • Nutrition involvement and modifications
   • Calorie restrictions
   • Exercises
   • Other lifestyle factors
2. Clinical interventions
   • Fecal microbiome transfer
   • Antibiotics and Probiotics
   • Pharmabiotics

Impacts of non-antibiotic drugs proceeding on the microbiota

We define "modulation" as manipulating one or many of the following targets: first, the relative spreading of bacteria species or strains; second, the actual quantity of bacteria; third, their metabolic activities; and fourth, their interaction with the host. More boards could be adapted— antigens, virulence, biofilms, and bacterial-like examples— but we've decided to stick to the ones listed above. While modulating the microbiome, care should be taken to avoid unintended consequences such as an overview or raise of pathogenic species, the broadcast of antibiotic resistance, or the generation of adverse host reactions (Claesson et al., 2012).

1. Lifestyle modification
• Microbiome and Diet

Diet is now abundantly clear as a key regulator of the gut microbiome, both in the long and short term; this makes perfect sense because bacteria rely on what we eat for the majority of their nourishment. Studies comparing communities and individuals with drastically varied eating habits indicate the long-term consequences of nutrition. These many variations are the result of all-time or, at the very least, very longstanding dietary habits. In short term, drastic dietary fluctuations, such as lowering the intake of fiber eliminating fermentable oligo-, di-, or monosaccharides, gluten, polyols (FODMAPs), or drastically boosting the intake of protein, can all affect microbiota composition. The microbiome also appears to have significant purposeful redundancy, which helps it to retain stability in dietary changes

• Calorie restrictions

An extreme dietary strategy: fasting, exemplifies the difficulties that dietary research faces. Although alterations in microbiome composition and diversity have always been observed in anorexia nervosa and its related eating disorders, separating cause from effect has proven difficult. It would be remarkable if increased fasting does not affect the gut microbiome; what has to be determined is if there are microbial characteristics particular to eating ailments that could show a major role in their pathogenesis. Given the growing interest in the gut microbiome's potential role in obesity, researchers are looking into the microbiome's involvement in several calorie-cutting measures. Fasting induces to change in the microbiota that has been linked to not only favorable metabolic effects but also to improvements in intestinal inflammation and also in central nervous system disorders. The microbiome's potential role in weight loss and metabolite benefits of bariatric surgery has been investigated.

- Exercise

Given the practically common association between physical exercises and food behaviors as a fragment of what is called a "healthy lifestyle," trying to measure the influence of exercises on the microbiota presents similar challenges. Professional athletes, on the one hand, ingest significantly more protein, which has a major influence on the makeup of the microbiome. The brain-gut axis is a recognized pathway for bidirectional transmission between two organs that are both impacted by exercise. This has been expanded to encompass the microbiota according to many lines of research. The vagus nerve, neuroendocrine mediators such as the hypothalamus-pituitary-adrenaline axis, and a range of neurotransmitters and the local hormones, some of that is also produced by microbiota, are hard-wired to facilitate information flow lengthwise the brain gut-microbe axis. This controls a variety of homeostatic functions, such as digestion, satiety, and hunger. Additionally, the vagus controls the tone of the gastrointestinal inflammatory response, which has been directly observed by pro-inflammatory responses following vagotomy in mice (O'Sullivan et al., 2015).

- Additional lifestyle factors

Additional lifestyle factors have been related to alterations in the microbiota, including cigarette smoking, alcohol consumption, and recreational drug use. Given the recognized connection between cigarette smoking and mouth cancer, the oral microbiome has also been of particular interest concerning its first of these. (Lee et al., 2018).

2. Clinical interventions

- Fecal microbiome transfer

Though it appears to have been used on an empiric basis for a few decades, fecal microbiota transplantation (FMT) only gained scientific acceptance in over a decade or two. The outcomes in recurrent Clostridioides difficile–associated illness (CDAD) have been particularly excellent, with cure rates of up to and above 90% reported. Limited clinical trial findings show that FMT may also be intricated in the development of the symptoms of hepatic encephalopathy, which has been known for some time to be significantly influenced by the small intestinal microbiota. Microbiome modification, particularly FMT, has promise for treating non-alcoholic steatohepatitis and its more non-alcoholic fatty liver diseases (NAFLD), which may be caused by the microbiome. Clinical trials are anticipated. Additionally, it is most important to keep in mind that FMT can transmit or affect viruses and other non-bacterial species found in the gut microbiome. FMT involves some risk. In addition to the theoretical possibility that the transmission of bacteriological signatures linked to disease states results in the development of disorders in the recipient, infectious agents can also be transmitted. For these reasons, it is necessary to regulate FMT. Numerous preparations and distribution techniques have been used, with variances in apparent efficacy; however, it is yet to understand which mechanisms of the transplanted fecal microbiome are genuinely required for efficacy (Cammarota et al., 2018).

- By Antibiotics

Antibiotics that are taken orally, or that undergo enterohepatic circulation and biliary excretion, will invariably influence gut microbiome to a lesser or greater amount, regardless of the mode of administration. These "innocent bystander" impacts might reduce host fighting to infections, allowing the CDAD or fungal outgrowth to occur, and certain populaces are particularly vulnerable. Antibiotics affect inflammation, metabolism, and cancer through their consequence on the microbiota; the overall impacts, whether beneficial or it is harmful, that are influenced by microbial, antibiotic, and host factors. These antibiotic effects' long-term ramifications for human health are being realized now. Initial and frequent contact with antibiotics in infancy, even in extremely modest dosages that we swallow throughout the food chain because of their usage in animal husbandry, appears to incline us to the expansion of inflammatory and metabolites illnesses later in our lifespan. It is certainly necessary to issue a call to arms to report the global use of antibiotics (Lewis et al., 2015).

• Probiotics and Prebiotics

Probiotics have been praised for millennia for a variety of health benefits; however, most of these claims have yet to be confirmed in high-quality clinical trials. Orally taken probiotics show to have more distant effects on the liver and CNS. Now we have a good grasp of how probiotics interact with their hosts to produce these of effects; for example, the molecular foundation of specific *Bifidobacteria* species' anti-inflammatory properties has been characterized in excessive detail in elegant in vitro and animal studies.

Cereals and plants like bananas, onions, chicory root, garlic, and Jerusalem artichokes may contain substances having prebiotic benefits, however, these substances are often present in little amounts and may not have prebiotic properties in these forms.

Prebiotics such as Galacto Oligosaccharides (GOSs), FOSs, and inulin chicory fiber, are more biologically dynamic and selective. Human milk oligosaccharides are crucial prebiotics that are given to infants in breast milk. They encourage the growth of *Bifidobacteria* and, as a result, have been connected to several health advantages. The infant's microbiota and immune system are also benefited from other breast milk ingredients. Prebiotic research currently focuses on creating highly selective "designer" prebiotics that is intended to affect only certain taxa of the gut microbiome.

Specified the enormous inter-distinct variation in the gut microbiota's makeup, it might be impossible to anticipate consistent outcomes from a particular microbial strategy in any illness condition. The development of "personalized bacteriotherapy" may be facilitated by efforts to clarify which bacteria or host characteristics influence responses. There is a lot of work to accomplish. Although this approach is appealing in theories and synbiotic preparation has had significant successes, synergy is not always feasible, and it is not always possible to determine the exact proportions of probiotics or prebiotics to any observable benefit. This is known as a synbiotic. (Quigley, 2019).

• Pharmabiotics

The word "pharmabiotic" is coined to refer to any materials with potential health benefits which can be extracted from microbiota, microbiota–host exchanges in the gut. This includes not only alive but also dead or altered organisms, also as bacterial harvests or metabolites. Natural antibiotics produced by bacteria, bacteriocins, genetically modified organisms bacteriophages, and short-chain fatty acids are some actual examples.

Although these and other technologies are still relatively new in terms of clinical use, they provide intriguing prospects for microbiota modification and may be critical to resolving the present antibiotic problem. CRISPR-based technologies, which are still evolving, have revolutionized genome deletion and have already used to build new antibacterial strategies (Shanahan, 2009).

• Effects of non-antibiotic drugs on the microbiota's

Interferences that alter the microbiota composition by modulating intrinsic protection systems against microbial colonization are expected to change the microbiota composition. Proton pump inhibitors (PPIs) have been related to a propensity to *Clostridioides difficile* infection, small intestinal bacteria overgrowth, and enteric infections, however, the evidence is mixed. Studies of human feces have shown that PPI users have a lower number of Clostridiales and a more number of, *Micrococcaceae, Actinomycetales and Streptococcaceae;* these alterations were before linked to increased vulnerability to this feared antibiotic complication. Similarly, drugs that alter motility and intestinal transport, of that which there are several, may also alter the microbiota composition. Numerous other drugs likely involve the microbiota with resultant improvement or decrease in effectiveness or initiation of side effects yet other fruitful field for forthcoming microbiota research. (A. Singh et al., 2018).

Conclusion

The function and structure of gut microbiota can be altered through dietary factors as various nutritional components (carbohydrates, vitamins, lipids, proteins, minerals, etc.), food additives, cooking, and processing, and these modifications are intimately linked to preserving bodily health. Long-term unhealthful eating patterns, like the western diet, play a significant role in many non-communicable diseases. A nutrition intervention program based on these fundamental concepts has been proven to be successful. For many years, research has shown how diet and gut flora interact. However, it is unavoidable that many factors other than diets such as age, heredity, smoking, physical activity, and the like, will also alter the alignment of the gut microbiome, making it difficult to pinpoint the precise role of diet in diseases.

References

Cammarota, https://doi.org/10.1136/gutjnl-2017-314049

Claesson, M. J.-P. W. (2012). Gut microbiota composition correlates with diet and health in the elderly. https://doi.org/10.1038/nature11319

Lee, S. H to Lee, J. H. (2018). https://doi.org/10.3390/jcm7090282

Lewis, J. D., to F. D. (2015). https://doi.org/10.1016/j.chom.2015.09.008

O'Sullivan, to P. D. (2015). Exercise and the microbiota. https://doi.org/10.1080/19490976.2015.1011875

Quigley, E. to W.B. Saunders. https://doi.org/10.1016/j.cgh.2018.09.028

Quigley, E. M. M., https://doi.org/10.12688/f1000research.20204.1

Shanahan, F. https://doi.org/10.1113/jphysiol.2009.174649

Singh, A., Proton Pump Inhibitors https://doi.org/10.1002/ncp.10181

Singh, to Liao, W. (2017). https://doi.org/10.1186/s12967-017-1175-y

Citrus indica: Characterization and Potential as a Nutraceutical

Upasana Deb,[*,a] *and Sheena Haorongbam*[b]

[a,b]North Eastern Hill University, India
E-mail: [*]upasanadeb123@gmail.com

Abstract

The study aims to introduce the fruit called *Citrus indica* or the Indian Wild Orange. It originated in The Nokrek Biosphere Reserve of Meghalaya and is referred to as the mother of Oranges. It is used by the Garos to make their traditional medicines, which makes it a suitable candidate for nutraceuticals. However, limited literature is available about the plant. With the loss of importance, the plants have started disappearing and the numbers have drastically reduced. Nowadays, among so many *Citrus* fruits, it is very difficult to identify the fruit. Therefore, with this paper, we aim to understand its morphology to make its identification easier. Also, biochemical characteristics have been studied which proves that the fruit is indeed unique and has the potential to become a nutraceutical. Therefore, more work and an increase of awareness are required on this plant to save it from getting extinct.

Keywords: *C. indica*, Biochemical, Morphology, Nutraceutical, Unique

Introduction

Citrus fruits are well known to the world, especially after the occurrence of Covid 19. People consume it for its rich vitamin C content and antioxidant properties. *Citrus* fruits belong to the family of Rutaceae and have many different species which vary morphologically as well as physicochemically (Kumar *et al.*, 2010). One such species of *Citrus* which is considered the progenitor of Citrus is *Citrus indica* (Singh, 1981). As the name suggests, it originated in India and was discovered by Tanaka (1928). The fruit was discovered in the foothills of Himalayas, in the Nokrek region of the West Garo Hills district of Meghalaya, India. It is a unique and rare plant known to be present only in the place of its occurrence, i.e., Nokrek Biosphere Reserve (Malik et al., 2006).

The plant was named the Indian Wild Orange by Tanaka as no vernacular names were present at that time (Bhattacharya and Dutta, 1956). However, today the local tribes of Garos call it "Memang Narang." According to them, the plant has medicinal properties which makes it a major constituent of their traditional medicine. It is used by the Garos for the treatment of smallpox, jaundice, and various other ailments like stomach ailments in humans and animals (Malik et al., 2006; Momin et al., 2016). But not much is known about the fruit, therefore with this paper, we tried to reveal its characteristics- morphological and physiochemical.

Materials and Methods

Morphology: For the morphological study, we took ten fruits and noted the measurements of the fruit. We cut the fruit transversely to see the segments of the fruit, seed size, etc. We also observed the number of sections and seeds in each fruit. The color of the peel of the fruit, inner segments, and of seed were also seen.

Biochemical Properties: These parameters were studied by Upadhaya, 2013. The tests and the methods applied are given in Table 1.

Table 1. Biochemical Tests conducted on the juice of *C.indica* by Upadhaya (2013)

S.No.	Tests	Methodology
1	Acidity	Titration with NaOH to pH 8.1 at 20 °C and 32 °C. Results were expressed in citric acid per liter.
2	Juice Content	Ten fruits were selected and weighed. After that, the juice was extracted from them. The juice content was then measured. Following that, the formula given below needs to be applied. % Juice content = (Juice Weight /Fruit Weight) x 100
3	Total Solid Content	AOAC, 1990
4	pH	Digital pH meter.
5	Carbohydrate	Estimated by Anthrone method using glucose as standard and anthrone as a reagent, absorbance at 630 nm
6	Protein	Estimated by Kjeldahl method Total Nitrogen (%) = Burette reading × 14 × 0.02 × 1000 / Sample weight
7	Dietary Fibre	Estimated gravimetrically % Crude fibre in ground sample = {Loss in weight on ignition (W2 – W1) – (W3 – W1)/ Weight of the samp
8	Microminerals: Copper, Iron, Manganese, Zinc, and Selenium	Done by Atomic Absorption Spectrophotometer
9	Ascorbic acid (mg/100g)	Estimated by 2,6-dichloroindophenol titrimetric method (AOAC, 1995) ascorbic acid (mg/100g) = {(concentration of working standard/ burette reading for blank) x (burette reading for sample/sample volume) x final volume of extractant}
10	Total Phenol Content (mg/100g)	Determined with the Folin-Ciocalteu spectrophotometric assay using gallic acid as standard and Folin-Ciocalteu as reagent, absorbance at 650 nm.
11	Total Flavonoid content (mg/100g)	Estimated by the Aluminium chloride spectrophotometric method, absorbance was measured at 500 nm.
12	Total Carotenoid content (mg/100g)	Sample was extracted using n-hexane-acetone-ethanol (v/v; 2:1:1) mixture extraction solvent. The mixture was then be centrifuged at 4000 rpm at 4°C. Supernatant was collected and made to the final volume with extraction solvent. Absorbance was measured at 450 nm and the result was expressed as β-carotene equivalents.

Results and Discussion

The morphological data proved that the fruit was indeed unique. The fruits of *Citrus indica* were of almost the same size, however, the shape of the fruits varied. Some fruits were wrinkled while some were smooth. Other features are represented in Table 2.

Table 2. Results of the Morphological Study of the fruit of *C.indica*

S.No	Fruit Features	Description
1	Color of the peel	Deep orange when ripe, green while unripe
2	Size	2.5 cm in diameter (average)
3	Segments	9 (average)
4	Color of the segment (pulp)	Lime yellow
5	Segment size	1.8 cm (average
6	Seed no.	Varies, sometimes one seed in one fruit while others had 11 seeds in one fruit
7	Seed size	0.9 cm (average)
8	Seed color	Ivory

Fig. 1. Different shapes of *C.indica* fruits

Fig. 2. Varying seed numbers in *C. indica* fruits

The results of Biochemical analysis of the juice of *C.indica* done by Upadhaya and Chaturvedi, 2013 are given in Table 3.

Table 3. Results of the biochemical analysis of the juice of *C. indica* by Upadhaya and Chaturvedi (2013).

S.No.	Tests	Results
1	Total acidity	40.69 g Citric acid/L
2	Total solid content	65.96 g/L
3	pH	2.51±0.31
4	Carbohydrate	21.22±3.00 (mg/L)
5	Protein	0.95mg/100 ml
6	Crude Fibre	2.54±0.31
7	Phosphorus	0.20±0.00 mg/100g
8	Potassium	6.36±0.26 (ppm)
9	Calcium	10.29±0.24 (ppm)
10	Sodium	0.101±0.02 (ppm)
11	Copper	0.033±0.00 (ppm)
12	Iron	0.237±0.04(ppm)
13	Manganese	0.067±0.03(ppm)
14	Zinc	0.523±0.03 (ppm)
15	Selenium	1.005±0.44 (ppm)
16	Ascorbic acid (mg/100ml)	695.64±14.43
17	Total Phenol Content (mg/100 ml)	53.33±0.58
18	Total Flavonoid Content (mg/100 ml)	3.29±0.46

Society is changing, and the heritage of traditional knowledge is getting lost. Even the Garos today have limited knowledge about their ethnic medicines and the way of preparation and ingredients. But it should not be forgotten that those medicines were purely natural. Without the interference of chemical in it, the medicines hardly had any side effects. In the long run, these medicines should not be forgotten. Nowadays, even the modern people of Garo Hills may get confused in identifying the ethnic plants especially *C. indica* as many varieties of *Citrus* are available and the knowledge to distinguish them is limited.

Understanding the morphological aspect of the fruit is very important for primary identification. Morphological study revealed many details of the plant, one was the variation in the fruit shape and the other was the variable seed count in each fruit. This data suggests that there's variation among the species of *C. indica* which needs to be studied. Only by knowing the right species of *C. indica* one can use it for the preparation of traditional medicine.

Biochemical studies revealed that the plant has the properties to become a nutraceutical. The ascorbic acid content of the fruit is much higher than the commercial fruit juices which had an ascorbic acid range of 24–43mg/100ml (Siopidou and Moshatou, 2000). So, people with lower immunity can build resistance to disease by taking this fruit juice. But in contrast to ascorbic acid, the phenol content was much lower than other *Citrus* fruits, which had an average range of 600–820 (µg GAE/mL) (Rekha et al., 2012).

Whatever data we have are quite old and had been generated through manual or old techniques. With time, there had been advancement in techniques. Some more tests for nutraceutical properties like vitamins, amino acid profiling, fatty acids, anti-nutritional components, etc need to be done especially by the modern techniques to classify it as a nutraceutical.

The study came up with many observations, the most important among them is that the fruit is of great traditional importance and its utilization is still a big secret. Low pH and extreme sourness may play an important role in its unpopularity among people. That is the reason people hardly care about the existence of the plant, and now it got endangered. So, the value addition of the fruit is required to make it more consumer-friendly.

If the plant gets extinct, it will be a big loss to the ecosystem and society, as the ancestor of oranges will be lost, so as its benefits which had the potential to boost the nutraceutical and pharmaceutical sectors.

Conclusion

More study is required on the plant, as the studies done so far are not enough to describe its full potential. It can be a very good nutraceutical and can also be used in the preparation of medicines. Value addition of the fruit juices may influence their flavor and increase their consumption. Thus, with more demands, people will start to recognize the plant and we may see more of it in future.

References

Bhattacharya S. C. and Dutta, S. (1956). Classifications of Citrus fruits of Assam. ICAR Sci. Monograph, 20(1), 1–110.

Kabasakalis, V., Siopidou, D. and Moshatou, E. (2000). Ascorbic acid content of commercial fruit juices and its rate of loss upon storage. Food chemistry. 70(3), 325–328.

Kumar, S., Jena, S. N. and Nair, N. K. (2010). ISSR polymorphism in Indian wild orange (*Citrus indica* Tanaka, Rutaceae) and related wild species in North-east India. Scientia Horticulturae, 123(3), 350–359.

Malik, S. K., Chaudhury, R., Dhariwal, O. P. and Kalia, R. K. (2006). Collection and characterization of *Citrus indica* Tanaka and *C. macroptera* Montr.: wild endangered species of northeastern India. Genetic resources and crop evolution. 53(7), 1485–1493.

Momin, K. C., Suresh, C. P., Momin, B. C., Singh, Y. S., and Singh, S. K. (2016). An ethnobotanical study of wild plants in Garo Hills region of Meghalaya and their usage. International Journal of Minor Fruits, Medicinal and Aromatic Plants, 2(1), 47–53.

Rekha, C., Poornima, G., Manasa, M., Abhipsa, V., Devi, J. P., Kumar, H. T. V., and Kekuda, T. R. P. (2012). Ascorbic acid, total phenol content and antioxidant activity of fresh juices of four ripe and unripe citrus fruits. Chemical Science Transactions. 1(2), 303–310.

Singh, B. (1981). Establishment of first gene sanctuary in India for Citrus in Garo Hills. Concept Publishing Company. Delhi, India.

Tanaka, T. (1928). On certain new Species of citrus. Studia Citrologica 2(1), 155–164.

Upadhaya, A. (2013). Biodiversity and nutritive value of semi wild and wild citrus species of Meghalaya. North Eastern Hill University, Shillong. Retrieved from: http://hdl.handle.net/10603/169749

Evaluating Impact of Clotrimazole on Inhibition of Candida Albicans and Surface Roughness in Permanent Soft-Liners

Miloni Bhatt,[*,a] *Dipti Shah,*[b] *and Kalpesh Vaishnav*[c]

[a]Karnavati School of Dentistry, Karnavati University, Gandhinagar, India
[b]Karnavati School of Dentistry, Karnavati University, Gandhinagar, India
[c]Karnavati School of Dentistry, Karnavati University, Gandhinagar, India
E-mail: [*]milonibhatt@karnavatiuniversity.edu.in

Abstract

Permanent soft liners used for the treatment of denture stomatitis have the drawback of increased porosity and surface roughness. This study aimed to analyze the change in the zone of inhibition and surface roughness with the addition of clotrimazole in soft liner. Long-term permanent soft liners Permasoft (groups 1 and 3) and Mollosil (group 2 and 4) were used to fabricate specimen discs (n=48/group) and blocks(n=8/group) with and without clotrimazole in an aseptic environment using a die. Samples were stored in salivary substitutes. Surface roughness tester -SJ-210 Mitutoyo Instrument was used on blocks to measure surface roughness. One-way ANOVA test, post hoc Tukey analysis, and Student t-test were performed. After 42 days decrease in the zone of inhibition was observed (p<0.001) for group 2 and 4 and a constant increase in surface roughness for all groups (p < 0.001) Zone of inhibition in control groups was nil. The addition of clotrimazole in soft liner inhibits the growth of candida Albicans in vitro and can be used for treating candidiasis. Surface roughness values increase within clinically acceptable limits.

Keywords: Zone of inhibition, Candida Albicans, Surface Roughness, Permanent Soft-liners

Introduction

The oral tissues provide adequate support to the removable dental prosthesis. Adaptation of the prosthesis is compromised due to the constant remodeling of alveolar ridges and oral tissues with time. Ill-fitting prosthesis along with predisposing systemic conditions like immunocompromised diseases, diabetes, vitamin deficiency, etc. in geriatric patients leads to the formation of sore spots and tissue trauma. These are the areas of force concentration or misfit of the denture[1]. For the fabrication of a new prosthesis, the abused mucosa needs to be conditioned with soft liners which are applied to the current prosthesis. This helps in regaining the adaptation of the prosthesis to the tissues. (Karin hermana Neppelenbroek et al., 2017) The soft liner which can be short-term or long-term and acrylic or silicone-based depending on the duration of use is applied between the impression surface of the prosthesis and oral mucosa. It acts as a shock absorber by providing a cushioning effect. Some known drawbacks of these materials are leaching out of plasticizers over time which makes them more porous in composition. This porosity provides a niche for the growth of opportunistic fungi such as candida Albicans causing candidiasis. Treatment of candidiasis includes systemic and topical use of antifungal agents. Topical agents are more effective on the fungi that invade superficial tissues, but they are less palatable leading to discomfort for the patients and also an uncertainty that the topical gents will be applied by the patients. Thus studies have been carried out by incorporating the antifungal agents directly into the soft liner (Bhatt et al., 2022). The incorporation of antifungal agents into temporary soft lining materials has been demonstrated to be effective and viable to extend their longevity and reduce biofilm accumulation.

The addition of antifungals into soft liners leads to alteration in the surface properties of the material such as hardness, roughness, water absorption, and elasticity. (Urban et al., 2014). Amongst the properties, the surface roughness of the material changes due to the leaching out of alcohol and plasticizers from the soft liner into the liquid medium. The rougher the surface of the material, the greater biofilm formation favoring the emergence/maintenance of oral pathologies (Bueno mg et al., 2015).

The literature lacks data regarding the incorporation of antifungal agents in long-term soft liners. (Grzegorz et al., 2014). This study aimed to examine the change in the zone of inhibition of candida albicans and the change in surface roughness by incorporation of clotrimazole into long-term denture soft liners in the in-vitro experimental setting. Based on this a null hypothesis was formulated stating "There will be no significant change on candida growth and no change in surface roughness between control samples and samples containing clotrimazole.'

Materials and Methods

The permanent long-term soft liners selected for this study were Acrylic based soft liner (Permasoft DENTSPLY) and Silicone based soft liner (Mollosil Detax). The material was manipulated according to the manufacturer's guidelines for Group 1- Permasoft control and Group 3-Mollosil control. The antifungal drug Clotrimazole was incorporated at the minimum inhibitory concentration of 5µg/ml. Thus 0.6mg Clotrimazole/60gm of powder was incorporated while manipulating material for Group 2-

Permasoft with Clotrimazole and Group 4-Mollosil with Clotrimazole. The material is transferred into stainless steel dies in an aseptic environment. Total n=48 discs (radius 2cm, height 0.2cm) and n=8 blocks (2cmx2cmx2cm) were fabricated for each group.

Fig. 1. Samples of disc and block

All samples were fabricated by a single operator to maintain uniformity. Samples were stored in 4 separate Petri dishes containing 10 ml salivary substitute (GC Dry Mouth Gel- GC) and incubated at 37°C in an incubator to simulate the conditions of the oral cavity.

Swabs were obtained from 0.5McFarland concentration candida Albicans (ATCC10231) strain and lawn culture was done on Petri dishes containing Saburaud dextrose agar medium.

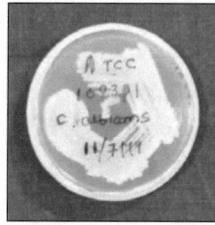

Fig. 2. ATCC Strain 10231

Samples were placed in the center of each petri dish. 8 such samples from each group were incubated at 37°C for 24 hrs. To measure the visible zone of inhibition created by the drug was in mm. "Figure 3'

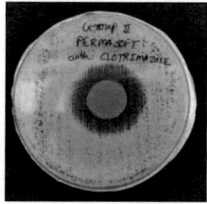

Fig. 3. Sample showing zone of inhibition

Surface roughness tester -SJ-210 Mitutoyo Instrument (JIS1994 Standard) was used to measure surface roughness on the blocks. 8 samples from each group were tested at a time. The stylus was passed on the surface and the Ra value displayed on the monitor was noted. "Figure 4."

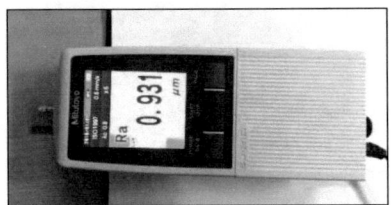

Fig. 4. Surface Roughness Value tested on roughness tester

The same procedure was repeated at time intervals of 1, 7,14,21,35, and 42 days and results of the zone of inhibition and surface roughness were noted down. The data obtained concerning zone of inhibition, surface hardness, and surface roughness was evaluated using a one-way analysis of variance (ANOVA) test, Tukey's post hoc analysis using SPSS software (SPSS 12.0; SPSS, INC, CHICAGO, III)

Results

The mean values of the zone of inhibition of 4 groups were measured in mm at time intervals and noted.

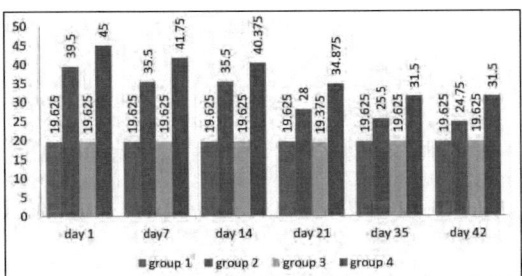

Graph 1. Zone of inhibition in mm at various time intervals

For Groups 1 and 3, no change in values of the zone of inhibition was observed (p>0.001) at time intervals of 1, 7,14,21,35, and 42 days. For Group 2 the value of the zone of inhibition decreased on each time interval (p<0.001) except

between days 7 and 14 where it remained constant (p value>0.01). For Group 4, the value of the zone of inhibition decreased constantly from day 1 to 35 (p<0.01) except between day 35 and 42 (p>0.001). The maximum zone of inhibition was observed on day 1 for groups 2 and 4. On one way analysis of variance (ANOVA) test for the zone of inhibition, the value decreased over time in between and within groups.(p<0.001) .'Table-1'

Table 1. Comparison of Zone of Inhibition (in mm) in Between Groups and Within Groups by One Way ANOVA

		Sum of Squares	df	Mean Square	F	ANOVA P VALUE	Significance
DAY 1	Between Groups	4216.125	3	1405.375	2861.855	<0.0001	S
	Within Groups	13.750	28	0.491			
	Total	4229.875	31				
DAY 7	Between Groups	3044.250	3	1014.750	3071.676	<0.0001	S
	Within Groups	9.250	28	0.330			
	Total	3053.500	31				
DAY 14	Between Groups	2777.844	3	925.948	1471.009	<0.0001	S
	Within Groups	17.625	28	0.629			
	Total	2795.469	31				
DAY 21	Between Groups	1329.344	3	443.115	848.356	<0.0001	S
	Within Groups	14.625	28	0.522			
	Total	1343.969	31				
DAY 35	Between Groups	774.125	3	258.042	614.908	<0.0001	S
	Within Groups	11.750	28	0.420			
	Total	785.875	31				
DAY 42	Between Groups	760.250	3	253.417	630.726	<0.0001	S
	Within Groups	11.250	28	0.402			
	Total	771.500	31				

From Tukey post hoc analysis, results for Group 2 and 4 showed significant changes in values at particular and at different time intervals suggesting a change in values of the zone of inhibition(p<0.001). Whereas change in the zone of inhibition was insignificant for Groups 1 and 3.

The mean values of surface roughness of 4 groups were measured in Ra value at all-time intervals and noted (Graph-2).

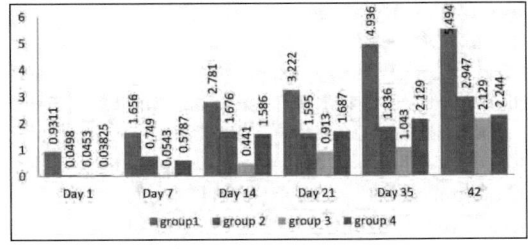

Graph 2. Surface Roughness at various times intevals

Using student t-tests for Groups 1, 2, and 4; a constant increase in the value of surface roughness was observed from day1 to day 42($p<0.0001$). Whereas for Group 3, a statistically significant increase in the value of surface roughness was observed from day 1 to 21 and 35 to 42($p<0.0001$). On analyzing results using the one-way analysis of variance (ANOVA) test, a significant increase in roughness was observed over time in between and within all groups from day 1 to 42($p<0.001$) "Table 2."

Table 2. Comparison of surface roughness (Ra value) in between groups and within groups by One way ANOVA

		Sum of Squares	df	Mean Square	F	ANOVA P VALUE	Significance
DAY 1	Between Groups	4.717	3	1.572			
	Within Groups	.000	28	0.000	1752316.373	<0.0001	S
	Total	4.717	31				
DAY 7	Between Groups	10.671	3	3.557			
	Within Groups	.000	28	0.000	2429162.764	<0.0001	S
	Total	10.671	31				
DAY 14	Between Groups	21.946	3	7.315			
	Within Groups	.001	28	0.000	335367.857	<0.0001	S
	Total	21.946	31				
DAY 21	Between Groups	22.813	3	7.604			
	Within Groups	.884	28	0.032	240.918	<0.0001	S
	Total	23.697	31				
DAY 35	Between Groups	69.085	3	23.028			
	Within Groups	.000	28	0.000	26184528.076	<0.0001	S
	Total	69.085	31				
DAY 42	Between Groups	59.120	3	19.707			
	Within Groups	.001	28	0.000	982042.571	<0.0001	S
	Total	59.120	31				

On comparing values using the Tukey post hoc analysis test, all groups had statistically significant p-value ($p<0.0001$) except for groups 2 and 4 on day 21.

Discussion

For this study, the addition of clotrimazole led to the formation of a zone of inhibition in candida growth and also altered the surface roughness of these materials. This proves that the null hypothesis formulated for this study was rejected. There is slow, prolonged release of the drug directly onto the affected tissues which gives the benefit of reduced drug dosage with maximum effect against candidiasis (Pachavaet et al., 2014). Clotrimazole and nystatin are the most commonly used drugs for topical application to treat oral candidiasis. Since nystatin is a short-term release drug, its use in

permanent soft liners serves lesser importance. It has been shown that the incorporation of drugs at commercially available concentrations in soft liners may affect their morphological structure and properties such as tensile strength, water sorption, modulus of elasticity, weight, hardness, and roughness.

Based on the statistical findings, for groups 1 and 3, it can be proved that long-term soft liners do not have antifungal properties of their own. These results are similar to Griiber et al. who showed that silicone and methacrylate soft denture liners would support the growth of C. albicans (Bueno et al., 2017). The drug-releasing and antifungal properties in Permasoft with clotrimazole (group 2) rapidly decreases within the first 7 days and then decreased gradually under the acrylic-based soft liner. Silicone soft liners are more stable in composition than acrylic liners because they do not contain a plasticizer. As there is no leaching out of products when they are stored in salivary substitutes, they retain their softness and elasticity over time. Due to this reason, the drug-releasing property of Mollosil with clotrimazole (group 4) is relatively more constant with gradual loss of zone of inhibition that reaches a plateau level at day 35. These results are similar to the studies done by Marta Radnai et al. (2010) for different soft liners and different antifungal drugs.

There is a development of resistant antifungal strains against the drug if the antifungal stays in contact with the mucosa for a longer time; which is not a desirable outcome. The antifungal activity of the drug diminishes after 42 days, this created a lesser chance of formation of resistant strains and a more effective treatment outcome. After 42 days, the soft liner with clotrimazole will function similarly to a soft liner control group and serves all the purpose except for the antifungal property. The drug-releasing or diffusing properties amongst the four groups can be compared as follows from minimum to maximum.

Mollosil control=Permasoft control < Permasoft with clotrimazole <Mollosil with clotrimazole.

The surface roughness is linked directly or indirectly to relevant factors related to removable dentures, such as retention, resistance to staining, microbial adhesion, the health of oral tissues, and patient comfort. In this study, the minimum to maximum values of surface roughness on day 1 was in the following order; Mollosil with clotrimazole<Mollosil control=Permasoft with clotrimazole<Permasoft control. While the minimum to maximum values of surface hardness at day 42 was Mollosil control <Mollosil with clotrimazole<Permasoft control < Permasoft with clotrimazole. There is a constant increase in the value of surface roughness from day 1 up to day 42 for all the groups, due to leaching out of the plasticizers and low molecular weight antifungal which makes the surface rough by the creation of pores and craters. These results are not similar to the study done on temporary soft liners and tissue conditioners by incorporation of drugs like miconazole, nystatin, and itraconazole by Mirian et al. (Bueno et al., 2017). As the surface roughness increases, the chances of absorption of the fluids, microbes adhesion and biofilm formation in material increases (Machado et al., 2009). The roughness value of Permasoft with clotrimazole is less than that of Permasoft control because when the bound drug molecules leach out, the polymerization chain reaction continues, making a stronger and uniform matrix. For Mollosil control and mollosil with clotrimazole groups, the increase in value of surface roughness is due to the release of plasticizers and the release of the drug particles.

Conclusion

Within the limitations of this in vitro study, it can be concluded that Permasoft and Mollosil do not have any antifungal properties of their own. When clotrimazole is added at the minimum inhibitory concentration in permasoft and mollosil, the zone of inhibition for mollosil with clotrimazole is more than permasoft with clotrimazole and surface roughness of mollosil with clotrimazole is lesser than permasoft with clotrimazole. This makes mollosil with clotrimazole a more effective material for use.

References

Neppelenbroek, Karin Hermana, Mello Lima, Jozely Francisca, Hotta, Juliana, Galitesi, Lucas Lulo, Fraga Almeida, Ana Lucia Pompéia, Urban, Vanessa Migliorini (2017). Effect of incorporation of antifungal agents on the ultimate tensile strength of temporary soft denture liners. Journal of Prosthodontics, 27(2), 177–81.

Bhatt, M., Shah, D., Vaishnav, K., Harsolia, Z., and Dangi, R. (2022). An in-vitro study of the effect of incorporation of antifungal drugs upon the growth of candida and surface hardness in permanent soft liners. International Journal of Health Sciences, 6(S2), 3141–3155. https://doi.org/10.53730/ijhs.v6nS2.5761

Urban, V. M., Thiago F., Bueno. Effect of the addition of antimicrobial agents on shore hardness and roughness of soft lining materials. Oral Diseases 2014, 5,21;57-65

Bueno, M. G., Urban V. M., Barb, G. S., Silva, W. J., Porto, V. C., Pinto, L., et al. (2015). Effect of antimicrobial agents incorporated into resilient denture relines on the candida albicans biofilm. Oral Diseases, 2015, 57–65.

Grzegorz et al. (2014). Long-term soft denture lining materials. Materials, 5816–42.

Pachava, et al. (2014). Comparative antifungal efficacy of denture soft liners with clotrimazole : an invitro study abstract : Indian Journal Of Dental Advancement, 1593–1595.

Bueno M. G., Juliana, Sousa, Hotta. (2017). Surface properties of temporary soft liners modified by minimum inhibitory concentrations of antifungals. Oral Diseases. 28, 158–64.

Radnai, M., Whiley, R., Friel, T., Wright, P.S. (2010). Effect of antifungal gels incorporated into a tissue conditioning material on the growth of candida albicans. J. of Dentistry, 292–6.

Machado, Breeding, Vergani, Elias. Hardness and surface roughness of relining and denture base acrylic resins after repeated disinfection procedures. J prosthetic dent 2009 102(2):115–22.

Malek, MohammedShakil S., and Bhatt, Viral (2022). Examine the comparison of CSFs for public and private sector's stakeholders: A SEM approach towards PPP in Indian road sector, International Journal of Construction Management, DOI: 10.1080/15623599.2022.2049490

Identification of Road Surface Deficiencies Using Convolutional Neural Network

Jaykumar Soni[a], Rajesh Gujar[,b]*

[a]Department of Civil Engineering, School of Technology, Pandit Deendayal Energy University, Gandhinagar, India
[b]Department of Civil Engineering, School of Technology, Pandit Deendayal Energy University, Gandhinagar, India
E-mail: [*]jay.sphd19@sot.pdpu.ac.in; rajesh.gujar@sot.pdpu.ac.in

Abstract

The road transport sector is a significant medium for cargo as well as passenger transport. It plays a significant role in the growth of a country. The maintained assets aid the social and economic development of the country. Despite the importance of the road network, many developing countries consider manual methods for pavement maintenance. Artificial intelligence (AI) approaches require the least human interference and hence give higher accuracy. Following the above statement, the present research presents a successful framework for the classification of road distresses. For the classification of various cracks and potholes, 3D images of defects and residual networks (ResNet) are used. Residual Networks are a type of convolutional neural network (CNN). The framework achieved acceptable accuracy for training and testing both. Future research can be conducted to achieve higher accuracy. The research helps government authorities plan and schedule maintenance strategies accordingly.

Keywords: Pavement, Cracks, Potholes, ResNet

Introduction

As one of the components of the infrastructure chain, the transportation sector is distinctly responsible for the socio-economic growth of a region or country Roadway transportation has been the most integral part of civilization, affecting daily life and the economy. It provides mobility and contributes to production (Gothane and Sarode, 2016). Consideration of maintenance activities is one of the most significant parts of these assets (Kordestani Ghalenoeei et al., 2021). The inferior pavement results in increased costs for public, and ultimately the increased average spending leads to higher rehabilitation costs (Rashid and Gupta, 2017). Out of many developing countries spending a significant amount of money on road construction, only a few nations place a high emphasis on road maintenance (Salih and Edum-Fotwe, 2016). A periodically maintained road costs approximately the 10th part of the total rehabilitation of the same for a similar lifespan (Sethi, Devesh Tiwari, Neha Dhiman, 2016). To plan and schedule the maintenance of any pavement stretch, firstly, the pavement status is needed to be evaluated (Coenen & Golroo, 2017). Traditionally, the trained officials visit the road stretch and monitor the pavement manually (She et al., 2020). The conventional methods are inaccurate, hazardous, and low productive as they are dominated by experience, fatigue, and authorities' mindset (Shtayat et al., 2020). Moreover, periodically collected data can improve the quality of modeling and analytics, resulting in extended facility service life and decreased wheel-of-life costs. As a result, automatic distress detection is becoming increasingly popular among transportation organizations. There are various methods available in practice, i.e., manual detection, sensor-based methods, smartphone sensor-based methods, remote sensor methods, and artificial intelligence (AI)-based methods. Manual detection methods are more prone to error, labor-intensive, low productive, time-consuming, and dangerous in several conditions. Furthermore, because different types of pavement distress have spectral similarities, it is difficult to obtain accurate results by relying solely on spectral features. Artificial Intelligence has been included in almost every field, and civil engineering is one. The proposed research has tried to include AI for pavement defect detection. Other methods can identify the defects, but Artificial Intelligence has been used for error-free, faster, and budget-friendly results; that is why adopting the AI approaches can be considered one of the most feasible methods.

For instance, a framework of CNN was trained and tested using various crack images which showed 99.2% accuracy. Later on, a two-step methodology was proposed using CNN to enhance classification accuracy. The dataset images were captured from the roads of China using a smartphone. It also showed acceptable precision (Liu et al., 2020). Recently, the authors theoretically evaluated that although the CNN requires a bit larger data, it can efficiently classify the pavement images as compared to artificial neural networks (ANN), fully convolutional neural networks (FCNN), and support vector machine (SVM) (Soni et al., 2022). For noise clearance and pre-processing of the image data-set, complex systems may be required for other algorithms but the research reportedly justifies the use of CNN over FCN, saying that FCN requires a complex computational process of weight distribution and hence the high configuration systems are needed. However, CNN shares the weight of all the junctions

more simply. Hence for the present study, CNN is majorly considered. The research provides a framework for the identification of road defects. Available literature majorly considers cracks for conducting the research. The potholes are one of the major defects that cause discomfort and unsafe conditions for users. In this article, potholes and cracks are simultaneously given the importance. Residual rntwork (ResNet), a CNN, is used as a core algorithm due to its skill of maintaining low error rate in deep networks. Deep networks are those having large number of layers for greater performance. The details of the data used for the development of the framework, data collection, training and testing are presented in the following sections of the article.

Data-set

3D images of defects were collected from the road stretches of Ahmedabad and some remote areas of it. By using 3D images more accurate information about the object is obtained resulting in better detection. The considered road stretches are shown below.

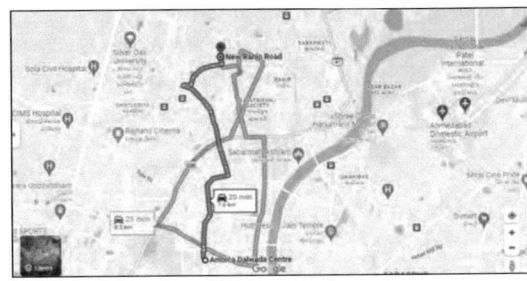

Fig. 1. Study Area-1 (Gota to New Ranip)

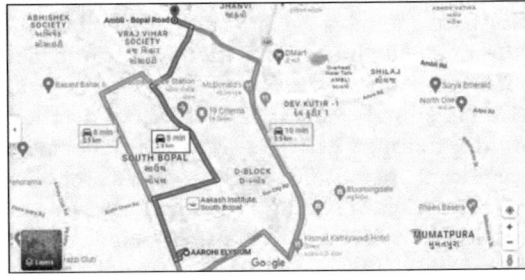

Fig. 2. Study Area-2 (Bopal Road)

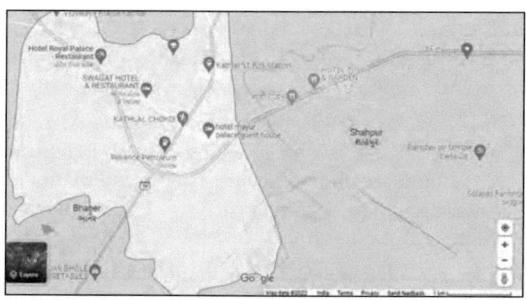

Fig. 3. Study Area-3 (Kathlal)

Study area comprises of total 19 Kms of roads. The images were collected using a smart phone having 48 MP camera and 2MP Depth sensor. The collected images were divided into 2 parts of training and testing considering the ration of 90:10.

Methodology

For the efficient detection of the defects, a 3-step methodology has been designed and used.

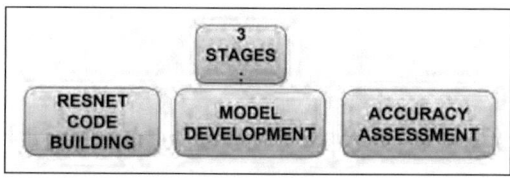

Fig. 4. Methodology

Firstly, the code for ResNet was built. TensorFlow is a python library used for the solution of classification, perception, understanding, discovering, and prediction. ResNet function was defined using different attributes. The attributes were name, layer, input shape, and class. The images of 64 pixels were obtained. Convolution operations are performed. Convolution is a mathematical operation used to extract features from an image, defined by an image kernel. The image kernel is nothing more than a small hypothetical matrix. Kernel is used to learn the different features of the original image. The kernel first moves horizontally, then shift down, and again moves horizontally. The sum of the dot product of the image pixel value and kernel pixel value gives the output matrix.

After each convolution operation, the image gets shrinked. Kernel will convert each 3 matrix into 1 1 matrix. At that time, padding is used to maintain the size of the matrix.

Training

As the training progressed, the number of filters also increased. Average pooling focused on the black-and-white portion of the images. Max pooling focused on the colored portion of the images. Colored part is the major bright part of the images. Softmax was used as an activation function. The activation function is the non-linear transformation that we do over the input signal. This transformed output is then sent to the next layer of neurons as input. Adam optimizer is used to shape the model in such a way that it gives the maximum accuracy. Total of 100 epochs was used to gain the highest accuracy. Once the model is developed. It is validated using the images reserved first for testing.

Testing

The images for the testing are first converted to the array. All the pre-processing operations like flipping, zooming, and rescaling were performed. The prepared image is then detected with the respective cracks. As a result, best matching category is shown as the identified crack. The achieved accuracy for training was 92% and the same for testing was 86%.

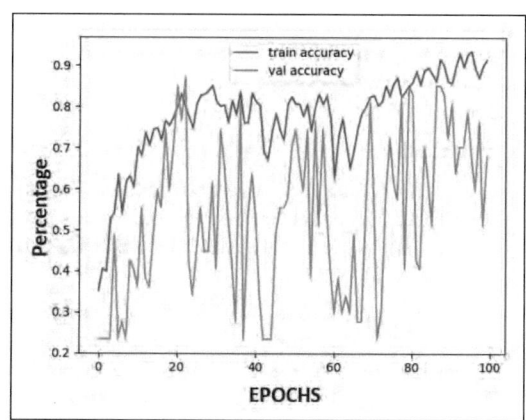

Fig. 5. Accuracy-Epochs

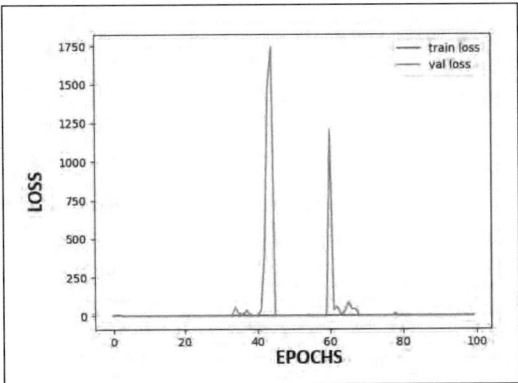

Fig. 6. Loss-Epochs

As shown in the above figures, maximum accuracy is achieved on 100 epochs. Loss is an indication of the bad precision of the model. The larger number of the loss, the worse the performance of the model. As shown in Fig. 9, the loss at 100 epochs is 0. Hence 100 epochs are the most suitable for the best performance of the model.

Conclusion

The present research presents a framework for the identification and classification of the defects of road pavement. The research suggests that automatic approaches especially, AI methods significantly help to classify the cracks and the potholes. A framework using ResNet (CNN) has been developed. SoftMax and Adam are used as an activation functions and an optimizer respectively. The training and testing accuracies are 92% and 86% respectively.

References

Coenen, T. B. J., and Golroo, A. (2017). A review on automated pavement distress detection methods. Cogent Engineering, 4(1), 1–23. https://doi.org/10.1080/23311916.2017.1374822

Gothane, S., and Sarode, M. V. (2016). Analyzing Factors, Construction of Dataset, Estimating Importance of Factor, and Generation of Association Rules for Indian Road Accident. Proceedings - 6th International Advanced Computing Conference, IACC 2016, 15–18. https://doi.org/10.1109/IACC.2016.13

Salih, Jamaa, Francis Edum-Fotwe, A. P. (2016). Investigating the Road Maintenance Performance in Developing Countries. 10(4), 444–448.

Sethi, K. C., Tiwari, Devesh, Neha Dhiman, A. K. S. (2016). A Review on Existing Indian Pavement Maintainance Management Systems. International Interdisciplinary Conference On Engineering Science & Management, December, 29–31.

Kordestani Ghalenoeei, N., Saghatforoush, E., Athari Nikooravan, H., and Preece, C. (2021). Evaluating solutions to facilitate the presence of operation and maintenance contractors in the pre-occupancy phases: a case study of road infrastructure projects. International Journal of Construction Management, 21(2), 140–152. https://doi.org/10.1080/15623599.2018.1512027

Liu, J., Yang, X., Lau, S., Wang, X., Luo, S., Lee, V. C. S., and Ding, L. (2020). Automated pavement crack detection and segmentation based on two-step convolutional neural network. Computer-Aided Civil and Infrastructure Engineering, 35(11), 1291–1305. https://doi.org/10.1111/mice.12622

Rashid, Z. Bin, and Gupta, R. (2017). Study of Defects in Flexible Pavement and Its Maintenance. International Journal of Recent Engineering Research and Development (IJRERD), 2(6), 30–37.

She, X., Zhang, H., Wang, Z., and Yan, J. (2020). Feasibility study of asphalt pavement pothole properties measurement using 3D line laser technology. International Journal of Transportation Science and Technology, xxxx. https://doi.org/10.1016/j.ijtst.2020.07.004

Shtayat, A., Moridpour, S., Best, B., Shroff, A., and Raol, D. (2020). A review of monitoring systems of pavement condition in paved and unpaved roads. Journal of Traffic and Transportation Engineering (English Edition), 7(5), 629–638. https://doi.org/10.1016/j.jtte.2020.03.004

Soni, J., Gujar, R., Shah, D., Parmar, P. (2022). A Review on Strategic Pavement Maintenance with Machine Learning Techniques. In P. Shah, J., Arkatkar, S.S., Jadhav (Ed.), Intelligent Infrastructure in Transportation and Management. Studies in Infrastructure and Control (pp. 147–157). Springer Singapore. https://doi.org/https://doi.org/10.1007/978-981-16-6936-1_12

Analysis of Road Surface Deformation Using Radar Image

Kishan Patel,[a] and Rajesh Gujar*,[a]

[a]Pandit Deendayal Energy University, India
E-mail: *Rajesh.Gujar@sot.pdpu.ac.in

Abstract

The status of the city's infrastructure can be identified through the road surface. The rough surface affects road safety and driving comfort. To minimize road hazards, pavement conditions must be periodically inspected for damaged surfaces. A radar image provides quick and efficient data collection. It offers non-destructive techniques with a large spatial coverage for assessing road conditions and classifying distress. The images collected from satellites, high-resolution cameras, and sensors should be analyzed to correlate surface distress. Recent research provides an opportunity for Synthetic Aperture Radar (SAR) and Interferometric Synthetic Aperture Radar (InSAR) based satellite images to monitor and manage pavement and infrastructure. Therefore, this study aims to highlight the analysis of radar images to detect and improve deteriorating roadway surfaces. The results showed the deficiencies on the surface that can be used to mitigate bad pavement conditions and allow road users to use good road infrastructure with safety and comfort.

Keywords: Road pavement, Synthetic Aperture Radar, Surface deformation

Introduction

Recently, surveillance of road surface conditions has become more critical. Well-maintained road surfaces boost health and comfort standards for road users. Therefore, continuous monitoring of road conditions is necessary to improve the transport system in terms of driving safety and comfort. The timely detection of issues and frequent data collecting are two major obstacles in pavement maintenance. The latter becomes more challenging in the case of manual inspection. Numerous studies and the experiences of transportation agencies demonstrate that the early identification of issues and the application of preventive measures lengthen the asset's useful life, save total maintenance costs, and preserve quality and safety (*AASHTO Guidelines for Pavement Management Systems*, n.d.). Urban towns and cities have the largest population settled in past years. It has become an important requirement to adopt more convenient tools for the analysis, planning, and management of road infrastructure. Typically, the cost of preserving a frequently maintained road is maybe less than three times of a deteriorated road which occurred due to lack of timely maintenance (Officials and Program., 2009). Therefore, radar image has become more reliable as there are no limits to data collection. Images collected from the satellite have significant potential for timely detecting, classifying, and analyzing pavement distress.

In the past two decades, numerous uses of Synthetic Aperture Radar (SAR) technology and Interferometric Synthetic Aperture Radar (InSAR) for extensive monitoring studies have been researched. (Ouchi, 2013). Most extensive geography is covered by remotely sensed satellite images. These data sets can be utilized for a wide range of, such as from the earth sciences to military reconnaissance. Although these data may cover broad regional or continental regions in a single image; data collection, quality, and use might be restricted by revisit durations, atmospheric interferences, and spatial resolution (Schnebele et al., 2015). This paper aims to utilize significant satellite remote sensing techniques in road surface detection through data collected to achieve a high level of accuracy in the assessment of the condition of a road surface. It also explores the requirement for an effective and automated pavement health monitoring system in the transportation industry. The use of radar images in transportation research is a growing and economically beneficial area of study.

Study Area

The different road networks with surface distresses are required for this research. Therefore, three different road networks of Ahmedabad city (Gujarat, India) were identified based on the importance and usage of the roads. The first study area is Science city road, with residential and commercial buildings. The second study area is the Gota-Ognaj road which passes through the residential and institutional buildings. The Priyakant Parikh Marg is selected as the third study area from the local urban area. Figure 2 shows the Google Earth images of all three study areas with red, blue, and green color outlines, respectively.

Data Collection

Mainly three types of datasets, such as radar (SAR) images, optical images, and digital images of selected study areas required. We have acquired the Sentinel-1A product for SAR images provided by European Space Agency (ESA) Copernicus Open Access Hub. A total of 12 Sentinel-1A images were acquired between April and September 2022. Landsat-9 images provided by USGS Earth Explorer were used for the optical dataset. Landsat-9 gives better optical sources compared to Sentinel-2 for the selected study. This was concluded after analyzing the optical features

of both sources. Whereas the high-resolution digital camera is used for the collection of digital images.

Methodology and Data Analysis

SAR image is a complex image. It needs to be processed to detect road surfaces amongst other geographic features such as water bodies, vegetation cover, open lands, buildings, etc. The transmitted and backscattered radiation of the illuminated scene is measured to create SAR images. A radar image has phase as well as amplitude in each pixel. Amplitude facilitates differentiating the surface characteristics by measuring the radiation backscattered by each pixel's object. The amplitude is more dependent on the ability of the surface to reflect away the radiation and its roughness (Ferretti et al., 2007). The Single looking SAR image needs to be converted into Multilooking first. After that, various image processing should be done, such as speckle filtering and backscattering that helps to remove dominant multiplicative and additive noise present in SAR image, respectively. The former noise was minimized by a Refined-Lee speckle filter with a 3 × 3 window (Yommy et al., 2015). Figure 1 shows the step-by-step procedure of SAR image processing.

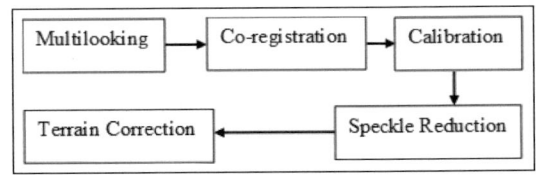

Fig. 1. Data processing steps for SAR imagery

The multi-look helps to improve radiometric resolution but reduces spatial resolution. Pixel value truly represents the radar backscatter of the reflecting surface. At the same time, co-registration superimposes the images with the same acquisition mode and orbit. Blurred surfaces and features can be reduced and/or removed by applying speckle filtering to the images. Digital Elevation Model (DEM) is used to correct SAR geometric distortion by geocoding the image for terrain correction. Generally, three types of terrain corrections give good results, namely sigma naught, beta naught, and gamma naught. Out of

them, sigma naught gives better-calibrated radar brightness measurements (Meyer et al., 2020).

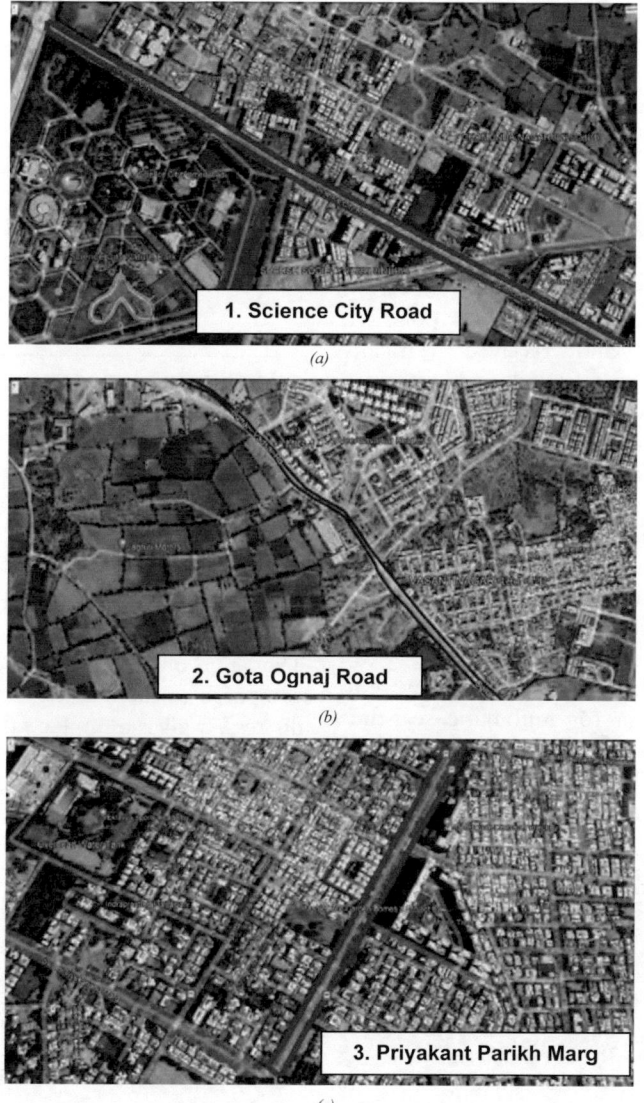

Fig. 2. Study area of road network selected for Ahmedabad city (a) Science City Road, (b) Gota-Ognaj Road and (c) Priyakant Parikh Marg

Detection of Road Surface

The road network typically appears in SAR data as vague lines. This is due to the road's smoothness relative to its surrounding structures, hence providing a mirror-like reflection resulting in low returns of a radar signal. With range-

oriented roads, the effect is more pronounced. Due to multiple bounce scatterings, roadways appear as very bright lines in the azimuth direction when certain configurations, such as highway borders, road rails, elevated roads, bridges, etc., are present. Since radars are side-looking sensors, the direction of looking greatly influences the geographic features of the acquired image (Hendry et al., 1988).

Significant SAR polarization configuration states based on the transmitted-received electromagnetic signals give different results after the post-processing of the images. It has two horizontal and vertical complex images of varying polarization combinations. The former has images with polarised signals transmitted and received horizontally (HH) and transmitted horizontally and received vertically (HV). In contrast, the latter has images with polarised signals transmitted and received vertically (VV) and transmitted vertically and received horizontally (VH). Only one image is required for any road network for the inspection date. The combination of any two polarization is dual-polarization, while the combination of all four implies quad-polarization. The scattering matrix can be converted into other polarisation bases after it has been acquired, such as the circular polarisation base, allowing the same polarimetric data to be read from several angles (Ouchi, 2013). All the images are then processed in software to develop an algorithm for automatic real-time distress detection on the surface. The highlighted road network of the whole Ahmedabad city extracted from the SAR image is shown in Fig. 3.

Results and Discussion

The surface deformation results were obtained by analyzing the thousands of different pixel values of the $\sigma 0_{VH}$ and $\sigma 0_{VV}$. Out of these values, four values are shown in Table 1.

Table 1. Pixel values of Sigma and Amplitude of Science city road

Sr. No.	$\sigma 0_{VH}$	$\sigma 0_{VV}$	Amplitude VH
1	0.024778	0.366377	599.121933
2	0.023525	0.181795	523.420579
3	0.025458	0.175113	306.649855
4	0.028796	0.273875	243.685022

Mathematical modeling converts these values into measurable quantities, say lengths. From these values, we can identify the deficiencies in the road surface. It should be verified with manually calculated values and the digital images of these deficiencies captured.

It can be seen from Fig. 4 that the detected value of surface distress can be verified with the digital image of that geo location of the pavement. The value obtained from the SAR image is then compared to the manually collected value of the distress. It gives a similar value with an error of 8 mm. The possible reason for this is due to taking the mean value of each pixel. In some cases, volume backscatter is produced due to other road facilities.

Fig. 3. Road detection of Ahmedabad City

Fig. 4. Road surface deficiency

Conclusion

This article proposes the use of SAR images to identify road surface deficiencies. It can be done by detecting the road surface from the satellite image and converting the pixel values into measurable quantities. Thus the exact location of the surface deformations can be identified accurately. This method can be used to determine and analyze the degree of deficiencies at the early stage. Therefore, the maintenance and repair work of the road surface can be done well in advance to improve the experience of the road surface for its users.

References

AASHTO Guidelines for Pavement Management Systems. (n.d.).

Ferretti, A., Monti-Guarnieri, A., Prati, C., Rocca, F., and Massonet, D. (2007). InSAR Principles - Guidelines for SAR Interferometry Processing and Interpretation. ESA Training Manual, 19.

Hendry, A., Quegan, S., and Wood, J. (1988). The visibility of linear features in SAR images. Remote Sensing. Proc. IGARSS'88 Symposium, Edinburgh, 1988. Vol. 3, 1517–1520.

Meyer, F. J., Ajadi, O. A., and Hoppe, E. J. (2020). Studying the applicability of X-Band SAR data to the network-scale mapping of pavement roughness on US roads. Remote Sensing, 12(9). https://doi.org/10.3390/RS12091507

Officials., A. A. of S. H. and T., and Program., R. I. (2009). Rough roads ahead : fix them now or pay for it later. AASHTO.

Ouchi, K. (2013). Recent Trend and Advance of Synthetic Aperture Radar with Selected Topics. 716–807. https://doi.org/10.3390/rs5020716

Schnebele, E., Tanyu, B. F., Cervone, G., and Waters, N. (2015). Review of remote sensing methodologies for pavement management and assessment. European Transport Research Review, 7(2). https://doi.org/10.1007/s12544-015-0156-6

Yommy, A. S., Liu, R., and Wu, A. S. (2015). SAR Image Despeckling Using Refined Lee Filter. 2015 7th International Conference on Intelligent Human-Machine Systems and Cybernetics, 2, 260–265. https://doi.org/10.1109/IHMSC.2015.236

Malek, MohammedShakil S., and Bhatt, Viral (2022). Examine the comparison of CSFs for public and private sector's stakeholders: a SEM approach towards PPP in Indian road sector, International Journal of Construction Management, DOI: 10.1080/15623599.2022.2049490

Malek, MohammedShakil S., and Gundaliya, P. J. (2020). Negative factors in implementing public–private partnership in Indian road projects, International Journal of Construction Management, DOI: 10.1080/15623599.2020.1857672

A Deep Dive into Motor Imagery EEG Classification with Transfer Learning Approach Using DenseNet-121 Model

Ayonija Pathre[*,a] *, Dr. S.Veenadhari*[b]

[a,b]Rabindranath Tagore University Bhopal, India.
E-mail: [*]ayo.pathre@gmail.com

Abstract

BCI based on motor imagery (MI) has been extensively used in exoskeleton rehabilitation. In realistic use, the poor signal-noise ratio of electroencephalograms (EEG) leads to low accuracy of identification in BCI. Numerous pieces of research have thus focused on improving feature extraction & classification methods. In this work, we have presented a novel technique for feature extraction and analysis for the single-trial MI EEG data dependent on a deep convolution neural network. This may be utilized for spatial frequency feature learning & MI EEG classification. Tentative research is Performed on 2 MI EEG datasets (BCI competition III dataset IVa & a self-collected right index finger MI dataset) to validate the efficiency of our algorithm in the evaluation of MI-based EEG using CNN (base approach) with MI-based EEG using Alexnet and MI base EEG using DenseNet-121 both proposed approach methods. Higher efficiency in the classification shows that their suggested approach is a useful MI-based BCI pattern recognition algorithm.

Keywords: Brain-Computer Interface (BCI), electroencephalogram (EEG), convolutional neural network (CNN), Densenet-121 Model, Transfer learning

Introduction

Since personal computing was introduced in the 1970s, engineers have incessantly worked towards reducing the communication gap between humans & computer technologies. This technique started with the growth of GUI, computers & mice this has contributed to more intuitive innovations, especially with growing artificial intelligence. The ultimate barrier between computers and humans is now being overcome by the use of BCIs that enable computers to be operated deliberately via brain signal activity monitoring. However, BCI may be built to utilize EEG signals in many ways for control; MI-BCIs, where users can see activities in their limbs to operate the system, get the most interest.

A BCI is a computer-based system that collects, analysis, & interprets brain signals into instructions sent to an output device to perform a requested response. Therefore, BCIs are not using the brain's normal peripheral nerve & muscle output pathways. This description restricts the word BCI purely for structures which is measure & then usage central nervous system signals (CNS) (Ma, 2018). Thus, for instance, a communications network triggered by voice or muscle is not a BCI. Moreover, the EEG equipment is not a BCI alone since it simply captures brain signals but does not produce a user-friendly output. It is a misinterpretation that BCIs are read-only devices. BCIs don't read the brains of unsuspecting or reluctant users to obtain information but allow users to act on the whole world using brain impulses rather than muscle. The employer creates brain signals after a training period. The BCI also interprets the signals after training & converts them into instructions to a user-specified external device. The high time resolution as well as the relative simplicity and efficiency of brain signals obtaining in comparison to the conventional including such fMRI & MEG are common for BCI EEG signals. These are easier to obtain than fMRI or MEG. EEG signals thus provide problems with the processing they may contain external noise or are susceptible to signal distortions since they are non-stationary. The stimuli evoke brain impulses that are subsequently detected by the BCI system to conclude the user's will. Random BCIs do not need external activation and performance levels based on activity generated by mental activity are conducted (S.U.Amin 2019).

The paper is structured as occurs. We first evaluate various states of art methods in the area of EEG in the BCI system. The next section describes the problem statement, proposed methodology, and dataset description. The fourth section presents the experimental results. The last sections define the conclusion and future scope of the proposed method.

Review of Literature

EEG signal processing is one of the major BCI systems' difficulties in building dependable interfaces for a range of BCI systems. There are several kinds of EEG signals utilized for BCI systems to control external devices and each kind is employed in different applications based on the field of view for research in the brain.

Deng et al. (2021) present a famous EEGNet deep learning model & compare it to the traditional algorithm FBCS Pattern. This paper then takes the view that EEGNet's 1-D convolution can be described by special DWT and also that EEGNet's deep convolution is comparable to CSP method. This study enhances EEGNet by using the TCSGL algorithm to increase its efficiency. Proposed methodology TSGL-EEGNet is calculated for 4-class MI

Classification tasks in BCI Competition IV2a & BCICIIIa datasets. accuracy (kappa), higher than EEGNET, C2CM, MB3DCN, SS-MEMDBF & FBCSP.

Jeong et al. (2021) suggest a multi-layer time pyramid pooling method to enhance motor imagery-based BCI efficiency. The approach presented includes multilayer multiscale pooling & fusion algorithms to capture different characteristics of an EEG signal, detailed and integrated into current CNNs. In human-computer interactions, the detection of a user's intentions is important. BCIs were recently widely discussed to help to detect & forecast users' intents more accurately.

A. Akrout. et al. (2020) Introduces a new technique for left/right-hand segmentation and analysis of both the foot and tongue movement by utilizing two methods of LD the ANN as well as the CNN. A broad selection of spatial and frequency domains is drawn from EEG data and trained in the classification tasks of the ANN and CNN networks. The EEG mental task signals are collected and categorized according to these designs. Furthermore, the suggested techniques are confirmed by and contrasted with the EEG datasets of BCICIV-2a. The findings indicate which CNN framework exceeds the ANN framework by 60,55%.

Jeong, J.-H. et al. (2020) Suggest a new deep-learning technique that may be utilized to study and classify the MI-EEG for spatial frequency. In particular, the CNN multilayer model is developed as said by MI-EEG signal spatial frequency features. EEG-PR is a significant constituent of the BCI system based on MI. CSP is used for feature extraction. However, previous information and sophisticated parameter modification are frequently needed to extract optimum CSP features. CNN is now a common DL method.

N. Lu, et al. (2020) This research has performed a thorough comparative investigation. CNN-based technique, RNN-based methodology, TCN-based technique, a parallel grouping of CNN & SRU, and cascaded CNN & SRU combinations have been built & associated created on comprehensive tests. The tests were carried out equally using the same dataset, the same data pre-processing, and then the same platform. Studies demonstrated TCN has the greatest performance & the parallel mixture of CNN & RNN has achieved 2nd best efficiency, which has encouraged us to investigate deep network STFL methods for future development. (Lu and Yin, 2018).

Material and Methodology

In existing work, the author used CNN with ReLU activation functions. Furthermore, the greatest results have been achieved when samples from the same recording session were trained on the network(Lu and Yin, 2019).

First of all, we transform raw EEG data into image representation and then use CNN. As sensor motor rhythm occurs typically in frequency ranges of 8–14 Hz (μ Rhythm) & 18–26 Hz (β Rhythm) for Finger Dataset, we apply bandpass filtering to get a signal of between 8–30 Hz since the original EEG data. Moreover, EEG data are typically utilized for pattern analysis from 0.5 to 2.5 seconds following visual cue presentation. In this research, the signals are also retrieved for further processing during this time. Raw EEG data is filtered in a range of 8-30 Hz for BCI III dataset IVa. It would be noted that this dataset has 118 electrodes. To minimize the load of future computation & eliminate the impact of duplicated channels, we collect EEG data from 49 channels for further analysis according to the research suggests. Every sample may be presented, after the preceding processing, as a size 49 to 3500 matrix, with 3500 sampling points and 49 electrodes. Subsequently, frequency domain decomposition, duration segmentation of 0.5 to 2.5 seconds, separation of energy & normalization are performed(www.bbci.de).

Step 1: From the 118 channels we select the particular 49 channels that have the main signal frequency, which was all done in MATLAB along with the selection process we also applied a bandpass filter on the data to remove the unwanted frequency. Step 2: Create a CSV file from the precise data (MATLAB file) to overcome the difficulty for the next work Step 3 Split the CSV into 250 different CSVs for a better process.

Thus, it makes 28 images per CSV and interpolates and images also. Step 4: Only proper images and their labels are saved on the other hand rejection of distorted images and their labels took place simultaneously. Step 5: While loading the images a Denoising operation was performed to remove the noise from the images. Step 6: We experimented with VGG16, VGG19 but didn't achieve the desired result. So, we decided to go with the Transfer Learning approach. Thus, end up using the DenseNet121. Step 7: Pass the images into the model (DenseNet121) for training applying 10-Fold Validation for improving accuracy and validation accuracy.

Results and Discussion

The software & hardware platforms of the suggested CNN (Densenet-121) model are Intel(R) Core (TM) i7-10700k 5.10 GHz CPU, 32 GB RAM DDR4, Jupyter notebook, Python 3.8, and TensorFlow 2.4.3 (GPU), GPU NVIDIA RTX 2060 SUPER cuDNN and CUDA ToolKit 11.2v. The Relu activation function is used to introduce non-linearity instead of Tanh. It speeds up 6 times at the same precision. Use dropout to respond with over-fitting instead of regularisation. The training time is, therefore, doubled by a 0.5 drop-out rate. Pooling overlaps to decrease the network size. It lowers the top 1 & top 5 error rates correspondingly by 0.4% & 0.3%.

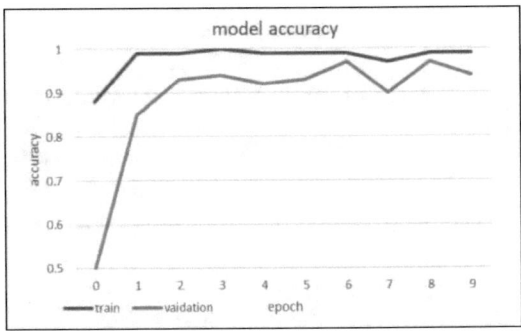

Fig. 1. Accuracy of classification of the validation set and training set in CNN (Densenet-121) AA subject training

The figure above shows the Epoch of AA subject Train and Validation Accuracy. Fig. 1 shows that, upon 1600 iterations, the classification accuracy of the training set exceeds the maximum value & remains constant, whereas the classification accuracy of the validation set has been maintained at maximum value, so that model achieves the best training effect after iterations & the trained model is assumed optimal classification optimal.

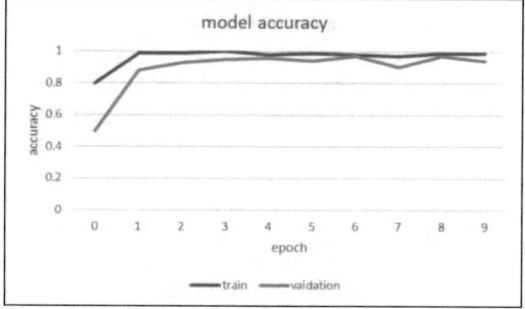

Fig. 2. Accuracy of classification of the validation set & training set in CNN (Densenet-121) AL subject training

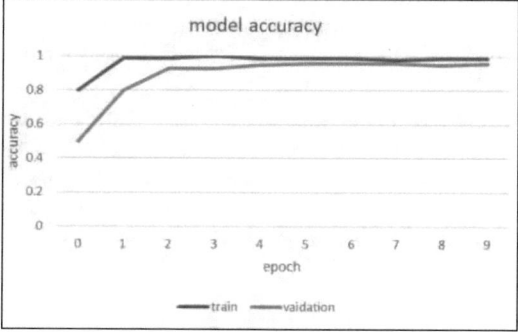

Fig. 3. Accuracy of classification of the validation set & training set in CNN (Densenet-121) AV subject training

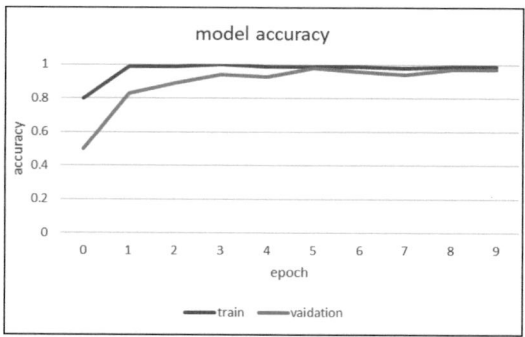

Fig. 4. Accuracy of classification of the validation set & training set in CNN(Densenet-121) AW subject training

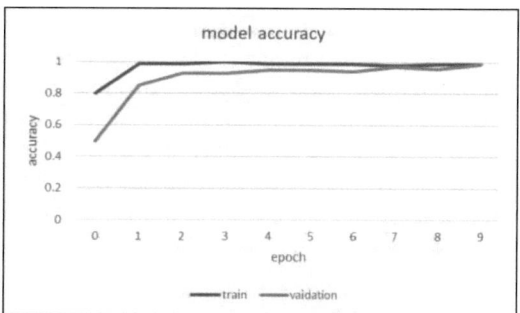

Fig. 5. Accuracy of classification of the validation set and training set in CNN (Densenet-121) AY subject training

The above shows the Image Plot for all Subject. All subjects in a given dataset, CNN (Densenet-121) models have identical architecture and hyperparameters. Therefore, some users have distinct features. Thus, the uniform parameter setting can result in a suboptimal solution for all subjects.

Table 1. Comparison of different approaches Accuracies in terms of Subjects

Subject	MI-based EEG using CNN	MI-based EEG using Alexnet	MI base EEG using DenseNet-121
aa	100	99	98
al	90	99	98
av	90	98	99
aw	80	98	99
ay	80	98	99

Fig. 6. Result Accuracy Graphs of Different Approaches

The above graph shows the comparative view of MI base EEG using CNN, MI-based EEG using Alexnet, and MI base EEG using DenseNet-121.

Conclusion

BCI provides a new path to information exchange between an external device as well as the brain through metabolic or EEG brain transformation to control messages for systems and platforms. The EEG receives data from time series with various variations recorded on the scalp using many sensors. We have suggested in this article TL using the Densenet-121 model method to reduce the training study total and enhance performance by learning a domain-independent kernel. It may be utilized for the study and categorization of MI-EEG spatial frequency. In conclusion, the suggested densenet-121 model is compared with MI base EEG using CNN, MI-based EEG utilizing Alexnet algorithms. The training period for the proposed model is also quite low. -We want to emphasize that the proposed classification technique is of major importance for real-life BCI systems.

References

Deng, X., Zhang, B., Yu, N., Liu, K., and Sun, K. (2021). Advanced TSGL-EEG Net for Motor Imagery EEG-Based Brain-Computer Interfaces. IEEE Access (9), 25118–25130.

K. -W. Ha and J. -W. Jeong,(2021) Temporal Pyramid Pooling for Decoding Motor-Imagery EEG Signals IEEE Access (9), 112–125

Akrout, A., Echtioui, A., Khemakhem, R., and Ghorbel, M. (2020). Artificial and Convolutional Neural Network of EEG-Based Motor imagery classification: A Comparative Study International Conference on Sciences and Techniques of Automatic Control and Computer Engineering, 46–50.

Miao, Minmin, Hu, Wenjun, Yin, Hongwei, Zhang, Ke (2020). Spatial-Frequency Feature Learning and Classification of Motor Imagery EEG Based on Deep Convolution Neural Network, Computational and Mathematical Methods in Medicine, 20(1981728), 13.

Lu, N., Yin, T., and Jing, X. (2018). Deep Learning Solutions for Motor Imagery Classification: A Comparison Study8th International Winter Conference on Brain-Computer Interface (BCI), 2020, (1–6).

Lu, N., Yin, T., and Jing, X. (2019). A Temporal Convolution Network Solution for EEG Motor Imagery Classification IEEE 19th International Conference on Bioinformatics and Bioengineering (BIBE), 796–799.

Amin, S. U., Alsulaiman, M., Muhammad, G., Bencherif, M. A., and Hossain, M. S. (2019). Multilevel Weighted Feature Fusion Using Convolutional Neural Networks for EEG Motor Imagery Classification in IEEE Acces (7), 18940–18950.

Ma, X., Qiu, S., Du, C., Xing, J., and He, H. (2018). Improving EEG-Based Motor Imagery Classification via Spatial and Temporal Recurrent Neural Networks, Annual International Conference of the IEEE Engineering in Medicine and Biology Society (EMBC), (40), 1903–1906.

Das, R., Maiorana, E., and Campisi, P. (2018). Motor Imagery for Eeg Biometrics Using Convolutional Neural Network IEEE International Conference on Acoustics, Speech and Signal Processing (ICASSP), 2062–2066. http://www.bbci.de/competition/iii/

Realistic Method of Domestic Power Management System

B. V. Manikandan,[a,*] K. Banumalar,[b,*] and R. Reethika[c]

[a]Senior Professor, Mepco Schlenk Engineering College(Autonomous), Sivakasi
[b]Professor, Mepco Schlenk Engineering College(Autonomous), Sivakasi
[c]Student, Mepco Schlenk Engineering College(Autonomous), Sivakasi
E-mail: *bvmani@mepcoeng.ac.in; *kbanumalar@mepcoeng.ac.in

Abstract

An effective domestic power management system is proposed which comprises multitasking such as smart metering, appliance control, net metering, communicating with utility, and easy fuse of call registry. Smart appliance control safeguards and controls individual appliances based on their operating conditions and as well as preferences. Future homes will sure to have PV panels on the rooftop and therefore power management through net metering becomes the viable option. This will help the domestic consumer to know about the status of return of supply and the cause of interruptions if any power interruption occurs on an unscheduled basis. Also, if any fuse goes out due to overloading or disturbance, either in the domestic premises or on the pole, fuse of call may be easily entered in the automatic log book registry. An effective domestic power management system will be a handy and effective tool for domestic consumers.

Keywords: Net metering, Wi-Fi module, Arduino, Energy meter, Power Management,

Introduction

The energy meter is an electrical measuring instrument, which is employed to record electricity consumed over a specified period in terms of units. It paves the way for billing by the power utility company. Since electrification is being done on large scale, the demand for this also has increased manifold (Alahakoon et al., 2016). A smart meter is a device that captures data such as electrical energy usage, voltage levels, current, and power factor. Smart meters pass this information on to the buyer for a clearer picture of usage models, as well as to the power company for system monitoring and user billing. Smart meters usually take energy readings in real-time, at regular intervals throughout the time horizon. Smart meters allow for a bidirectional connection between the meter and the central system.

Compared to automatic meter reading (AMR), advanced metering infrastructure (AMI) enables both-way communication between the meter and the supplier (Ali et al., 2021; Sahana et al., 2015). Wireless or standard wired connections, like power line carriers, can link the meter to the network (PLC). Wireless communication options in common use include cellular communications, and Wi-Fi (readily available). Since the inception of electricity deregulation and market-driven pricing throughout the planet, utilities are trying to find a way to ensure that consumption and production are in tandem (Ali et al., 2021). Electrical and gas recoding meters are not smart, they merely track overall usage and do not tell us when the energy has been used. Smart meter permits near-real-time monitoring of power consumption. This helps power utilities to charge various tariffs for usage depending on the usage time of day. It also makes it easier for companies to create more realistic money-flow models. Utilities' labor costs are minimized because smart meters do not require manpower even for taking reading remotely.

Background Study

AMI (advanced metering infrastructure) helps to have communication between the customer and the utility (two-way) (Gulezar Shamim and Mohd. Rihan, 2017). This helps the grid in making recommendations to consumers so that they can better manage their load utilization. Other AMI capabilities include automatic meter billing, controlling power usage, demand-based pricing, and preventing theft (Banumalar et al., 2018). A smart home controller, sometimes known as a futuristic smart meter, is a device able to do net energy metering, control of device smartly, and enable communication with both suppliers and customers (Das et al., 2015; Saravana Kumar et al., 2021).

Tania Tony et al. 2016 describe a smart house controller (SHC) with net metering and device control capabilities. The meter in question can conduct both net metering and smart appliance control, however, it's made up of two microcontrollers. The authors of [Vishnu S et al, 2016] propose using Wi-Fi to implement and develop a low-cost single-phase static energy meter with Iota ideas. Using the Wi-Fi module, the system will be able to interact between the meters. As a result, energy use may be tracked. Rather than using approximation-based methodologies, the billing is based on actual consumption. Vishnu et al. (2016) and Manikandan et al. (2011) proposed one digital tele-watt-hour meter System (DTS), which is a totally complete electrical energy recording instrument. Compared to ordinary household meters, the device has two major advantages out of many. The first is distant data communication, and the second is a time of use (TOU) metering.

Proposed System

Energy metering is done efficiently with the help of smart energy meters. The proposed unit will do net metering, control of home appliances, theft of energy detection, and two-directional communication between utility and user. Net metering is done with suitable Arduino programming which receives input from the current transformer and potential transformer. The readings calculated, i.e. voltage, current, frequency, and Power, are displayed using LCD. The meter values are sent to both the user and the utility using the Wi-Fi module. The calculated power values are compared with the maximum power required for the corresponding loads. If the measured value exceeds the maximum value, the utility alerts the user regarding the same. Then the home control system is enabled which controls the loads. Thus efficient energy management is done. In this system, there is an additional feature for power outage queries. If there occurs an unpredicted power outage, the user can communicate with the utility by pressing a button installed in the In-meter. By pressing that user can get a message with the details of the user who have asked for the query. After getting the message, the utility sends the answer to the user about the reason for power outage.

Figure 1 shows the block diagram for the proposed system. This block diagram depicts the interrelation between the components of the proposed system and the task such as,

- Net metering
- Sensing
- Appliance control
- Bi-directional communication

Wi-Fi module plays a major role in bidirectional communication. It can either run an application or delegate all Wi-Fi networking duties to another processor. With the help of the Wi-Fi module, the user and utility receive an alert. Figures 2 and 3 show the proposed system's flow chart, which depicts the proposed system's flow. This depicts the utility and the user gets alerted during power fluctuations. It also depicts how energy management is done effectively and efficiently.

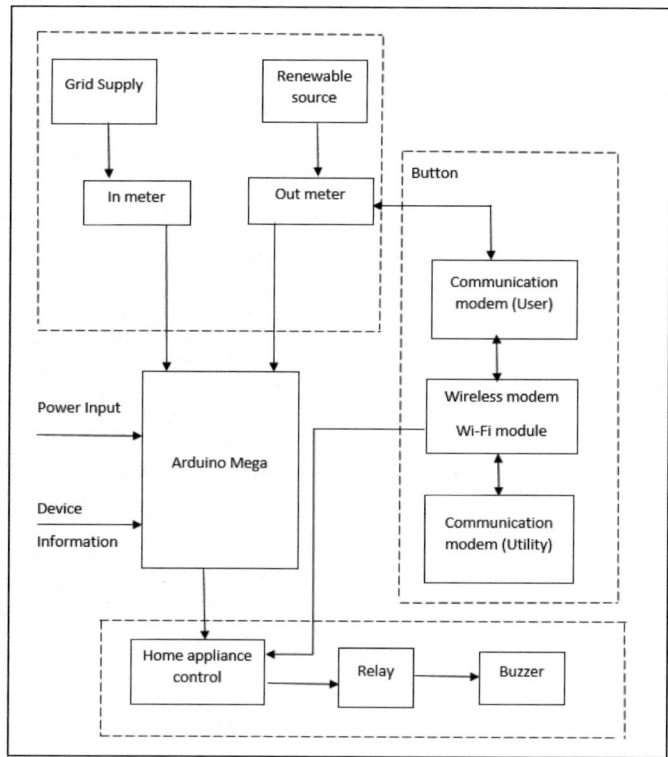

Fig. 1. Block Diagram of Proposed system

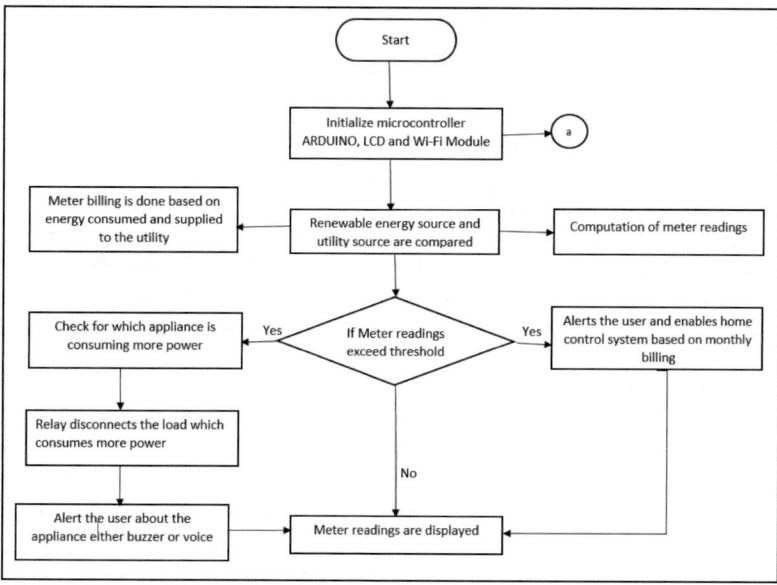

Fig. 2. Flow Chart of Proposed system

In this flowchart, first, the system gets initialized. After initializing, the system compares the renewable energy source and utility source for net metering. Based on this meter billings are done. If the net-metered values exceed the maximum load value, then the home automation system is turned on and alerts the user via a buzzer.

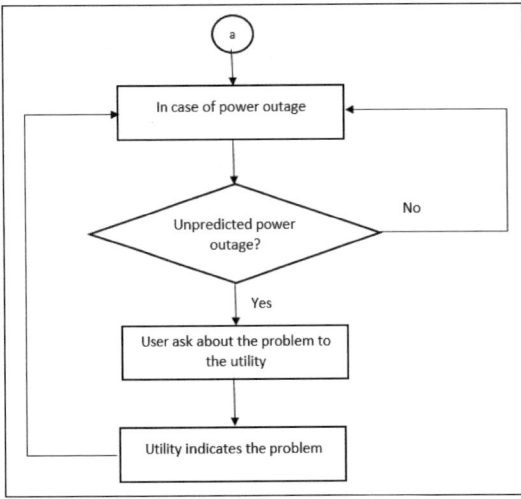

Fig. 3. Flow chart

Thereby the user can save energy. The net-metered values are displayed on the LCD. After initializing, if the in-meter does not show or read any values, the user has to check for a power outage. If there is a power outage, the user has to check whether it is predicted or unpredicted. If it is unpredicted, then there is a button on the user side. By pressing that user can ask the reason for the power outage of the utility. The utility sends the answer to the user's query and the answer is displayed on the LCD.

Hardware Implementation

Figure 4 shows the hardware implementation for the project. Two single-phase energy meters are connected to the user. One is used for measuring how much power is consumed from the utility and the other is to measure the amount of power sent to the utility. Opto-coupler is connected between the two energy meters for isolation. The voltage

and current values are measured using voltage and current transformers.

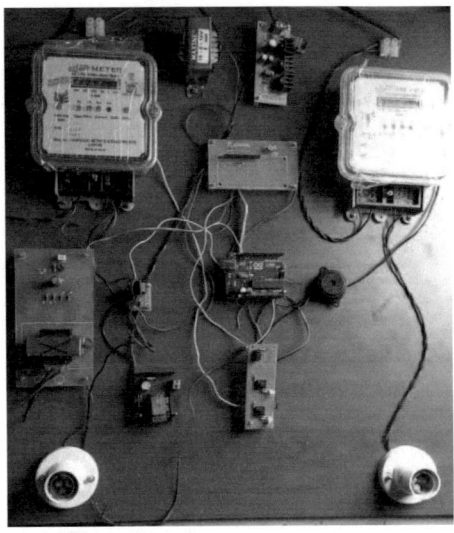

Fig. 4. Hardware implementation

The transformers are connected to the relay circuit which is connected to the loads through intermediate circuit connections. The relay circuit is used to control the appliances individually. The driven relay operates as a switch that can either open or close according to the requirement. If the voltage value exceeds the maximum value, the relay circuit disconnects the appliances connected to the circuit. The energy meters are also connected to the grid side for demonstration purposes. The measured meter values are sent to the microprocessor unit thereby net metering is done. The net values are displayed on the LCD. If the meter values exceed the threshold, the buzzer alerts the user. To communicate with the user about the system's fault, a Wi-Fi module is used.

Result

The system was first implemented with net metering alone. This project was come up with different objectives. The first objective is to do net metering by using in and out energy meters. Net metering was achieved based on renewable energy sources and energy from the grid. The second objective is appliance control. The appliance control was achieved by comparing the energy consumed by the appliances with the maximum load value based on that; appliances are controlled using a home control system. If any appliance exceeds the maximum rated value, the home control system disconnects the load thereby preventing appliances from being damaged. The third objective is bi-directional communication. Bidirectional communication was achieved by sending meter data to both user and utility. If any problem occurs in the meter, the user gets an alert message via a buzzer or through a mobile phone. Likewise, if there occurs a fuse change in the home, it will be automatically updated on the Utility side.

Conclusion

Effective domestic power management system has been proposed and demonstrated successfully. The first objective is appliance control. The appliance control was achieved by comparing the energy consumed by the appliances with the maximum load value based on that; appliances are controlled using a home control system. If any appliance exceeds the maximum rated value, the home control system disconnects the load thereby preventing appliances from being damaged. The second objective is to do net metering by using in and out energy meters. Net metering was achieved based on renewable energy sources and energy from the grid. The third objective is bi-directional communication. Bi-directional communication was achieved by sending meter data to both the user and the utility. If any problem occurs in the meter, the user gets an alert message via a buzzer or through a mobile phone. Likewise, if there occurs a fuse change in the home, it will be automatically updated on the Utility side. Thus, through the validation of all listed objectives, it has been established that this new power management system will be assisting consumers to deal with power management more efficiently.

References

Alahakoon, Damminda, and Yu, Xinghuo (2016). Smart electricity meter data intelligence for future energy systems: A survey, IEEE Transactions on Industrial Informatics, 12(1), 425–436.

Sahana, M. N., Anjana, S., Ankith, S., Natarajan, K., and Shobha, K. R. (2015). Home energy management leveraging open IoT protocol stack, Recent Advances in Intelligent Computational Systems (RAICS).

Gulezar Shamim and Mohd. Rihan. (2017). A technical review on smart grids in India. 4th IEEE International Conference on Electrical Computer and Electronics (UPCON) Uttar Pradesh Section, 642–648.

Banumalar, K., Manikandan, B. V., Chandrasekaran, K., and Arul Jeyaraj, K. (2018). Optimal placement of phasor measurement units using clustered gravitational search algorithm. Journal of Intelligent & Fuzzy Systems 34(6), 4315–4330.

Manikandan BV, Raja SC, \Venkatesh P and Mandala M (2011), "Comparative study of two congestion management methods for the restructured power systems", Journal of electrical engineering and technology 6 (3), 302-310.

Das, Himshekhar, and Saikia, L. C. (2015), "GSM enabled smart energy meter and automation of home appliances." International Conference on Energy, Power, and Environment: Towards Sustainable Growth (ICEPE), IEEE.

Ali, A. N., Premkumar, K., Vishnupriya, M., Manikandan, B. V., and Thamizhselvan, T., (2021). Design and development of realistic PV emulator adaptable to the maximum power point tracking algorithm and battery charging controller, Solar Energy, 220, 473–490.

Saravanakumar, R., Banumalar, K., Manikandan, B.V and Chandrasekaran, K (2021). Realistic Method for Placement of Phasor Measurement Units through Optimization Problem Formulation with Conflicting Objectives. Electric Power Components and Systems 49(4–5), 474–487.

Tony, Tania, Sivraj, P., and Sasi, K. K. (2016) "Net energy meter with appliance control and bi-directional communication capability" International Conference on Advances in Computing, communications and Informatics (ICACCI), IEEE

Babu, Vishnu S., Ashwin Kumar, U., Priyadharshini, R., Premkumar, Krithika, and Nithin, S, (2016). An intelligent controller for smart home. International Conference on Advances in Computing, Communications and Informatics (ICACCI).

Performance Evaluation of High Voltage Gain Bidirectional DC-DC Converter for Modern Electric Vehicles

Ms. K. S. Krishna Veni,[a,*] Dr.S. T. Jaya Christa,[b] Dr. N. Rathina Prabha,[c] and Ms. M. Aiswarya[d]

[a]Assistant Professor, MepcoSchlenk Engineering College, Sivakasi, Tamilnadu, India
[b]Associate Professor, MepcoSchlenk Engineering College, Sivakasi, Tamilnadu, India
[c]Associate Professor, MepcoSchlenk Engineering College, Sivakasi, Tamilnadu, India
[d]PG Scholar, MepcoSchlenkEngineering College,Sivakasi,Tamilnadu, India
E-mail: *krishnaveni.ks@mepcoeng.ac.in

Abstract

In this research, the primary objective is to design a DC-DC converter with a high voltage transfer ratio that can operate in buck-boost mode by utilizing a particular switching strategy to enable a bidirectional flow of power. This converter is a core part of the energy management system, and it is receiving more attention in the automobile industry. EVs, which operate on batteries, has recently gained popularity as an emerging technology. Initially, DC-DC converters were used in EVs. But to utilize the available power generated during regenerative braking operation, a bidirectional flow of current is required through which the battery will be charged during regenerative braking operation and the battery will discharge during the motoring mode of the electric vehicle. This research investigates the working of a new high gain Bidirectional DC-DC Converter (BDC) which is suitable for EVs.

Keywords: High voltage gain DC-DC converter, electric vehicles, non-isolated, bidirectional power flow

Introduction

In modern life, the dependence on electrical products has increased due to the advancement in technology. Utilization of energy resources has become extremely without polluting the environment is important to achieve sustainable development in society. But due to the fast depletion of fossil fuels and the continuous increase in pollutant factors across the globe, the whole world is looking towards the use of renewable energy resources. The conventional source of electricity as of now is fossil fuels which cause severe environmental impacts.

For improved environmental protection with minimal economic compromise, the employment of renewable energy resources has become a very popular choice. As a result of the rising demand for clean energy sources, researchers are concentrating on designing an appropriate converter with a flow of power in both directions.

A highly efficient and high voltage transfer ratio converter is discussed in Deepak Raviet al. (2018); that circuit is efficient, but the voltage gain is low when the flow of current is from High Voltage (HV) side to Low Voltage(LV) side. For obtaining a high gain, a coupled inductor is used in a transformer-based isolated topology (Wu and Ke, 2021). Whereas, Non-isolated BDC topologies (Bhajana et al., 2019) are easy to implement and control. The Non-isolated BDC is preferred because of its simpler design, reduced size of the converter, and minimal cost. The drawbacks observed in it are limited gain and high switching stress. To increase the gain and decrease voltage stress various topologies (Lee and Do, 2019) have been proposed. Coupled-inductor bidirectional converters are presented by Afzal et al. (2021) for increasing the gain. A unidirectional DC-DC converter suitable for DC microgrid applications is proposed by Rathina Prabha et al. (2020); Gnanavadivel et al. (2021). The quadratic converter battery side voltage has a high ripple current and it reduces the functioning of the battery. An aquadratic converter with a coupled inductor is introduced to decrease the current ripples. The coupled inductance and leakage inductance in the circuit increase the switching stress and the voltage transfer ratio is similar to a quadratic converter. For Regenerative Braking in EVs, a Takagi suzeno sliding-modeis proposed (M.Kumar et al., 2021)and that article describes theoperation of Bidirectional Buck-Boost (BBB) converter-based PMDC machine drive. A high gain converter in both modes is proposed by N. Elsayad et al. (2019). From the literature survey, it is inferred that further studies related to high-gain BDC are required.

Conventional System

The non-isolated topologies do not have a transformer and it is preferred for low-power applications. These topologies can be used when the size and the weight are important concerns in design of converters. The isolated BDC topology consists of a transformer and these convert the DC to AC and the AC voltage is passed through a high-frequency AC transformer and it is rectified to DC. The conventional half-bridge BDC converter topology is showcased in Fig. 1. The Half Bridge BidirectionalDC-DCConverter (HBBDC) consists of two source sides high

and low voltage sides, two switches S1&S2 with Diode D1 & D2, and two capacitors. The switches are driven by the PWM signals with the time period T and the duty cycle. This operates in two modes one is step-up and down mode. During boost mode, the battery will discharge and during the buck mode power flow in a reverse direction and the battery will be charged. The simulated diagram of HBBDC is shown in Fig. 2.

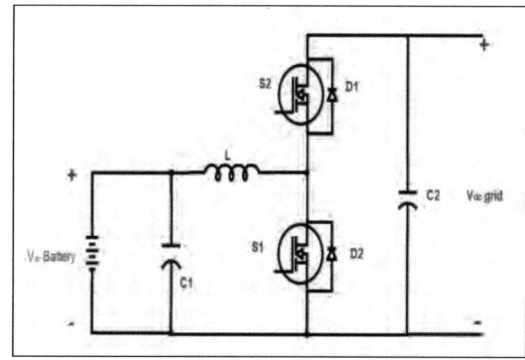

Fig. 1. Half Bridge BDC converter (HBBDC)

In the boost mode, there are two intervals. During the Interval 1 S1 is ON and during the Interval 2 S1 is OFF. If the switch S1 conducts, then the diode D1 will get reverse biased, the inductor gets charged and its current will increase until the removal of a gate pulse. When the switch S1 does not conduct, a sudden change in current will not happen immediately so a reversal of voltage polarity occurs and D1 will be forward-biased. So, the inductor charges the output capacitance hence the voltage is stepped up. In the step-down mode also, we have two intervals. During interval 1, the switch S2 conducts and diode D2 is reverse biased. Therefore, the inductor gets charged and the output capacitor gets charged. During Interval 2, the switch S2 is in OFF state and the Diode D2 is forward biased. So, the inductor will discharge through the freewheeling diode and hence the voltage will be stepped down.

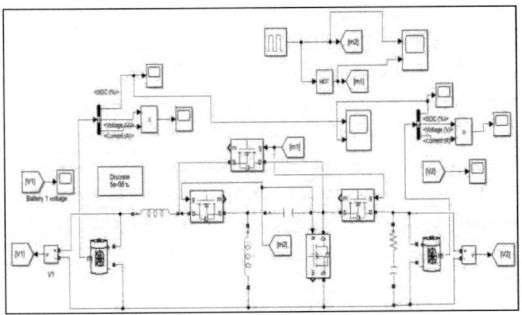

Fig. 2. Simulation diagram of an HBBDC converter

MATLAB simulation is carried out for a frequency of 20kHz, LV side is taken as 24V and HV side is taken as 200V. The designed values of the inductor and capacitor are 528μH and 5.2μF respectively.

Proposed System

The proposed Modified Half Bridge BDC (MHBDC) is showcased in Fig. 3. A voltage multiplier circuit is introduced in the converter to obtain high voltage gain. This circuit consists of four switches, three capacitors, and two inductors namely L1 & L2.

a

b

Fig. 3. Circuit diagram of (a) Modified HBBDC converter and (b) High gain cell

The four switches are driven by the PWM signals. Both modes of operation are obtained based on the pulses given to the switches. This converter can operate in both modes and also power flow is bidirectional.

In the buck mode, switch S1 conducts and the remaining switches will be in OFF state, but the body diodes of the remaining three switches will be in ON state. In this mode, during the first interval the switch S1 conducts, therefore inductor L1 is in the charging state and the inductor L2 will be in discharging state.

a

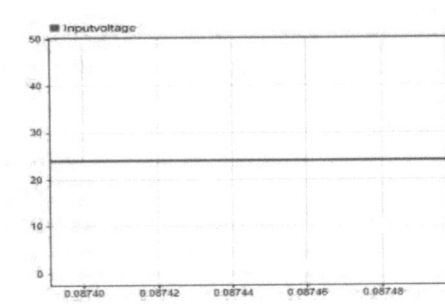

b

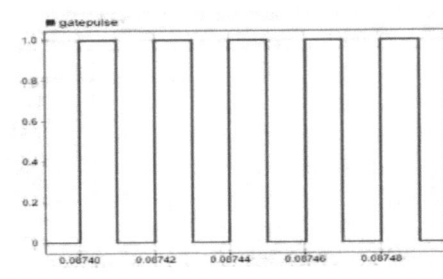

c

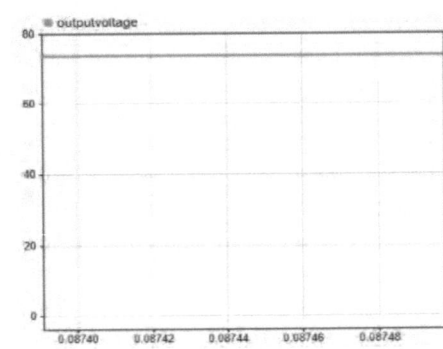

d

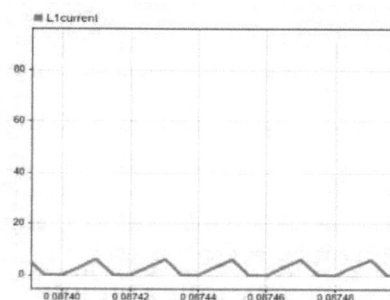

b

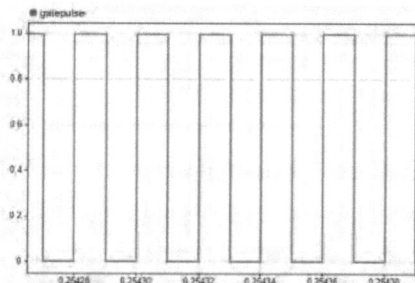

e

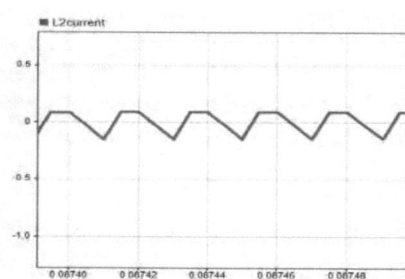

c

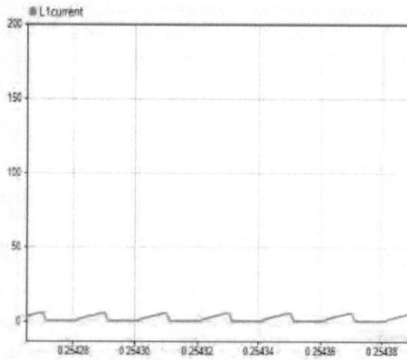

Fig. 4. Simulation waveforms of the proposed converter in boost mode (a) input voltage (b) gate pulse (c) output voltage (d)Inductor current L1 (e) Inductor current L2

During the second interval, the switch S1 does not conduct, therefore the inductor L2 will be in a charging state and the inductor L1 is in discharging state. So, the voltage is boosted up. The simulation output of the proposed system during boost operation is shown in Fig. 4.

d

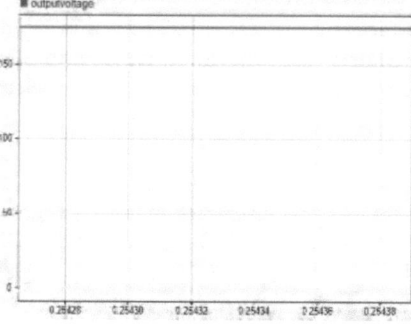

a

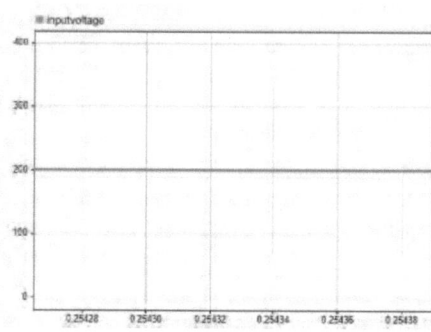

e

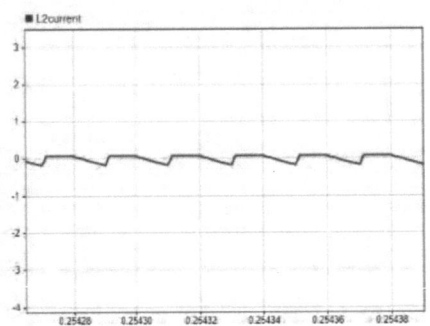

Fig. 5. Simulation output of the proposed converter in buck mode (a) input voltage (b) gate pulse (c) output voltage (d) Inductor currents L1 (e) Inductor current L2.

During step-down operation switch S1 will be in OFF state and S2, S3 and S4 are in ON state. In this mode, during the first interval the switch S2, S3 and S4 are in ON state, therefore inductor L1 is in the charging state and inductor L2 is in discharging state. During the second interval, switch S2, S3 and S4 are in OFF state and so, the voltage is stepped down in this mode. The simulation output of the proposed system in buck mode is shown in Fig. 5. The HBBDC and the modified HBBDC converter output voltage for various duty cycles are simulated and they have been tabulated in Table 1. From this tabulation, it is evident that the modified topology results are comparatively better than the conventional topology.

Table 1. Comparative analysis of conventional and the modified topology

Duty Cycle	OutputVoltage Boost Mode		Output Voltage Buck mode	
	Half Bridge	Modified Half Bridge	Half Bridge	Modified Half Bridge
0.1	18.33	29.11	26.5	78.7
0.2	36.67	39.17	29.77	124.1
0.3	55	50.43	33.95	150.2
0.4	73.33	62.08	39.46	165.7
0.5	91.67	73.83	47.08	175.3
0.6	110	85.74	58.2	181.6
0.7	128.8	97.61	75.82	185.7
0.8	147.2	117.4	106.7	188.7
0.9	165.4	219.6	159.8	190.8

The maximum voltage in LV side is taken as 24V and at the HV side is 200V. In the proposed system during step up mode, input of 24V and the output voltage of 73.88V is got for a duty cycle of 0.5. In the buck mode, for the input of 200V, output of 175.4V is obtained for a duty cycle of 0.5. Therefore, during the regenerative braking reversal of power flow will happen and the voltage will be stepped down.

Conclusion

In this article, a modified HBBDC converter is proposed and its performance is analyzed by simulating it in the MATLAB environment. The proposed non-isolated converter can function in both the buck mode and boost mode. From the obtained results it is inferred that the modified HBBDC converter provides high voltage gain than that of the conventional HBBDC converter.

References

Afzal, R., Tang, Y., Tong, H., and Y. Guo, (2021). A high step-up integrated coupled inductor-capacitor DC-DC converter, IEEE Access, 9, 11080–11090.

Bhajana, V. V. S. K., Drabek, P., and Aylapogu, P. K. (2019). Design and implementation of a zero voltage transition bidirectional DC-DC converter for DC traction vehicles, International Transactions on Electrical Energy Systems, 29(5), 1–14.

Ravi, Deepak, Reddy, BandiMallikarjuna, Shimi, S. L., Samuel, Paulson. (2018). Bidirectional DC to DC Converters: An Overview of Various Topologies, Switching Schemes and Control Techniques, International Conference on Recent Trends in Engineering & Sciences, 7(4).

Elsayad, N., Moradisizkoohi, H., and Mohammed, O. A. (2019). Design and implementation of a new transformerless bidirectional DC-DC converter with wide conversion ratios, IEEE Trans. Ind. Electron., 66(9), 7067–7077.

Gnanavadivel, J., Shunmathi, M., Muthu Thiruvengadam, P., Narthana, S. (2021). Analysis of DC-DC Converter with High Step-up Gain for Alternative Energy Sources, International Journal of Engineering Trends and Technology. 69(18), 162–168.

Kumar, M., Kumar, K., and Chaudhary, K. (2021). Modified non-isolated bidirectional DC–DC converter for regenerative braking for electric vehicle applications, Proceedings of the Symposium on Power Electronic and Renewable Energy Systems Control Singapore, Lecture Notes in Electrical Engineering, 77–88.

Lee, S. W., and Do, H.-L. (2019). Quadratic boost DC-DC converter with high voltage gain and reduced voltage stresses, IEEE Transactions on Power Electronics, 34(3): 2397–2404.

Naresh, S. V. K., and Peddapati, S. (2021). New family of transformerless quadratic buck-boost converters with wide conversion ratio, International Transactions on Electrical Energy Systems, 31(11), 1–21.

RathinaPrabha, N., Gnanavadivel, J., Krishna Veni, K. S. (2020). Performance Investigation of Single Switch Dual Output DC-DC SEPIC Converter for PV Applications, International Journal of Advanced Science and Technology, 2(9), 4676–4683.

Wu, Y., and Ke, Y.-T. (2021). A novel bidirectional isolated DC-DC converter with high voltage gain and wide input voltage, IEEE Transactions on Power Electronics 36(7), 7973–7985.

Zagreb Index In Various Signed Graphs

P. Ambika,[a,*] V. R. Vinothini,[b,*] and N. Suganya Baby[c,*]

[a]Department of Mathematics, Bishop Heber College, Trichy, India
[b]Department of Mathematics, Bannari Amman Institute of Technology, Erode, India
[c]Department of Mathematics, Kongu Engineering College, Erode, India
E-mail: *ambika.ma@bhc.edu.in; *VINOTHINIVR @bitsathy.ac.in; *suganyababy.sh@kongu.edu)

Aim

The aim of this research, we proposed and studied grade-based topological indices (TI) such as positive Zagreb Noteindex $(ZI)^+$ index) and negative Zagreb index $((ZI)^-$ index). Further investigated the $(ZI)^+$ index) and $(ZI)^-$ index in various signed graphs is imitative. Additionally describe the results by illustrations.

Keywords: Graphs, Topological index(TI), and Zagreb index

Introduction

The positive grade of the node u in the signed graph is defined by the amount of positive edges are occur in the node u then it is symbolized by $d_+(u)$. The negative grade of the node u in the signed graph is defined by amount of negative edges are occur in the vertex u, which is symbolized by $d_-(u)$.

The maximum positive grade of signed graph Σ is maximum positive grade along the points in Σ it is represented by $\Delta_+(G)$. The maximum negative grade of the signed graph Σ is maximum negative grade along the points in Σ it is represented by $\Delta_-(G)$.

Note that the sum of positive grade and negative grade of a node in $u \in \Sigma$ is the grade of node in the underlying graph $G = (V, E)$.

The positive grade of an edge uv in a signed graph is defined by quantity of positive arcs are adjacent to the edge uv and it is symbolized by $d_+(uv)$. The negative grade of the edge uv in the signed graph is defined by number of negative edges adjacent to the edge uv and it is represented by $d_-(uv)$.

The minimum positive grade of the signed graph Σ is minimum positive grade along the edges of Σ it is represented by $\delta_{E+}(G)$. The minimum negative grade of the signed graph Σ is minimum negative grade along the edges of Σ it is represented by $\delta_{E-}(G)$.

$$TI(G) = \sum_{pq \in G} F\big(d(p), d(q)\big)$$

Zagreb index is defined as

$$M_r(G) = \sum_{p \in V(G)} \big(d_p\big)^r$$

In this investigation, we explain the Zagreb index of signed graphs. Further, we investigate some properties and bounds of the Zagreb index of signed graphs.

Zagreb Index in Signed Graph

In this segment, we define the positive and negative Zagreb index in signed graphs and examine the properties and limits of the Zagreb index in signed graphs.

Definition 1: The first positive Zagreb (ZI_1^+) index and second positive Zagreb (ZI_2^+) index of the signed graphs is defined as $ZI_1^+(\Sigma) = \sum_{u \in V(G)} \left(d_u^+\right)^2$ and $ZI_2^+(\Sigma) = \sum_{uv \in E(G)} \left(d_u^+ d_v^+\right).$

Definition 2: The first negative Zagreb (ZI_1^-) index and second negative Zagreb (ZI_2^-) index of the signed graphs is defined as $ZI_1^-(\Sigma) = \sum_{u \in V(G)} \left(d_u^-\right)^2$ and $ZI_2^-(\Sigma) = \sum_{uv \in E(G)} \left(d_u^- d_v^-\right).$

Theorem 1: In a positive K regular signed graph by means of m vertices, then the $ZI_1^+(\Sigma)$ index and $ZI_1^-(\Sigma)$ index is $ZI_r^+(\Sigma) = m(k)^2 = ZI_1^-(\Sigma).$

Proof: Take on Σ be a positive K-regular signed graph of m points. Therefore we get $d^+(v_i) = K$, for every $v_i \in \Sigma$. The positive Zagreb $ZI_1^+(\Sigma)$ index of signed graphs is defined as $ZI_1^+(\Sigma) = \sum_{p \in V(G)} \left(d_p^+\right)^2.$

$$ZI_1^+(\Sigma) = \sum_{p \in V(G)} (k)^2$$
$$= (k)^2 + (k)^2 + \dots |V| times$$
$$ZI_1^+(\Sigma) = |V|(k)^2 = m(k)^2$$

Take on Σ be a negative K regular signed graph with m points. Therefore we get $d^+(v_i) = K$, for every $v_i \in \Sigma$. The negative Zagreb (ZI_1^-) index of signed graphs is defined as $ZI_1^-(\Sigma) = \sum_{p \in V(G)} \left(d_p^-\right)^r.$

$$ZI_1^-(\Sigma) = \sum_{p \in V(G)} (k)^2$$
$$= (k)^2 + (k)^2 + \dots |V| times$$
$$ZI_1^-(\Sigma) = |V|(k)^2 = m(k)^2$$

Hence $ZI_r^+(\Sigma) = m(k)^2 = ZI_1^-(\Sigma)$

Theorem 2: In a positive K regular signed graph by m points, then the $ZI_2^+(\Sigma)$ index and $ZI_2^-(\Sigma)$ index are $ZI_2^+(\Sigma) = \dfrac{m \cdot k^3}{2} = ZI_2^-(\Sigma).$

Proof: Take on Σ be a positive K regular signed graph of m points. Therefore we get $d^+(v_i) = K$, for every $v_i \in \Sigma$. There are $\left(\dfrac{m \cdot k}{2}\right)$ edges in Σ since Σ is regular. The second positive Zagreb $ZI_2^+(\Sigma)$ index of signed graphs is defined as $ZI_2^+(\Sigma) = \sum_{uv \in E(G)} \left(d_u^+ d_v^+\right).$

$$ZI_2^+(\Sigma) = \sum_{uv \in E(G)} \left(d_u^+ d_v^+\right). = \sum_{uv \in E(G)} (k \cdot k)$$
$$= k^2 + k^2 + \dots + |E| times = k^2 \cdot \left(\dfrac{m \cdot k}{2}\right)$$
$$ZI_2^+(\Sigma) = \dfrac{m \cdot k^3}{2}$$

Suppose Σ stand a negative K regular signed graph with m points. Therefore we get $d^+(v_i) = K$, for every $v_i \in \Sigma$. There is $\left(\dfrac{m \cdot k}{2}\right)$ edges in Σ since Σ is regular. The negative Zagreb (ZI_1^-) index of signed graphs is defined as $ZI_2^-(\Sigma) = \sum_{uv \in E(G)} \left(d_u^- d_v^-\right).$

$$ZI_2^-(\Sigma) = \sum_{uv \in E(G)} \left(d_u^- d_v^-\right). = \sum_{uv \in E(G)} (k \cdot k)$$
$$= k^2 + k^2 + \dots + |E| times = k^2 \cdot \left(\dfrac{m \cdot k}{2}\right)$$
$$ZI_2^-(\Sigma) = \dfrac{m \cdot k^3}{2}$$

Hence
$$ZI_2^+(\Sigma) = \dfrac{m \cdot k^3}{2} = ZI_2^-(\Sigma).$$

Illustration 1: Positive and negative 2-regular signed graph $\Sigma(G,\sigma)$.

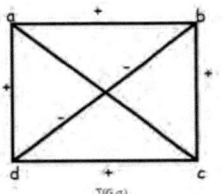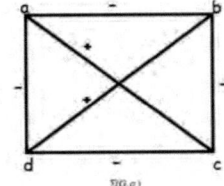

Fig. 1. Positive and negative 2-regular signed graph.

In a positive 2-regular signed graph the positive grade of the every vertices in $\Sigma(G,\sigma)$ is 2. The order and size $\Sigma(G,\sigma)$ is $O(\Sigma)=m=4, S(G)=6$. The $ZI_r^{+}(\Sigma)$ index $ZI_1^{+}(\Sigma)=(4)(2)^2 =|V|(k)^2 =16$. In a negative 2-regular signed graph $\Sigma(G,\sigma)$ the negative grade of the every node in $\Sigma(G,\sigma)$ is 2. The order $\Sigma(G,\sigma)$ is $O(\Sigma)=m=4, S(G)=6$. The $ZI_1^{-}(\Sigma)$ index $ZI_1^{-}(\Sigma)=(4)(2)^2 =|V|(k)^2 =16$.

Theorem 3: In a positive and negative complete signed graph of m vertices, then the $ZI_1^{+}(\Sigma)$ index and $ZI_1^{-}(\Sigma)$ index is

$$ZI_1^{+}(\Sigma)=m(m-1)^2 = ZI_1^{-}(\Sigma).$$

Proof: $\Sigma(G,\sigma)$ be a positive complete signed graph. It suggests that the induced subgraph $<V^+>$ is a complete graph of order $O(G)=m$. It suggests the positive grade of all points in G is $(m-1)$ *and* m number of points in the induced subgraph $<V^+>$. Hence an (n-1) regular graph here is $\dfrac{m(m-1)}{2}$ arcs in a complete graph of m points. The positive Zagreb (ZI_1^{+}) index of signed graphs is defined as

$$ZI_1^{+}(\Sigma)=\sum_{p\in V}\left(d_p^{+}\right)^2 =\sum_{p\in V}(m-1)^2$$
$$=(m-1)^2 +(m-1)^2 +....|V|\,times$$
$$ZI_1^{+}(\Sigma)=m(m-1)^2$$

Let $\Sigma(G,\sigma)$ be a negative complete signed graph. This suggests that the induced subgraph $<V^->$ is a complete graph of order $O(G)=m$. This suggests the negative grade of all vertices in G is $(m-1)$ *and* m number of points in the induced subgraph $<V^->$. The negative Zagreb (ZI_1^{-}) index of signed graphs is defined as

$$ZI_1^{-}(\Sigma)=\sum_{p\in V}\left(d_p^{-}\right)^2 =\sum_{p\in V}(m-1)^2$$
$$=(m-1)^2 +(m-1)^2 +....|V|\,times$$
$$ZI_1^{-}(\Sigma)=m(m-1)^2$$

Hence $ZI_1^{+}(\Sigma)=m(m-1)^2 = ZI_1^{-}(\Sigma)$.

Theorem 4: For a positive and negative complete signed graph of m points, then $ZI_1^{+}(\Sigma)$ index and $ZI_1^{-}(\Sigma)$ index is

$$ZI_2^{+}(\Sigma)=\dfrac{\left(m\cdot(m-1)^3\right)}{2}= ZI_2^{-}(\Sigma).$$

Proof: Take on $\Sigma(G,\sigma)$ be a positive complete signed graph this suggests the induced subgraph $<V^+>$ is a complete graph of order $O(G)=m$. This suggests the positive grade of all points in G is (m-1) and m number of points in the induced subgraph $<\quad>$. In an (n-1) regular graph there is $\dfrac{m(m-1)}{2}$ arcs in a complete graph of n points.

The positive Zagreb (ZI_2^{+}) index of signed graphs is defined as

$$ZI_2^{+}(\Sigma)=\sum_{uv\in E}\left(d_u^{+}d_v^{+}\right) =\sum_{uv\in E}(m-1)^2$$
$$=(m-1)^2 +(m-1)^2 +....|E|\,times$$
$$ZI_2^{+}(\Sigma)=(m-1)^2 \cdot\left(\dfrac{m\cdot(m-1)}{2}\right)$$
$$ZI_2^{+}(\Sigma)=\dfrac{\left(m\cdot(m-1)^3\right)}{2}$$

Let $\Sigma(G,\sigma)$ be a negative complete signed graph it suggests the induced subgraph $<V^->$ is a complete graph of order $O(G)=n$. This suggests the negative grade of all points in G is (n-1) and n number of points in the induced subgraph $<V^->$. The negative Zagreb (ZI_2^-) index of signed graphs is defined as

$$ZI_2^-(\Sigma)=\sum_{uv\in E}\left(d_u^- d_v^-\right)=\sum_{uv\in E}(m-1)^2$$

$$=(m-1)^2+(m-1)^2+....|E|times$$

$$ZI_2^-(\Sigma)=(m-1)^2\cdot\left(\frac{m\cdot(m-1)}{2}\right)$$

$$ZI_2^-(\Sigma)=\frac{\left(m\cdot(m-1)^3\right)}{2}$$

Hence $ZI_2^+(\Sigma)=\dfrac{\left(m\cdot(m-1)\right)^2}{2}=ZI_2^-(\Sigma)$.

Illustration 2: Positive and negative complete signed graph $\Sigma(G,\sigma)$.

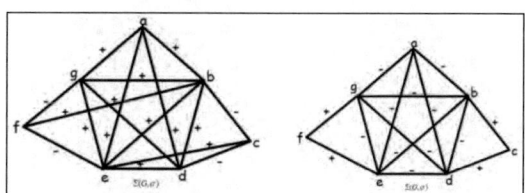

Figure 2. $\Sigma(G,\sigma)$.

In a positive complete signed graph $\Sigma(G,\sigma)$ the positive grade of the every vertices in $<V^+>$ is 4. The order $\Sigma(G,\sigma)$ is $O(\Sigma)=m=7, S(\Sigma)=14$. The Zagreb (ZI_1^+) index $ZI_1^+(\Sigma)=7(7-1)^2=7\cdot36=252$ and

$$ZI_2^+(\Sigma)=\frac{m\cdot(m-1)^3}{2}=\frac{7\cdot216}{2}=756.$$

Similarly the Zagreb (ZI_1^-) index $ZI_1^-(\Sigma)=7(7-1)^2=7\cdot36=252$ and

$$ZI_2^-(\Sigma)=\frac{m\cdot(m-1)^3}{2}=\frac{7\cdot216}{2}=756.$$

Theorem 5: In a positive signed graph $K_{m,n}$, then the Zagreb (ZI_1^+) and Zagreb (ZI_2^+) index is $ZI_1^+(\Sigma)=mn(m+n)$ and $ZI_2^+(\Sigma)=m^2n^2$.

Proof: $\Sigma(G,\sigma)$ be a positive signed graph of node sets $V_m^+ \& V_n^+$ respectively. This implies the induced subgraph $<V_{m,n}^+>$ is a positive complete bipartite graph of vertex sets $V_m \& V_n$. This suggests the positive grade of all points in V_m and V_n is $n \& m$ respectively such that $d^+(v_i)=n, \forall v_i\in V_m$ and $d^+(v_j)=m, \forall v_j\in V_n$, here is mn arcs in a positive complete bipartite signed graph $K_{m,n}$ of (m,n) vertices. Therefore Zagreb (ZI_1^+) index

$$ZI_1^+(\Sigma)=\sum_{p\in V}\left(d_p^+\right)^2=\sum_{p\in V_m}\left(d_p^+\right)^2+\sum_{q\in V_n}\left(d_p^+\right)^2$$

$$=\sum_{p\in V_m}(n)^2+\sum_{q\in V_n}(m)^2=\left(n^2+n^2+...+|V_m|times\right)+$$

$$\left(m^2+m^2+...+|V_n|times\right)$$

$$=\left(n^2+n^2+...+m\,times\right)+\left(m^2+m^2+...+n\,times\right)$$

$$=m\cdot n^2+n\cdot m^2$$

$$ZI_1^+(\Sigma)=mn(m+n)$$

Similarly

$$ZI_2^+(\Sigma)=\sum_{uv\in E}\left(d_u^+ d_v^+\right)=\sum_{\substack{uv\in E\\u\in V_m\,and\\v\in V_n}}\left(d_u^+ d_v^+\right)=\sum_{\substack{uv\in E\\u\in V_m\,and\\v\in V_n}}(mn)$$

$$=mn+mn+...+|K_{mn}|times=mn(mn)$$

$$ZI_2^+(\Sigma)=m^2n^2$$

Hence the Zagreb (ZI_1^+) and Zagreb (ZI_2^+) are $ZI_1^+(\Sigma)=mn(m+n)$ and $ZI_2^+(\Sigma)=m^2n^2$.

Theorem 6: For a negative signed graph $K_{m,n}$, then the Zagreb (ZI_1^-) and Zagreb (ZI_2^-) are $ZI_1^-(\Sigma)=mn(m+n)$ and $ZI_2^-(\Sigma)=m^2n^2$.

Proof: $\Sigma(G,\sigma)$ be a negative signed graph of points sets V_m^- & V_n^- respectively. This infers the induced sub graph $<V_{m,n}^-$ > is a positive complete bipartite graph of vertex sets V_m & V_n. This infers the positive grade of every points in V_m and V_n are n & m respectively such that $d^-(v_i)=n, \forall v_i \in V_m$ and $d^-(v_j)=m, \forall v_j \in V_n$, there is mn arcs in a positive signed graph $K_{m,n}$ of (m,n) vertices. Therefore Zagreb (ZI_1^-) index

$$ZI_1^-(\Sigma)=\sum_{p\in V}\left(d_p^-\right)^2 = \sum_{p\in V_m}\left(d_p^-\right)^2 + \sum_{q\in V_n}\left(d_p^-\right)^2$$

$$= \sum_{p\in V_m}(n)^2 + \sum_{q\in V_n}(m)^2 = \left(n^2+n^2+...+\left|V_m\right| times\right)+$$

$$\left(m^2+m^2+...+\left|V_n\right| times\right)$$

$$=\left(n^2+n^2+...+m\,times\right)+\left(m^2+m^2+...+n\,times\right)$$

$$=m\cdot n^2 + n\cdot m^2$$

$$ZI_1^-(\Sigma)=mn(m+n)$$

Similarly

$$ZI_2^-(\Sigma)=\sum_{uv\in E}\left(d_u^- d_v^-\right)=\sum_{\substack{uv\in E \\ u\in V_m \, and \\ v\in V_n}}\left(d_u^- d_v^-\right)$$

$$=\sum_{\substack{uv\in E \\ u\in V_m \, and \\ v\in V_n}}(mn)=mn+mn+...+\left|K_{mn}\right| times$$

$$=mn(mn)$$

$$ZI_2^-(\Sigma)=m^2n^2$$

Hence the Zagreb (ZI_1^-) and Zagreb (ZI_2^-) are $ZI_1^-(\Sigma)=mn(m+n)$ and $ZI_2^-(\Sigma)=m^2n^2$.

Illustration 3: Positive complete bipartite signed graph $\Sigma(G,\sigma)$.

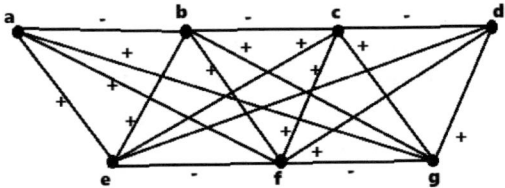

Figure 3. $\Sigma(G,\sigma)$.

In a positive signed graph $\Sigma(G,\sigma)$ the positive grade of every vertices in V_1^+ and V_2^+ are m=4 and n=3 respectively. The order $\Sigma(G,\sigma)$ is $O(\Sigma)=7$. There are 12 edges in positive complete bipartite signed graph $\Sigma(G,\sigma)$. The Zagreb (ZI_1^+) and Zagreb (ZI_2^+) are $ZI_1^+(\Sigma)=84$ and $ZI^+(\Sigma)=144$.

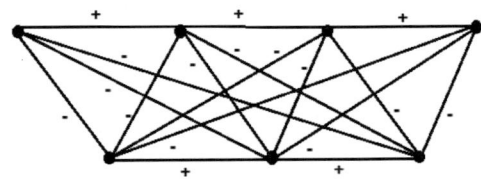

Figure 4: $\Sigma(G,\sigma)$.

In a Negative complete bipartite signed graph $\Sigma(G,\sigma)$ the Negative grade of every vertex in V_1^+ and V_2^+ are m=4 and n=3 respectively. The order $\Sigma(G,\sigma)$ is $O(\Sigma)=7$. There is 12 edges in Negative complete bipartite signed graph $\Sigma(G,\sigma)$. The Zagreb (ZI_1^-) and Zagreb (ZI_2^-) are $ZI_1^-(\Sigma)=84$ and $ZI_2^-(\Sigma)=144$.

Conclusion

In this investigation, we explain the Zagreb index of signed graphs. Further, we investigate some properties and bounds of the Zagreb index of signed graphs. In the future, the various topological indices in signed graphs are investigated and apply the indices various applications.

References

Abirami, M., Ganesh, Y.J., and Kalaivani K. Atom-Bond Connectivity Index in various Graphs, ECS Transactions, 107(1), 12121–12129.

Cartwright, D., and Harary, F. (1956). Structural balance a generalization of Heider's theory, Psychological Review, 1956, 63(5), 277–293.

Ch. Das, Kinkar, Gutman, Ivan., and Boris Furtula (2012). On atom–bond connectivity index, Filomat. 26(4), 733–738.

Zaslavsky, Thomas (1981). Characterizations of signed graphs. Journal of Graph Theory. 5(4). 401–406.

Domination-Chromatic Number in Operations on Fuzzy Graphs

K. Prabhavathi,[a] A. Sekar,[b] B. Usha,[c] and A. L. Nachammai[d]

[a]Department of Mathematics, Bannari Amman Institute of Technology, Erode, Tamil Nadu
[b]Department of Mathematics, Sri Ramakrishna Engineering College, Coimbatore, Tamil Nadu
[c,d]Department of Mathematics, Kongu Engineering College, Erode, Tamil Nadu
E-mail: PRABHAVATHIK@bitsathy.ac.in; sekar.arumugam@srec.ac.in; usha_b.sh@kongu.edu

Abstract

In this effort, we emerged the concept of a Dom-Chro set and Dom-Chro number of an FG. Further study the Dom-Chro number of various graphs like complete, complete bipartite FG, union, join, and Cartesian also some bounds of the Dom-Chro number are investigated.

Keywords: Domination number, chromatic number, Dom-Chro number.

Introduction

For a given χ-coloring of an FG G, a dominating set $S \subseteq V(G)$ is thought selected dom-chro set, if it covers at least one vertex from each color of G. The dom-chro number $\chi_{fd}(G)$ is the minimum cardinality engaged above all the dom-chro sets of FG $G(\sigma, \mu)$.

Let $G_1(\sigma_1, \mu_1)$ and $G_2(\sigma_2, \mu_2)$ be FG on $V_1 \& V_2$ respectively with $V_1 \cap V_2 = \varphi$. The union of $G_1 \& G_2$ denoted by $G_1 \cup G_2$ is the FG G on $V_1 \cup V_2$ demarcated by $G = (G_1 \cup G_2) = ((\sigma_1 \cup \sigma_2'), (\mu_1 \cup \mu_2'))$ wherever

$$(\sigma_1 \cup \sigma_2')(x) = \begin{cases} \sigma_1(x), if\ x \in V_1 \\ \sigma_2'(x), if\ x \in V_2 \end{cases}$$

$$(\mu_1 \cup \mu_2)(xy) = \begin{cases} \mu_1(xy), if\ xy \in E_1 \\ \mu_2(xy), if\ xy \in E_2 \\ 0\ otherwise \end{cases}$$

The join of $G_1 \& G_2$ is the FG $G_1 + G_2$ on $V_1 \cup V_2$ is demarcated by $G = (G_1 + G_2) = ((\sigma_1 + \sigma_2), (\mu_1 + \mu_2))$ wherever

$$(\sigma_1 + \sigma_2)(x) = \begin{cases} \sigma_1(x), if\ x \in V_1 \\ \sigma_2(x), if\ x \in V_2 \end{cases}$$

$$(\mu_1 + \mu_2)(xy) = \begin{cases} \mu_1(xy), if\ xy \in E_1 \\ \mu_2(xy), if\ xy \in E_2 \\ \mu_1(x) \wedge \mu_2(y), if\ x \in V_1 \& y \in V_2 \end{cases}$$

The Cartesian product of $G_1 \& G_2$ represented by $G_1 \times G_2$ is the FG $G_1 \times G_2$ on $V_1 \times V_2$ demarcated by $G = (G_1 \times G_2) = ((\sigma_1 \times \sigma_2), (\mu_1 \times \mu_2))$ wherever $(\sigma_1 \times \sigma_2)(x_1, x_2) = \sigma_1(x_1) \wedge \sigma_2(x_2)$ and

$$(\mu_1 \times \mu_2)(x_1 x_2)(y_1 y_2) = \begin{cases} \sigma_1(x_1) \wedge \mu_2(x_2 y_2), if\ x_1 = y_1 \\ \sigma_2(x_2) \wedge \mu_1(x_1 y_1), if\ x_2 = y_2 \\ 0, \quad otherwise \end{cases}$$

In this effort, we emerged the concept of a Dom-Chro set and Dom-Chro number of an FG. Further study the Dom-Chro number of various graphs like complete, complete bipartite FG and join Cartesian also some bounds of the Dom-Chro number are investigated.

Domination Chromatic Number of FG

In this segment, we develop the concept of domination chromatic (Dom-Chro) numbers in FG and then investigate the bounds of Dom-Chro numbers in different FG.

Definition 1. For a given χ-coloring of an FG G, a dominating set $S \subseteq V(G)$ is thought to be dom-chro set, if it covers at least one vertex from each color of G. The dom-chro number $\chi_{ifd}(G)$ is the minimum cardinality taken over all the dom-chro sets of FG $G(\sigma,\mu)$.

Theorem 1: In a complete FG $G(V,E)$, then $\chi_{fd}(G) = O(G)$.

Proof: Let $G(V, E)$ be a complete FG. Therefore around is a strong arc among every couple of nodes in V clearly $G(V, E)$ n-color FG. Let D is a $\gamma(G)$ set of the complete FG $G(V, E)$. Clearly, $D = \{v_1 | v_1 = \Delta_N(G)\}$ this implies D did not cover all color in $G(V, E)$ This implies Vis the dom-chro set and it covers all the colors in complete FG $G(V, E)$ Hence $\chi_{fd}(G) = |V| \Rightarrow \chi_{id}(G) = O(G)$.

Illustration 1

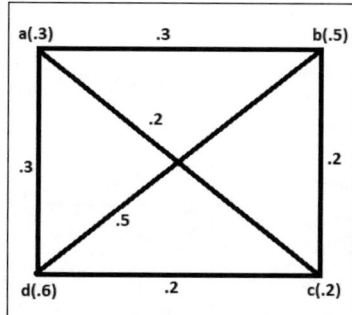

Fig. 1. Complete FG $G(V, E)$

In Fig. 1, the dom-chro set and number of the complete FG $G(V, E)$ are $\{a,b,c,d\} = V$ & $\chi_{fd}(G) = 1.6 = O(G)$ respectively.

Theorem 2: For a complete bipartite FG, $G(V, E)$, then $\chi_{ifd}(G) = \gamma_{if}(G)$.

Proof: Assume $G(V, E)$ be a complete bipartite FG. Consequently the vertex V_1 & V_2 disjoint sets clearly $G(V, E)$ 2-color FG. Let D be a $\gamma(G)$ set of the complete bipartite FG $G(V, E)$. Clearly $D = \{v_1, v_2 | v_1 \in V_1 \& v_2 \in V_2\}$ this implies v_1 & v_2 having a different color since $v_1 \in V_1 \& v_2 \in V_2$. Therefore the set D covers all the colors in complete bipartite FG $G(V, E)$. Hence $\chi_{fd}(G) = |D| \Rightarrow \chi_{fd}(G) = \gamma_f(G)$.

Illustration 2

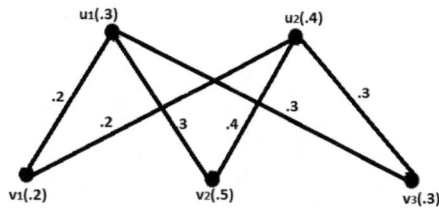

Fig 2: Complete bipartite FG

In fig 2, the dom-chro set and number of the complete bipartite FG $G(V, E)$ are $\{u_1, v_1\}$ & $\chi_{fd}(G) = 0.5 = \gamma_{if}(G)$ respectively.

Theorem 3: In union FG $(G_1 \cup G_2)$ the dom-chro number $\chi_{fd}(G_1 \cup G_2) = \chi_{fd}(G_1) + \chi_{fd}(G_2)$.

Proof: Assume $G_1(\sigma_1,\mu_1)$ and $G_2(\sigma_2,\mu_2)$ FG with dom-chro number $\chi_{fd}(G_1)$ and $\chi_{fd}(G_2)$ respectively. Let $D_c \subseteq V(G)$ be a minimal dom-chro number of an FG $G_1 \cup G_2$. In $G_1 \cup G_2$ edges are in the following form

(i). $Đ \in {}_1$

(ii). $xy \in G_2$

Note that there is a strong arc among nodes in G_1 and G_2. This implies minimal dom-chro sets D_{C_1} and D_{C_2} of G_1 and G_2 are dominating set of $G_1 \cup G_2$. Therefore the different set of colours used in G_1 and G_2. Clearly D_{C_1} and D_{C_2} are not a dom-chro set of $G_1 \cup G_2$. Since some nodes colour are in $G_1 \cup G_2$ remain not covered by D_{C_1} and D_{C_2}. The set $D_{C_1} \cup D_{C_2}$ cover all the colours in $G_1 \cup G_2$. Hence the set $D_C = \left(D_{C_1} \cup D_{C_2} \right)$ is a dom-chro set of $G_1 \cup G_2$.

$$D_C = \left(D_{C_1} \cup D_{C_2} \right)$$
$$\left| D_C \right| = \left| D_{C1} \right| + \left| D_{C_2} \right|$$
$$\chi_{fd}(G_1 \cup G_2) = \chi_{fd}(G_1) + \chi_{fd}(G_2)$$

llustration 3

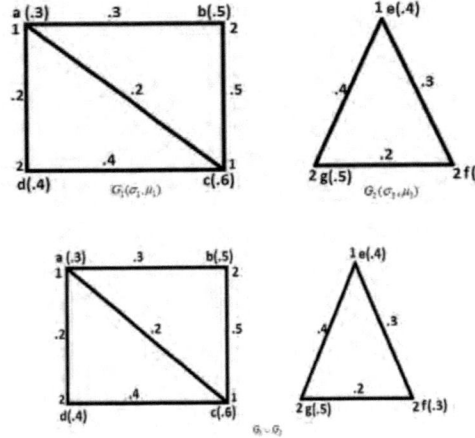

Fig. 3. $G_1 \cup G_2$

In Fig. 3, the dom-chro set of the FG's G_1 and G_2 are $\{a,d\} \& \{e,f\}$ respectively. Note the dom-chro number of the FG's G_1 and G_2 are $\chi_{fd}(G_1) = 0.7 \& \chi_{fd}(G_2) = 0.7$. The dom-chro set and number of the FG $G_1 \cup G_2$ are $\{a,d,e,g\} \& \chi_{fd}(G_1 \cup G_2) = 1.4$ respectively.

Theorem 4: In join FG $G_1 + G_2$ the dom-chro number $\chi_{fd}(G_1 + G_2) = \chi_{fd}(G_1) + \chi_{fd}(G_2)$.

Proof: Let $G_1(V_1, E_1)$ and $G_2(V_2, E_2)$ be an FG by way of dom-chro number $\chi_{fd}(G_1)$ and $\chi_{fd}(G_2)$

respectively. Let $D_C \subseteq V(G)$ be a minimal dom-chro number of an FG $G_1 + G_2$. In $G_1 + G_2$ edges are in the following form

i. $xy \in G_1$
ii. $xy \in G_2$
iii. $xy \in G_1 + G_2$

Note that there is an effective arc among nodes in G_1 and G_2. This implies minimal dom-chro sets D_{C_1} and D_{C_2} of G_1 and G_2 are dominating set of $G_1 + G_2$. Therefore the different set of colors used in G_1 and G_2. Clearly D_{C_1} and D_{C_2} are not a dom-chro set of $G_1 + G_2$. Since some of the node color in $G_1 + G_2$ remain not covered by D_{C_1} and D_{C_2}. The set $D_{C_1} \cup D_{C_2}$ cover all the colors in $G_1 + G_2$. Hence the set $D_C = \left(D_{C_1} \cup D_{C_2} \right)$ is a dom-chro set of $G_1 + G_2$.

$$D_C = \left(D_{C1} \cup D_{C_2} \right)$$
$$\left| D_C \right| = \left| D_{C1} \right| + \left| D_{C_2} \right|$$
$$\chi_{fd}(G_1 + G_2) = \chi_{fd}(G_1) + \chi_{fd}(G_2)$$

Illustration 4

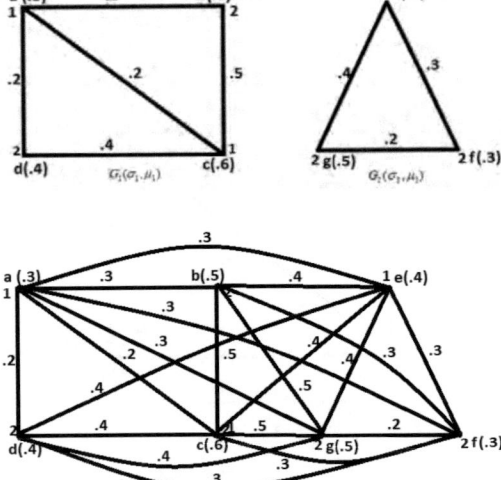

Fig. 4. The join of FG G_1 and G_2

In Fig. 4, the dom-chro set of the FG's G_1 and G_2 are $\{a,d\} \& \{e,f\}$ respectively. Note the dom-chro number of the FG's G_1 and G_2 are $\chi_{fd}(G_1) = 0.7 \& \chi_{fd}(G_2) = 0.7$. The dom-chro set and number of the FG $G_1 \cup G_2$ are $\{a,d,e,g\} \& \chi_{fd}(G_1 \cup G_2) = 1.4$ respectively.

Theorem 5: In a Cartesian product of FG $(G_1 \times G_2)$ the dom-chro number $\chi_{fd}(G_1 \times G_2) = \gamma_f(G_1 \times G_2)$

Proof: Assume $G_1(\sigma_1, \mu_1)$ and $G_2(\sigma_2, \mu_2)$ be a FG with γ set D_1 and D_2 respectively. Let $D_c \subseteq V(G)$ be a minimal dom-chro number of a FG $G_1 \times G_2$. In $G_1 \times G_2$ edges are in the following form

   i. $(x_1 y_1)(x_1 y_2) \in G_1 \times G_2$

   ii. $(x_1 y_1)(x_2 y_1) \in G_1 \times G_2$

Let D be a minimum γ_f set of $G_1 \times G_2$. Take on D is not a minimal dom-chro set of $G_1 \times G_2$. There is a vertex color of $G_1 \times G_2$ are not covered by the set D. This implies there is a vertex $(x_1 y_1)$ in $G_1 \times G_2$ is not dominated by the set D. This is contradict to our assumption D be a γ set of $G_1 \times G_2$. Hence D is a minimal dom-chro set of $G_1 \times G_2$. Therefore we get $D_c = D \Rightarrow |D_c| = |D| \Rightarrow \chi_{fd}(G_1 \times G_2) = \gamma_f(G_1 \times G_2)$.

Illustration 5

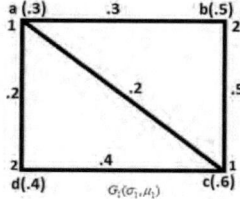 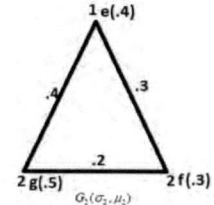

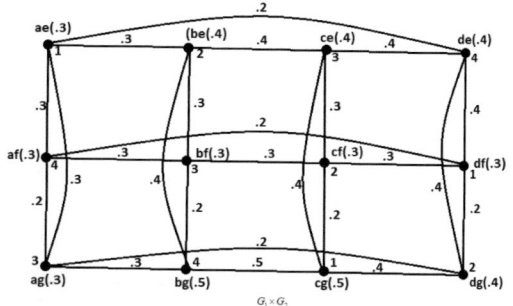

Fig. 5. The Cartesian product of G_1 and G_2

In Fig. 5, the dom-chro set and number of the FG $G_1 \times G_2$ are $\{ae, be, ce, de\} \& \chi_{fd}(G_1 \times G_2) = 1.1 = \gamma_f(G_1 \times G_2)$ respectively.

Theorem 6: In direct product of FG $(G_1 * G_2)$ the dom-chro number $\chi_{fd}(G_1 * G_2) = \gamma_f(G_1 * G_2)$.

Proof: Assume $G_1(\sigma_1, \mu_1)$ and $G_2(\sigma_2, \mu_2)$ be a FG with γ D_1 and D_2 respectively. Let $D_c \subseteq V(G)$ be a minimal dom-chro number of a FG $G_1 * G_2$. In $G_1 * G_2$ edges are in the following form $(x_1 y_1)(x_2 y_2) \in G_1 \times G_2$

Let D be a minimum γ_f set of $G_1 * G_2$. Take on D is not a minimal dom-chro set of $G_1 * G_2$. There is a vertex colour of $G_1 * G_2$ are not covered by the set D. This implies there is a vertex $(x_1 y_1)$ in $G_1 * G_2$ is not a dominated by the set D. This is contradict to our assumption D be a γ set of $G_1 * G_2$. Hence D is a minimal dom-chro set of $G_1 * G_2$. Therefore we get $D_c = D \Rightarrow |D_c| = |D| \Rightarrow \chi_{fd}(G_1 * G_2) = \gamma_f(G_1 * G_2)$.

Illustration 6

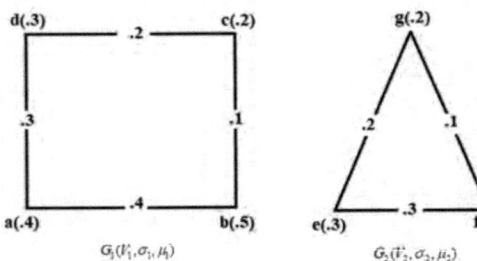

$G_1(V_1', \sigma_1, \mu_1)$ $G_2(V_2', \sigma_2, \mu_2)$

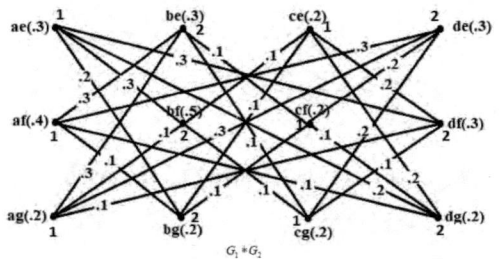

$G_1 * G_2$

Fig. 6. $G_1 * G_2$

In fig. 5, the dom-chro set and number of the FG $G_1 * G_2$ are *{ae,be,ce,de}* & $\chi_{fd}(G_1 \times G_2) = 1.1 = \gamma_f(G_1 * G_2)$ respectively.

Conclusion

In this effort, we emerged the concept of a Dom-Chro set and Dom-Chro number of a FG. Further study the Dom-Chro number of a various graphs like complete, complete bipartite FG, union, join and Cartesian also some bounds of the Dom-Chro number are investigated.

References

Ambika, P., Vinoth Kumar, N., Hellan Priya, J. DOM-CHRO Number In Operations On Intuitionistic Fuzzy Graphs, Journal of algebraic statistics, 13(2), 1058–1062.

Mordeson, N., and Nair, P. S. (2000). Fuzzy Graphs & Hypergraphs, Physica Verlag.

Prabakaran, P., Vinoth Kumar, N., and Preethi, N. (2021). Regular Domination in Various Fuzzy Graphs, Journal of Physics: Conference Series, 1947 (2021), 012054.

Rosenfeld, A, Zadeh L. A., Fu and Shimura M., (1975). Fuzzy sets & their applications (1975), 77–95.

Special Dio - Triples Involving Primes

Manju Somanath,[a,*] V. Sangeetha,[b] and T. Anupreethi[c]

[a]Assistant Professor & Research Advisor, Department of Mathematics, National College, (Autonomous, Affiliated to Bharathidasan University), Tiruchirappalli, Tamil Nadu, India.
[b]Assistant Professor, Department of Mathematics, National College, (Autonomous, Affiliated to Bharathidasan University), Tiruchirappalli, Tamil Nadu, India
[c]Research Scholar, Department of Mathematics, National College (Autonomous, Affiliated to Bharathidasan University), Tiruchirappalli, Tamil Nadu, India.
E-mail: *anupreethitamil@gmail.com(manjuajil@yahoo.com)

Abstract

The Construction of Dio - triple involving twin primes and cousin primes with properties D(2) and D(5) respectively are discussed in this paper. The recurrence relations are derived using the concept of pellian equations.

Keywords: Twin prime, Cousin prime, Dio - triple, Pell equation, Recurrence relations

Introduction

Diophantus of Alexandria noticed that the numbers $\frac{1}{16}, \frac{33}{16}, \frac{68}{16}, \frac{105}{16}$ have the property that the product of any two of these numbers is one less than the square of a rational number. Sets of integers with a similar property have been of interest for many years and a sequence $(a_1, \ldots. a_m)$ of non-negative integers, is said to be Diophantine m-tuple if each $a_i a_j + 1 (i \neq j)$ is the square of an integer. Fermat was the first to find a Diophantine 4- tuple, namely (1,3,8,120), and Baker and Davenport [2] showed that if (1,3,8,d) is a Diophantine 4- tuple, then d=120. More generally, for $k \geq 2$, the Diophantine triple $(k–1, k+1, 4k)$ extends uniquely to the Diophantine 4-tuple $\left(k-1, k+1, 4k, 16k^3 - 4k \right)$ and $k=2$ gives Fermat's example. No Diophantine 5- tuple is

known; however, $\left(\dfrac{777480}{8288641}, 1, 3, 8, 120 \right)$ is a Diophantine 5- tuple of rational numbers[1]. Many parametric families of Diophantine triples and 4- tuples are known [1,3,4]; for example,

$$\{(a,b,c): ab+1=q^2, c=a+b+2q\}, \qquad \ldots\ldots(1)$$

is a family of Diophantine triples for $0<a(b,q)0$, and

$$\{(a,b,c,d): ab+1=x2, bc+1=y2, ca+1=z2, d=a+b+c+2abc+2xyz\} \qquad \ldots\ldots(2)$$

is a family of Diophantine 4-tuples. Note that *(1)* shows how to pass from a Diophantine pair *(a,b)* to a Diophantine triple *(a,b,c)* and (2) shows how to pass from a Diophantine triple *(a,b,c)* to a Diophantine 4- tuple *(a,b,c,d)* Let us briefly consider the problem of extending a Diophantine pair *(a,b)* where $ab+1=q^2$, to a Diophantine triple *(a,b,c)*. We shall assume that *a* and *b* are coprime, and we require an integer *c* such that $ac+1=X^2$ and $bc+1=Y^2$ for some integers X and Y. If such a *c* exists, then X and Y must be solutions of the Diophantine equation $bX^2 = aY^2 + b - a$. Conversely, if X and Y are solutions of this equation, then a divides X^2-1 (because a and b are coprime) and so we can define c by $ac+1=X^2$ (equivalently, $bc+1=Y^2$ and(a,b,c) is a Diophantine triple. As the solutions of Diophantine equations behave erratically with respect to the coefficients, it, therefore, seems unlikely that we can parametrize all Diophantine triples of the form *(a,b,c)*. In this argument, the Diophantine triples in (1) correspond to the solutions $X=a+q$ and $Y=b+q$ Moreover, this argument enables one to make a reasonably efficient search for those values c for which *(a,b,c)* is a Diophantine triple. As an exceptional case, we encounter the formation of triples using two particular families of primes; twin primes and cousin primes. In this paper, we made an attempt to form a triple from the above said two prime numbers with respective properties D(2) and D(5).

Definition: 1 Twin Prime

A twin prime is a prime number that is either 2 less or 2 more than another prime number. In other words, a twin prime is a prime that has a prime gap of two. Sometimes the term "twin prime" is used for a pair of twin primes.

For example, (41, 43) are twin primes, since the difference between the two numbers 43-41=2. The first few twin prime pairs are (3,5), (5,7), (11,13), (17,19), (29,31).

Definition: 2 Cousin Prime

Cousin primes are prime numbers that differ by four that is, pair of cousin primes of the form (p,p+4). The first few cousin prime pairs are (3,7), (7,11), (13,17), (19,23), (37,41).

General Discussion and Results

Twin prime

Consider the pair $(TP_A, TP_B) = (p, p+2)$ of twin primes that are bound to satisfy the property $D(2): TP_A * TP_B + TP_A + TP_B + 2$ is a perfect square. Let Q_1 be a non-zero integer such that the equations $(TP_A + 1)Q_1 + TP_A + 2 = \sigma^2$;

$(TP_B + 1)Q_1 + TP_B + 2 = \tau^2$ are satisfied. These equations can be reduced to a pellian equation

$$\Omega^2 - (TP_A + 1)(TP_B + 1)\lambda^2 = 1 \qquad \ldots\ldots(3)$$

with basic solution *(p+2,1)* by applying the linear transformations $\tau = \Omega + (TP_B + 1)\lambda$ and $\sigma = \Omega + (TP_A + 1)\lambda$. Solving the pellian equation, we attain a Diophantine triple with the above-mentioned property as $(TP_A, TP_B, Q_1) = (p, p+2, 4p+7)$. Approaching for the extension of the pairs (TP_A, Q_1) and (TP_B, Q_1) we arrived at Diophantine triples

$$(TP_A, Q_1, Q_{A1}) = (p, 4p+7, 9p+14) \text{ and}$$
$$(TP_B, Q_1, Q_{B1}) = (p+2, 4p+7, 9p+20).$$

Keeping the first variable as TP_A and forming various pairs with second variables Q_{A1}, Q_{A2} .... we have reached a general expression for a triple satisfying the property D(2) as (TP_A, Q_N, Q_{N+1}) where $Q_N = (N+1)^2 p + N^2 + 4N + 2, N = 1, 2, 3$ .... Similarly, a series of triples can be generated with TP_B as the first variable and this has a general expression (TP_B, Q_N, Q_{N+1}) where $Q_N = (N+1)^2 p + 3N^2 + 4N, \ N = 1, 2, 3, \ldots$.

For further insights into the extension of triples, we consider the following theorems.

Theorem 1

Let $\Omega^2 - (TP_A + 1)(TP_B + 1)\lambda^2 = 1$ be the Diophantine equation, where (TP_A, TP_B) be a pair of twin primes. Then

a. The fundamental solution of $\Omega^2 - (p^2 + 4p + 3)\lambda^2 = 1$ is $(\Omega_1, \lambda_1) = (p+2, 1)$.

b. Define the string (Ω_n, λ_n), where
$$\begin{pmatrix} \Omega_n \\ \lambda_n \end{pmatrix} = \begin{pmatrix} p+2 & p^2+4p+3 \\ 1 & p+2 \end{pmatrix}^n \begin{pmatrix} 1 \\ 0 \end{pmatrix} \text{ for } n \geq 1.$$

Then (Ω_n, λ_n) is a solution of

$$\Omega^2 - (p^2 + 4p + 3)\lambda^2 = 1.$$

c. The points (Ω_n, λ_n) satisfy the recurrence relations $\Omega_n - (2\Omega_1)\Omega_{n-1} + \Omega_{n-2} = 0$;

$\lambda_n - (2\Omega_1)\lambda_{n-1} + \lambda_{n-2} = 0$, for $n \geq 4$.

Proof:

(a)

Since $\Omega_1^2 - (p^2 + 4p + 3)\lambda_1^2$

$$= (p+2)^2 - (p^2 + 4p + 3)(1)^2$$

$$= 1$$

is satisfied for the values

$(\Omega_1, \lambda_1) = (p+2, 1)$, (a) is proved.

b. The approach of Mathematical induction is employed here. Consider

$$\begin{pmatrix} \Omega_n \\ \lambda_n \end{pmatrix} = \begin{pmatrix} p+2 & p^2 + 4p + 3 \\ 1 & p+2 \end{pmatrix}^n \begin{pmatrix} 1 \\ 0 \end{pmatrix}.$$

Let n=1. We get $(\Omega_1, \lambda_1) = (p+2, 1)$ which is a solution of (3). Assume that the result is true for n-1, (ie) $\Omega_{n-1}^2 - (p^2 + 4p + 3)\lambda_{n-1}^2 = 1$. The general solution (Ω_n, λ_n) can be expressed as

$$\begin{pmatrix} \Omega_n \\ \lambda_n \end{pmatrix} = \begin{pmatrix} p+2 & p^2 + 4p + 3 \\ 1 & p+2 \end{pmatrix}\begin{pmatrix} \Omega_{n-1} \\ \lambda_{n-1} \end{pmatrix} \quad ....(4)$$

To check for its solvability, we have

$\Omega_n^2 - (p^2 + 4p + 3)\lambda_n^2$

$$= \left[(p+2)(\Omega_{n-1}) + (p^2 + 4p + 3)(\lambda_{n-1})\right]^2 -$$

$$(p^2 + 4p + 3)\left[\Omega_{n-1} + (p+2)\lambda_{n-1}\right]^2$$

$$= \Omega_{n-1}^2 - (p^2 + 4p + 3)\lambda_{n-1}^2$$

$$= 1$$

Thus the result is valid for n, so that (Ω_n, λ_n) is a solution of (3)

c. We have to prove that Ω_n and λ_n satisfy the recurrence relations.

From (4), for the particular values of n=1,2,3,4, the solution set is found to be

n	Ω_n	λ_n
1	$p+2$	1
2	$2p^2 + 8p + 7$	$2p + 4$
3	$4p^3 + 24p^2 + 45p + 26$	$4p^2 + 16p + 15$
4	$8p^4 + 64p^3 + 184p^2 + 224p + 112$	$8p^3 + 48p^2 + 92p + 56$

The solution set satisfies the recurrence relations $\Omega_n - (2\Omega_1)\Omega_{n-1} + \Omega_{n-2} = 0$;

$\lambda_n - (2\Omega_1)\lambda_{n-1} + \lambda_{n-2} = 0$, for $n \geq 4$.

To Check the Extension to Quadruple

Let R be any non-zero integer such that

$$pR + R + p + 2 = l^2 \qquad .... (5)$$

$$(p+2)R + R + p + 4 = m^2 \qquad ....(6)$$

$$(4p+7)R + R + 4p + 9 = n^2 \qquad ....(7)$$

Eliminating "R" from (6) & (7) and applying the transformations $m = \Omega + (p+3)\lambda$; $n = \Omega + (4p+8)\lambda$, the pellian equation $\Omega^2 - (4p+8)(p+3)\lambda^2 = 1$ is generated with initial solution $(2p+5, 1)$. This on substitution yields the values for $m = 3p + 8$ and $R = 9p + 20$.

The LHS of (5) can be expanded by these values and thus obtain as $9p^2 + 30p + 22$ which is never a perfect square for all values of P. Thus the possibility of extension to a quadruple with property $D(2)$ can be discarded.

Cousin Prime

To check the extendability of the pair of cousin primes $(CP_A, CP_B) = (p, p+4)$ satisfying the property $D(5): CP_A * CP_B + CP_A + CP_B + 5$, we follow the methods as above and thus form a system of equations, $(CP_A + 1)Q_2 + CP_A + 5 = \sigma^2$;

$\left(CP_B+1\right)Q_2+CP_B+5=\tau^2$ where Q_2 is any non-zero integer. Hence the Diophantine triple $\left(p, p+4, 4p+11\right)$ is deduced.

A series of triplets

$$\left(CP_A, Q_2, Q_{A1}\right), \left(CP_B, Q_2, Q_{B1}\right) \cdots\cdots$$

are thus derived giving a general interpretation

$$\left(CP_A, Q_N, Q_{N+1}\right);$$

$$Q_N = \left(N+1\right)^2 p + N^2 + 6N + 4, N = 1, 2, 3$$

....and

$$\left(CP_B, Q_N, Q_{N+1}\right);$$

$$Q_N = \left(N+1\right)^2 p + 5N^2 + 6N,$$

$$N = 1, 2, 3 \dots.$$

Remark

The existence of a quadruple considering the cousin primes can be dismissed:

For,

Let R be any non-zero integer such that

$$pR + R + p + 5 = l^2 \ \dots \ (8)$$

$$\left(p+4\right)R + R + p + 9 = m^2 \ \dots.. \ (9)$$

$$\left(4p+11\right)R + R + 4p + 16 = n^2 \ \dots. \ (10)$$

Eliminating "R" from (9) & (10) and applying the transformations $m = \Omega + \left(p+5\right)\lambda$; $n = \Omega + \left(4p+12\right)\lambda$, the pellian equation $\Omega^2 - \left(4p+12\right)\left(p+5\right)\lambda^2 = 4$ is generated with

initial solution $\left(2p+8,1\right)$. This on substitution yields the values for $m = 3p+13$ and $R = 9p+32$.

The LHS of (8) can be expanded by these values and thus obtained as $9p^2 + 42p + 37$ which is never a perfect square for all values of P. Thus the possibility of extension to a quadruple with property $D(5)$ can be discarded.

Conclusion

Considering the pair of twin primes and cousin primes separately, we made an attempt for the extension of triples with property $D(2)$ and $D(5)$ respectively. One may search for other primes and other properties for a possible extension to triples and quadruples.

References

Arkin, J., Hoggat, V. E., and Straus, E. G., (1979). On Euler's solution of a problem of Diophantus, Fibonacci Quart.17 pp. 333–339.

Baker, A., and Davenport, H., (1969). The equations $3x^2 - 2 = y^2$ and $8x^2 - 7 = z^2$, Quart.J.Math. (Oxford ser.92) 20, pp. 129–137.

Dujella, A., (1996). Generalized Fibonacci numbers and the problem of Diophantus, Fibonnaci Quart. 34, pp. 25–30.

On The Non-isolated Resolving Number of Product Graphs and Some Families of Graphs

R. Ananthakumar

Faculty, MEPCO Schlenk Engineering College(Autonomous), Sivakasi, Tamil Nadu, India. Pin Code: 626005
E-mail: ananthakumar1983@gmail.com, ananthakumar@mepcoeng.ac.in

Abstract

For set $W = \{W_1, W_2, ..., W_k\}$ (order is imposed) of nodes and a node v in a connected graph G. $code(v) = \left(d(v, w_1), d(v, w_2), ..., d(v, w_k)\right)$ is the symbol of v (k-vector) with respect to W, where $d(x,y)$ represents the distance between the nodes x and y. If distinct nodes in $V(G)$ have distinct codes, then W is said to be a revolving set of G. The resolving set with minimum number of nodes of G is the dimension of G ($dim(G)$). A revolving set W is a non-isolated resolving set if the sub-graph induced by W has no K_1 components. The minimum number of nodes of a non-isolated resolving set of G is called the non-isolated resolving number of a graph G and is denoted by $nr(G)$. Here we investigate the value of $nr(G)$ of graphs obtained from Corona, Lexicographic, and join of any two graphs. Also, non-isolated resolving numbers for some families of graphs are discussed.

Keywords: Resolving, Non-isolated resolving, $nr(G)$

Introduction

We use the notations n and m to denote the order and size of a graph respectively. Also, we assume graph G means, simple and connected. Here we introduce the following notations for convenience.

R-set for Resolving set
NR-set for Non-isolated resolving set.
$d(u,v)$ for the usual distance between nodes u and v in a connected graph G.

Notice that, in each connected graph G and each ordered set $W = \{W_1, W_2, ..., W_k\}$ of nodes of G, the ith coordinate of $code$ (W_i) is 0 and the ith coordinate of all other node representations is positive, implying that, while investigating whether distinct nodes receive distinct codes with respect to an ordered subset W of $V(G)$ for G, we consider only the nodes of $V(G)–W$. In general, the idea NR-set is existing for all graphs G, with an order at least three since $V(G)–\{v\}$ is an NR-set, where the node v is any antipodal node of G.

The idea of metric dimension of graph was initially studied by P.J. Slater (1976) and (1988) and Harary and Melter (1976). Various resolving concepts are created by merging resolving concepts with other graph-theoretic structures like connectivity, independence, or acyclic (Saenpholphat and Zhang, 2004). In Jeya Bala Chitra, S. Arumugam (2015) and J. Paulraj Joseph and N. Shunmugapriya, 2015) the idea of total R- set was studied. Here, we study NR-sets for product graphs and few families of graphs.

We refer C. Hernando, M. Mora, I. M. Pelayo and C. Seara (2010) for the concepts of twin nodes and twin graph.

Main Results

Here we found the value of $nr(G)$ where G is product graphs such as Lexicographic, Corona, comb, and join of two graphs, also upper bound for unicyclic graphs are discussed.

Now we define the notion called distinct neighbor number of a graph, and it is useful in finding non-isolated resolving number of product graphs.

Definition 1: Let G be any non-complete graph. A set $S \subseteq V(G)$ is said to be a *distinct neighbor set* of graph if for each couple $x, y \in V - S$, there exist $z \in S$ s.t $z \in N(x)$ and $z \notin N(y)$, or $z \in N(y)$ and $z \notin N(x)$. The least number of nodes in a distinct neighbor set of G is called *distinct neighbor number* for graph G and it is represented by $dn(G)$.

The distinct neighbor set exists for all non-complete graphs, for let a-b-c be any geodesic of length two joining a and b in G, then $V - \{a, b\}$ become a distinct neighbor set, since $bc \in E$ and $ac \notin E$. Hence $dn(G) \leq n - 2$. Also the property of a set being distinct neighbor set is super hereditary, that is if S is any distinct neighbor set then $S \cup \{v\}$ is also distinct neighbor set for all $v \in V - S$.

We now present distinct neighbor numbers of some standard graphs.

Example 2

a. For Path P_n

$$dn(P_n) = \begin{cases} 2 + 2k \text{ if } n \equiv 0 \,(\text{mod} 5) \\ 3 + 2k \text{ if } n \equiv 1 \text{ or } 2 \text{ or } 3 \,(\text{mod } 5) \\ 4 + 2k \text{ if } n \equiv 4 \,(\text{mod} 5) \end{cases}$$

b. For cycle C_n,

$$dn(C_n) = \begin{cases} 2 + 2k \text{ if } n \equiv 0 \text{ or } 1 \,(\text{mod} 5) \\ 3 + 2k \text{ if } n \equiv 2 \text{ or } 3 \,(\text{mod } 5) \\ 4 + 2k \text{ if } n \equiv 4 \,(\text{mod } 5) \end{cases}$$

Where $k = \dfrac{n-5}{5}$, for all $n \geq 5$ also $dn(P_3) = 1$ and $dn(P_4) = 2$.

c. For Star $K_{1,n}$, $dn(K_{1,n}) = n - 1$

d. For complete bipartite graph K_{n_1, n_2}, $dn(K_{n_1, n_2}) = n_1 + n_2 - 2, n_1, n_2 \geq 3$.

e. $dn(K_n - e) = n - 2$, for all $n \geq 4$, where K_n is a complete graph.

Example 3: Let G, H be any connected graphs and let order of G be n_1. Let $G \lhd_o H$ be comb product of G and H. Then

a. $nr(G \lhd_o C_n) = 2 \cdot n_1$.

b. $nr(G \lhd_o K_n) = n_1 \cdot (n-2)$.

c. $nr(G \lhd_o K_{m,n}) = n_1 \cdot (m + n - 3)$.

d. $nr(G \lhd_o P_n) = 2 \cdot n_1$, if the node o is not a leaf of path P_n.

Theorem 4: Let the unicyclic graph be G with cycle $C : (v_1, v_2, \ldots, v_k, v_1)$ and let $T_1, T_2, \ldots, T_t$ be the trees of G identified to nodes $v_{i1}, v_{i2}, \ldots, v_{it}$ of cycle C respectively. Then $nr(G) \leq 2 + 2t + \sum\limits_{i=1}^{t} nr(T_j')$, where T_j' is the tree induced by the nodes $V(T_j) \cup \{v_{ij}\}$ in G.

Proof. First we construct a NR-set as follows. Set $W = \{v_1, v_2\} \cup \left(\bigcup_{j=1}^{t} W_j'\right) \cup \left(\bigcup_{j=1}^{t} U_j\right)$, where w_j' is a minimum order NR-set of tree T_j' and $U_j = \{v_{ij}, w_j\}$ where w_j is a non-pendant node of T_j adjacent to v_{ij} if exist, else T_j' become P_2 or K_{1,n_1} In this case, let $V(K_{1,n_1}) = \{v_{ij}, u_{j1}, u_{j2}, \ldots, u_{jn}\}$ and without loss of generality assume $W_j' = \{v_{ij}, u_{j1}, u_{j2}, \ldots, u_{j(n-1)}\}$, now choose $U_j = \{u_{jn}\}$. Then $|W| \leq 2 + 2t + \sum\limits_{j=1}^{t} nr(T_j')$. The nodes v_1 and v_2 are adjacent in cycle, implies v_1 and v_2 are adjacent in <W>, also each W_j' are NR-sets, and the nodes in U_j are adjacent, we have <W> has no isolates in G.

Now any couple of nodes in cycle C are resolved by either v_1 or v_2. And any couple of nodes in $V(T_j')-W_j'$ is resolved by some node of W_j' only since $d_{T_j'}(x,y)=d_G(x,y) \ \forall x,y \in V(T_j')$, $(CP_4+1)Q_2$ Now let x in $V(T_i)-W_i$ and y in $V(T_m)-W_m$, or y in $V(C)-\{v_1,v_2\}$. Suppose both the nodes v_1 and v_2 are not resolved the couple *(x,y)*. Then we assume that the geodesics joining x and v_1, y and v_1 contains $v_{il} \in V(C)$ as a common node and $d(x,v_{il})=d(y,v_{il})$. In this case $d(x,w_l)+2=d(y,w_l)$, implies $w_j \in U_l$ resolves the couple *(x,y)* implies W resolves all pair of nodes. Hence $nr(G) \le 2+2t+\sum_{j=1}^{t} nr(T_j')$.

Theorem.5 Let G be any nontrivial graph and H be any non-complete graph. Then $nr(G*H)=n.$ $dn(H)$, $G*H$ is the Lexicographic of G and H, where $n=|V(G)|$.

Proof. Let $V(G) = \{v_1, v_2, ..., v_n\}$ and $V(G) = \{u_1, u_2, ..., u_n\}$, $H_i = \langle\{(v_i,u_j):1 \le j \le n_1\}\rangle$ be the duplicates of H in $G*H$ corresponding to v_i. Then for any two distinct nodes $(v_i,u_r),(v_j,u_s)$ in $G*H$, we have

$$d_{G*H}((v_i,u_r),(v_j,u_s)) = \begin{cases} d_G(v_i,v_j) & if \ i \ne j \\ 1 \ if \ i=j \ and \ u_r u_s \in E(H) \\ 2 & otherwise \end{cases}$$

Let W' be any R-set. Now for all couples of nodes x,y in $V(H_i)$, we have $d(x,z) = d(y,z)$ for all $z \in V(G*H)-V(H_i)$, no node in $V(G*H)-V(H_i)$ resolves any couple of nodes in $V(H_i)$ which implies that nodes from $V(H_i) \cap W'$ must resolve all the couples of nodes in $V(H_i)-W'$. Let $W_i' = V(H_i) \cap W'$ and let $x,y \in V(H_i)-W'$. Then $d(x,y) = 1$ or 2. Since some nodes say z of W_i' resolves the couple *(x,y)*, implies that xz in E an $yz \notin E$ or $yz \in E$ and

$xz \notin E$ is the only possibility. This shows that W_i' must be a distinct neighbor set of H_i, and hence $|W'| \ge |\bigcup_{i=1}^{n} W_i'| = n \cdot dn(H)$. In $G*H$ the induced graph $\langle \bigcup_{i=1}^{n} S_i \rangle$ is connected, where S_i is any minimum cardinality of the distinct neighbor set of H_i corresponding to S in $V(H)$ for all $i=1,2,3,...n$ implies that $nr(G*H) \ge n \cdot dn(H)$.

Let W_i be any *dn*-set of H_i (the copy of H) and let $W = \bigcup_{i=1}^{n} W_i$. Then $|W|=n.dn(H)$. Now we investigate that W resolves nodes of $G*H$. Let G^* be the twin graph of G and $V(G^*) = \{v_1^*,v_2^*,v_3^*...,v_n^*\}$ say. Let $x=(v_i,u_r), y=(v_j,u_r) \in V(G*H)-W$. Suppose $v_i^* \ne v_j^*$. Then there exists a node $v_k \in V(G)$ such that $d(v_i,v_k) \ne d(v_j,v_k)$, else the nodes v_i,v_j belongs to same distance similar equivalence class. Now $d(x,z) \ne d(y,z)$, for all z in $V(H_k) \cap W$, where H_i is the copy corresponding to v_k in $V(G)$ and hence $code(x) \ne code(y)$. Now we assume that $v_i^* = v_j^*$.

Case 1. $\langle v_i^* \rangle \cong K_{n_1}^c$ for some n_1.

Let $x=(v_i,u_r) \in V(H_i)-W$ and $y=(v_j,u_s) \in V(H_j)-W$. If $r=s$, then by definition of *dn*-set, there exists $z \in W_i$ s.t $xz \in E(H_i)$ and hence $d(x,z) = 1$ but $d(y,z) = 2$ in $G*H$ as $\langle v_i^* \rangle$ is independent, $code(x) \ne code(y)$. If $r \ne s$, then as W_i is *dn*-set of $V(H_i)$, there exists $z=(v_i,u_l) \in W_i$ with the one of the property, (i) $z \in N(x) \cap W_i$ but $z \notin N(v_i,u_s) \cap W_i$, implies that $d(x,z)=1$ but $d(y,z)=2$ or (ii) $z \in N(v_i,u_s) \cap W_i$ but $z \notin N(x) \cap W_i$, implies that $d((v_j,u_l),y)=1$ but $d((v_j,u_l),x)=2$ and hence $code(x) \ne code(y)$.

Case 2. $\langle v_i^* \rangle \cong K_{n_1}$ for some n_1.

Let $x=(v_i,u_r) \in V(H_i)-W$ and $y=(v_j,u_s) \in V(H_j)-W$. If $r=s$, then by definition of *dn*-set, there exists $z \in W_i$ such that $z \notin N(x)$ and hence $d(y,z)=1$ but $d(x,z)=2$ in $G*H$ as $\langle v_i^* \rangle$ is complete implies $code(x) \neq code(y)$. If $r \neq s$, then as W_i is dn-set of $V(H_i)$, there exist $z \in N(x) - W_i$ such that $xz \notin E$ and hence $d(y,z)=1$ but $d(x,z)=2$ as $\langle v_i^* \rangle$ is complete. Hence W is an R-set, which implies $nr(G*H) = n \cdot dn(H)$.

Remark 6 From the proof of above Theorem.5 we can conclude the followings:

(i) Let G be a graph of order n and $H = \bigcup_{i=1}^{k} H_i$ with each component H_i is not a complete graph. Then $nr(G*H) = n \cdot \sum_{i=1}^{k} dn(H_i)$.

(ii) Let $G_1, G_2, \dots, G_k$ be connected non-complete graphs. Then $nr(G_1 + G_2 + \cdots + G_k) = \sum_{i=1}^{k} dn(G_i)$, where $G_1 + G_2 + \dots + G_k$ is the sequential join of $G_1, G_2, \dots, G_k$.

Theorem 7: Let G be a graph of order n_1 and H be any non-complete connected graph. In this case $nr(G \circ H)$ is $n_1 \cdot dn(H)$ or $n_1 + (n_1 \cdot dn(H))$, where $G \circ H$ is the corona of G and H.

Proof. Let $V(G) = \{v_1, v_2, \dots, v_{n_1}\}$ and let $H_1, H_2, \dots, H_{n_1}$ be the duplicates of H such that v_i is adjacent to all the nodes of H_i for all $i = 1, 2, \dots, n_1$ respectively. Let W be any R-set of G. In $G \circ H$, for every couple of nodes x,y in $V(H_i)$, $d(x,u) = d(y,u)$ for all $u \in V(G \circ H) - V(H_i)$. Also $d(w,z) = 1$ or 2 for all $w,z \in V(H_i)$ for $i = 1, 2, \dots, n_1$ (since $w - v_i - z$ is a path of length two). As W resolves vertices of G, for each couple (x,y) of nodes $x, y \in V(H_i) - W$ there exists a $z \in V(H_i) \cap W$ such that $d(x,z)$ is one and $d(y,z)$ is two, or $d(x,z)$ is two and $d(y,z)$ is one. That is in H_i, the set $V(H_i) \cap W$ is a distinct neighbor set for each $i = 1, 2, \dots, n_1$.

Let $W' = \left(\bigcup_{i=1}^{n_1} S_i \right) \cup V(G)$, where S_i is any minimum order distinct neighbor set of H_i. Then

S_i resolves all couples of nodes in $V(H_i) - S_i$, $y \in V(H_j) - S_j$, $d(x,v_i) = 1$ and $d(y,v_j) \geq 2$. Hence $nr(G \circ H) \leq n_1 + (n_1 \cdot dn(H))$.

Suppose H admits a distinct neighbor set S in which $<S>$ has no isolates, then $\bigcup_{i=1}^{n} S_i$ is a minimum order NR-set of $G \circ H$, where S_i is a minimum order of distinct neighbor set of H_i corresponding to S in $V(H)$ for all $i = 1, 2, \dots, n$ implies that $nr(G \circ H) = n_1 \cdot dn(H)$.

On the other hand, assume that all the minimum cardinality neighbor set S of H has isolated in $<S>$, then $W' = \left(\bigcup_{i=1}^{n} S_i \right) \cup V(G)$ is a minimum order NR-set of $G \circ H$, as $\langle W' \rangle$ is connected and hence $nr(G \circ H) = n_1 + (n_1 \cdot dn(H))$.

Corollary.8 From Theorem.7 we conclude the followings:

i. Suppose H is any of graph given in Example. 2, in this case $nr(G \circ H)$ can be computed immediately using Theorem.7 and Example.2

ii. Also, by the arguments given in proof of Theorem 7, we have $nr(H + K_1) \leq dn(H) + 1$, where H is a non-complete graph.

iii. In the above theorem if we allow complete components $H_{k+1}, H_{k+2} \dots, H_t$, to H, then one can check the value of $nr(G \circ H)$ is

$$n_1 \cdot \sum_{i=1}^{k} dn(H_i) + \sum_{i=k+1}^{t} \left(|V(H_i)| - 1 \right) \text{ or}$$

$$n_1 \cdot \sum_{i=1}^{k} dn(H_i) + \sum_{i=k+1}^{t} \left(|V(H_i)| - 1 \right) + n_1.$$

Remark 9 From (ii) of Corollary.8 and (a) of Example 2 we have

a. For Path P_n

$$nr(P_n + K_1) = \begin{cases} 2 + 2k + 1 & if \ n \equiv 0 \ (\mathrm{mod}\,5) \\ 3 + 2k + 1 & if \ n \equiv 1 \ or \ 2 \ or \ 3 \ (mod\,5) \\ 4 + 2k + 1 & if \ n \equiv 4 \ (\mathrm{mod}\,5) \end{cases}$$

b. For cycle C_n,

$$nr(C_n + K_1) = \begin{cases} 2 + 2k + 1 & if\ n \equiv 0\ or\ 1\ (mod\ 5) \\ 3 + 2k + 1 & if\ n \equiv 2\ or\ 3\ (mod\ 5) \\ 4 + 2k + 1 & if\ n \equiv 4\ (mod\ 5) \end{cases}$$

Where $k = \dfrac{n-5}{5}$, for all $n \geq 5$

But in (Jeya Bala Chitra, S. Arumugam (2015)) it is given that $nr(P_n + K_1) = \dfrac{n}{2}$, which is wrong, because if we choose

$$W = \{u, v_1, v_5\} \cup \left(\bigcup_{i=1}^{k} \{v_{5i+2}, v_{5i+5}\} \right) \subseteq V(P_n + K_1)$$

$$= (V(P_n) = \{v_1, v_2, \dots, v_n\}) \cup (V(K_1) = \{u\}).$$

Then one can check W is a R-set and $<W>$ is connected (here graph induced by W, $<W>$ is a star) in $P_n + K_1$, and $nr(P_n + K_1) < \dfrac{n}{2}$. Suppose if $n = 20000$, then $\dfrac{20000}{2} = 10000$, but from (a.) of this remark $nr(P_{20000} + K_1) = 8000$.

The following are some problems for future research.

Problem.1 Characterize graphs G, which admits a minimum order distinct neighbor set S with $<S>$ has no isolates in G.

Problem.2 For given integer k, characterize graphs G with $dn(G) = k$.

Conclusion: Exited by the parameter non-isolated resolving sets studied in Jeya Bala Chitra, S. Arumugam (2015) and J. Paulraj Joseph and N. Shunmugapriya (2015), in this article $nr(G)$ of some classes of graphs and graphs obtained from product is found. In this article the value of $nr(P_n + K_1)$ was improved compared to this value that appeared in Jeya Bala Chitra, S. Arumugam (2015). Study of necessary and sufficient conditions for graphs G with $nr(G) \in \{2, n-2, n-3\}$, are attractive for future research. Also, the study of $dn(G)$ of the general graph will be useful for finding the exact value of a non-isolated resolving number of product graphs. Also, it is a good start to study how the removal or addition of vertex or edge of a graph G affects the value of $nr(G)$, and compare $nr(G)$ with other dimension parameters of graph. Also one can investigate this value for special class of graphs, graphs that admit exactly one NR-set.

Acknowledgment

The Author is grateful to his college, and college management for the support, and thankful to Research Project funding agency DAE-NBHM, Mumbai.

References

Slater, P. J. (1988). Dominating and reference sets in graphs. Journal of Mathematical Physical science. 22, 445–455.

Slater, P. J. (1976). Leaves of trees. Congress Numeruntum. 14, 549–559.

Harary, F., and Melter (1976). On metric dimensions of graphs. Ars Combinatorica. 2, 191–195.

Hernando, C., Mora, M., Pelayo, I. M., and Seara, C. (2010). Extremal Graph Theory for Metric Dimension and Diameter. Electronic Journal of. Combinatorics. 17: $\#$R30.

Saenpholphat, V. and Zhang, P. (2004). Conditional Resolvability in Graphs: A Survey. International Journal of Mathematics. (38), 1997–2017.

Chitra, Jeya Bala, Arumugam, S. (2015).

Resolving Sets Without Isolated nodes. Procedia Computer.Science (74), 38–42.

Paulraj Joseph, J. and Shunmugapriya, N. (2015). Total Resolving Number of Graphs. International journal of pure and applied mathematics. 3(57), 323–343.

Entropy and Distance Measures of Bipolar Pythagorean Neutrosophic Soft Set

S. Anitha,[a] and A. Francina Shalini[b]

[a]Research Scholar, Nirmala College for Women, Coimbatore, India
[b]Department of Mathematics, Nirmala College for Women, Coimbatore, India
E-mail: anitharaj.ps@gmail.com; francshalu@gmail.com

Abstract

In this paper, the idea of entropy and distance measures are defined for the Bipolar Pythagorean neutrosophic soft sets (BPNSS). Examples are given to show the feasibility and discussion was provided for the proposed methods to find the accurate measure.

Keywords: Bipolar Pythagorean neutrosophic soft sets, Pythagorean fuzzy sets, Distance measure, Entropy

Introduction

Considering the imprecision in decision-making, Zadeh [23] introduced the idea of fuzzy set and studied membership function. The concept of neutrosophy was introduced by F. Smarandache [17,18,19] in which he developed the degree of indeterminacy. Bipolar-valued fuzzy sets, an extension of fuzzy set and their operation was coined in 2000 by Lee[13]. Deli et al. [7] developed bipolar neutrosophic sets and study their application in decision-making. The concept of soft set theory as a new mathematical tool was initiated by Molodtsov[15] in 1999 and presented the fundamental results of soft sets. Bipolar soft sets and bipolar fuzzy soft sets are studied by Aslam et al.[2]. Neutrosophic soft sets were studied by some authors in [4,5,6,12,13,16]. To deal with vagueness considering the membership grade and non-membership grade satisfying the conditions $\mu + \nu \leq 1$ or s $\mu + \nu \geq 1$, and also, it follows that $\mu^2 + \nu^2 + \pi^2 = 1$, Pythagorean fuzzy set and subsets were discussed by Xindong [21] and Yagar [22]. The notion of bipolar neutrosophic soft set was introduced by Ali et al.[3] in 2017. The concept of the Pythagorean neutrosophic set was introduced by Jansi [10]. The concept of bipolar Pythagorean neutrosophic soft set was developed by Anitha [1].

In order to find the distances accurately various methods were proposed to measure the distances and some of them have ideal effect in its classification. Many authors discussed and applied the entropy, distance and similarity measures in the literature [8,9,20]. The commonly used distance measures are Hamming distance[20], Euclidean distance[20] and so on. In this paper, we propose an entropy and distance measure between bipolar Pythagorean neutrosophic soft sets and numerical examples were given. The organization of the paper is as follow. In Section 2, some preliminary concepts are introduced. In Section 3, we proposed entropy measures between bipolar Pythagorean neutrosophic soft sets. In Section 4, generalizations of Hamming distance, Euclidean distance, and their normalized versions for the bipolar Pythagorean neutrosophic soft sets are given. In Section 5, numerical examples and discussion were provided for the proposed methods. Finally, in Section 6, the conclusion and future works are stated.

Preliminaries

Definition: 1[11] Let U be a universe. A *bipolar fuzzy set* A in U is defined as: $A = \left\{ \left(u, T^+\left(u\right), T^-\left(u\right) \right) : u \in U \right\}$ where $T^+ : X \to [0,1]$ and $T^- : X \to [-1,0]$. The positive membership degree $T^+(u)$ denotes the truth membership corresponding to a bipolar fuzzy set A and the negative membership of an element $u \in U$ to some implicit counter-property corresponding to a bipolar fuzzy set A.

Definition: 2[14] Let X be a non-empty set. A *bipolar Pythagorean fuzzy set* (BPFS)

$$A = \left\{ \left(x, T_A^P, F_A^P, T_A^N, F_A^N \right) : x \in X \right\}$$

where $T_A^P : X \to [0,1]$, $F_A^P : X \to [0,1]$,

$T_A^N : X \to [0,1]$, $F_A^N : X \to [0,1]$ are the mappings

such that $0 \leq \left(\left(T_A^P(x) \right)^2 + \left(F_A^P(x) \right)^2 \right) \leq 1$ and $-1 \leq$

$-\left(\left(T_A^N(x) \right)^2 + \left(F_A^N(x) \right)^2 \right) \leq 0$ and

Definition: 3[1] Let U be a universe and E be a set of parameters. A *bipolar Pythagorean neutrosophic soft set (BPNSS)*

$$\mathbb{A} = \left\{ \left(e, \left\{ \begin{pmatrix} u, T^+(u), I^+(u), F^+(u), \\ T^-(u), I^-(u), F^-(u) \end{pmatrix} : u \in U \right\} \right) : e \in E \right\}$$

where $T^+ : X \to [0,1]$, $I^+ : X \to [0,1]$, $F^+ : X \to [0,1]$,

$T^- : X \to [-1,0]$, $I^- : X \to [-1,0]$, $F^- : X \to [-1,0]$

are the mappings such that

$0 \leq \left(\left(T^+(u) \right)^2 + \left(I^+(u) \right)^2 + \left(F^+(u) \right)^2 \right) \leq 2$ and

$-2 \leq -\left(\left(T^-(u) \right)^2 + \left(I^-(u) \right)^2 + \left(F^-(u) \right)^2 \right) \leq 0$.

Entropy Measure of BPNSS

Definition 1 Let U be a universe and E be a subset of a parameter A. Let $B_1. = (G_A, E)$. and $B_2. = (H_A, E)$ two BPNSS. The mapping $E : \text{BPNSS}(U) \to \mathbb{R}^+ \cup \{0\}$ tated as bipolar Pythagorean neutrosophic soft entropy if it satisfies:

i. E(B) = 0 if and only if B ∈ IFSS(U)

ii. E(B) is maximum if and only if

$T^{2^+}{}_{G_A(e)}(\text{u}) = I^{2^+}{}_{G_A(e)}(\text{u}) = F^{2^+}{}_{G_A(e)}(\text{u})$ and

$T^{2^-}{}_{G_A(e)}(\text{u}) = I^{2^-}{}_{G_A(e)}(\text{u}) = F^{2^-}{}_{G_A(e)}(\text{u})$ for

all $e \in E$ and $u \in U$.

iii. $E(B) = E(B^c)$ for all $B \in \text{BPNSS}(U)$

iv. $E(B_1) \leq E(B_2)$ if $\mathcal{B}_2 \subseteq \mathcal{B}_1$

Definition 2 For the BPNSS B. The bipolar Pythagorean neutrosophic soft entropy of E is defined as follows:

$$E(B) = 1 - \frac{1}{2mn} \sum_{i=1}^{m} \sum_{j=1}^{n} \begin{bmatrix} \left(T^{2^+}{}_{B(e_j)}(u_i) + F^{2^+}{}_{B(e_j)}(u_i) \right). \\ \left| I^{2^+}{}_{B(e_j)}(u_i) - I^{2^+}{}_{B^c(e_j)}(u_i) \right| \\ -\left(T^{2^-}{}_{B(e_j)}(u_i) + F^{2^-}{}_{B(e_j)}(u_i) \right). \\ \left| I^{2^-}{}_{B(e_j)}(u_i) - I^{2^-}{}_{B^c(e_j)}(u_i) \right| \end{bmatrix}$$

Example 3 Let $U = \{u_1, u_2, u_3, u_4\}$ be a universe set and $E = \{e_1, e_2, e_3\}$ be a set of parameters. Let $A = \{e_1, e_2\}$ be a subset of E.

1. Define $B_1 = (G_A, E) = \{e_1, G_A(e_1), e_2, G_A(e_2)\}$ where,

$$G_A(e_1) = \begin{cases} u_1, 0.50, 0, 0.50, -0.50, 0, -0.50, \\ u_2, 0.60, 0, 0.40, -0.20, 0, -0.80, \\ u_3, 0.10, 0, 0.90, -0.60, 0, -0.40, \\ u_4, 0.30, 0, 0.70, -0.40, 0, -0.60 \end{cases}$$

$$G_A(e_2) = \begin{cases} u_1, 0.30, 0, 0.70, -0.70, 0, -0.30, \\ u_2, 0.80, 0, 0.20, -0.20, 0, -0.80, \\ u_3, 0.20, 0, 0.80, -0.10, 0, -0.90, \\ u_4, 0.30, 0, 0.70, -0.40, 0, -0.60 \end{cases}$$

B_1 becomes intuitionistic fuzzy soft set since all the indeterminacy values are zero. Therefore by definition 3.2 E(B) = 0.

2. $B_2 = (G_A, E) = \{e_1, G_A(e_1), e_2, G_A(e_2)\}$ where,

$$G_A(e_1) = \begin{cases} u_1, 0.50, 0.50, 0.50, -0.50, -0.50, -0.50, \\ u_2, 0.40, 0.40, 0.40, -0.20, -0.20, -0.20, \\ u_3, 0.60, 0.60, 0.60, -0.40, -0.40, -0.40, \\ u_4, 0.30, 0.30, 0.30, -0.60, -0.60, -0.60 \end{cases}$$

$$G_A(e_2) = \begin{cases} u_1, 0.20, 0.20, 0.20, -0.70, -0.70, -0.70, \\ u_2, 0.80, 0.80, 0.80, -0.20, -0.20, -0.20, \\ u_3, 0.50, 0.50, 0.50, -0.10, -0.10, -0.10, \\ u_4, 0.40, 0.40, 0.40, -0.30, -0.30, -0.30 \end{cases}$$

E(B) = 1 that is maximum by definition 3.2, since all the membership values are equal.

3. Define $B_3 = (G_A, E) = \{e_1, G_A(e_1), e_2, G_A(e_2)\}$

where,

$$G_A(e_1) = \begin{cases} u_1, 0.50, 0.60, 0.40, -0.50, -0.90, -0.20, \\ u_2, 0.60, 0.80, 0.30, -0.60, -0.70, -0.20, \\ u_3, 0.70, 0.70, 0.60, -0.40, -0.60, -0.20, \\ u_4, 0.30, 0.80, 0.80, -0.50, -0.50, -0.30 \end{cases}$$

$$G_A(e_2) = \begin{cases} u_1, 0.70, 0.40, 0.30, -0.20, -0.30, -0.70, \\ u_2, 0.50, 0.60, 0.50, -0.10, -0.20, -0.80, \\ u_3, 0.70, 0.30, 0.40, -0.40, -0.50, -0.40, \\ u_4, 0.60, 0.60, 0.20, -0.60, -0.70, -0.50 \end{cases}$$

Then, $(B_3)^c = (G^c_A, E) = \{e_1, G^c_A(e_1), e_2, G^c_A(e_2)\}$

where,

$$G^c_A(e_1) = \begin{cases} u_1, 0.40, 0.40, 0.50, -0.20, -0.10, -0.50, \\ u_2, 0.30, 0.20, 0.60, -0.20, -0.30, -0.60, \\ u_3, 0.60, 0.30, 0.70, -0.20, -0.40, -0.40, \\ u_4, 0.80, 0.20, 0.30, -0.30, -0.50, -0.50 \end{cases}$$

$$G^c_A(e_2) = \begin{cases} u_1, 0.30, 0.60, 0.70, -0.70, -0.70, -0.20, \\ u_2, 0.50, 0.40, 0.50, -0.80, -0.80, -0.10, \\ u_3, 0.40, 0.70, 0.70, -0.40, -0.50, -0.40, \\ u_4, 0.20, 0.40, 0.60, -0.50, -0.30, -0.60 \end{cases}$$

Since the complement of truth membership becomes false membership and false membership becomes truth membership, the sum of indeterminacy and its complement is one, therefore E(B) = E(B^c) by definition 3.2

4. Let $B_1 = (G_A, E) = \{e, G_A(e) : e \in E\}$ and

$$B_2 = (H_A, E) = \{e, H_A(e) : e \in E\}$$

where,

$$G_A(e_1) = \left\{ e_1, \begin{cases} (u_1, 0.6, 0.5, 0.4, -0.6, -0.7, -0.1), \\ (u_2, 0.9, 0.5, 0.5, -0.5, -0.7, -0.2) \end{cases} \right\}$$

$$G_A(e_2) = \left(e_2, \begin{cases} (u_1, 0.9, 0.5, 0.6, -0.9, -0.7, -0.2), \\ (u_2, 0.9, 0.5, 0.6, -0.8, -0.7, -0.1) \end{cases} \right)$$

$$H_A(e_1) = \left(e_1, \left\{ \begin{cases} (u_1, 0.5, 0.4, 0.3, -0.6, -0.2, -0.4), \\ (u_2, 0.6, 0.3, 0.2, -0.5, -0.3, -0.2) \end{cases} \right\} \right)$$

$$H_A(e_2) = \left(e_2, \begin{cases} (u_1, 0.6, 0.4, 0.2, -0.5, -0.1, -0.1), \\ (u_2, 0.7, 0.6, 0.3, -0.4, -0.2, -0.3) \end{cases} \right)$$

Here $B_1 \subseteq B_2$. By definition 3.2, $E(B_2) \leq E(B_1)$ if $B_1 \subseteq B_2$ since the entropy values of B_1 and B_2 are $E(B_1) = 0.830$ and $E(B_2)$ 0.6725.

Methods to Measure the distance of Bipolar Pythagorean Neutrosophic Soft Sets (BPNSS)

Distance measure is a term that describes the difference between sets. Here, we introduce novel distance measures for BPNSS and study some of its mathematical properties.

Definition 1 Consider two BPNSS B_1 and B_2. Let d be a mapping and it satisfies the following conditions:

    i. $d(B_1, B_2) \geq 0$

   ii. $d(B_1, B_2) = d(B_2, B_1)$

  iii. $d(B_1, B_2) = 0$ if $B_1 = B_2$

  iv. $d(B_1, B_2) + d(B_2, B_3) \geq d(B_1, B_3)$

The distance measure between two bipolar Pythagorean neutrosophic soft sets is denoted by $d(B_1, B_2)$.

Definition 2 Let U be the universe and E be the parameter then hamming distance between any two BPNSS B_1 and B_2 is defined as

$$d^H_{BPNSS}(B_1, B_2) = \sum_{j=1}^{n} \sum_{i=1}^{m} \frac{\begin{array}{c}|\Delta_{ij}T(u)| + |\nabla_{ij}T(u)| + |\Delta_{ij}I(u)| + \\ |\nabla_{ij}I(u)| + |\Delta_{ij}F(u)| + |\nabla_{ij}F(u)|\end{array}}{6}$$

where $\Delta_{ij}T(u) = T^{2^+}_{B_1(e_j)}(u_i) - T^{2^+}_{B_2(e_j)}(u_i)$,

$$\nabla_{ij}T(u) = T^{2^-}_{B_1(e_j)}(u_i) - T^{2^-}_{B_2(e_j)}(u_i)$$

Theorem 3 Let B_1, B_2 and B_3 be three BPNSS defined in the B universe of discourse U. The introduced hamming distance satisfies the following properties:

    i. $d^H_{BPNSS}(B_1, B_2) \geq 0$

   ii. $d^H_{BPNSS}(B_1, B_2) = d^H_{BPNSS}(B_2, B_1)$

  iii. $d^H_{BPNSS}(B_1, B_2) = 0$ if $B_1 = B_2$

  iv. $d^H_{BPNSS}(B_1, B_2) + d^H_{BPNSS}(B_2, B_3) \geq d^H_{BPNSS}(B_1, B_3)$ (for any B_3)

Proof

i. Since $\left|\Delta_{ij}T(u)\right|, \left|\nabla_{ij}T(u)\right|, \left|\Delta_{ij}I(u)\right|, \left|\nabla_{ij}I(u)\right|,$

$\left|\Delta_{ij}F(u)\right|, \left|\nabla_{ij}F(u)\right|$ are all positive,

$d_{BPNSS}^{H}(\mathcal{B}_1, \mathcal{B}_2) \geq 0$.

ii. $\left|\Delta_{ij}T(u)\right|$ is same for both $d_{BPNSS}^{H}(\mathcal{B}_1, \mathcal{B}_2)$ and

$d_{BPNSS}^{H}(\mathcal{B}_2, \mathcal{B}_1)$ since $\left|T^{2^+}{}_{\mathcal{B}_1(e_j)}(u_i) - T^{2^+}{}_{\mathcal{B}_2(e_j)}(u_i)\right|$

$= \left|T^{2^+}{}_{\mathcal{B}_2(e_j)}(u_i) - T^{2^+}{}_{\mathcal{B}_1(e_j)}(u_i)\right|$ and this is true

for all membership degrees. Therefore,

$d_{BPNSS}^{H}(\mathcal{B}_1, \mathcal{B}_2) = d_{BPNSS}^{H}(\mathcal{B}_2, \mathcal{B}_1)$.

iii. Since $\nabla_{ij}T(u) = T^{2^+}{}_{\mathcal{B}_1(e_j)}(u_i) - T^{2^+}{}_{\mathcal{B}_2(e_j)}(u_i) = $

0 and $\nabla_{ij}T(u) = T^{2^-}{}_{\mathcal{B}_1(e_j)}(u_i) - T^{2^-}{}_{\mathcal{B}_2(e_j)}(u_i) = 0$

for $\mathcal{B}_1 = \mathcal{B}_2$, then $d_{BPNSS}^{E}(\mathcal{B}_1, \mathcal{B}_2) = 0$ if $\mathcal{B}_1 = \mathcal{B}_2$.

iv. Consider, $d_{BPNSS}^{E}(\mathcal{B}_1, \mathcal{B}_2) = $

$$\sum_{j=1}^{n}\sum_{i=1}^{m} \frac{\left|\Delta_{ij}T_1(u)\right| + \left|\nabla_{ij}T_1(u)\right| + \left|\Delta_{ij}I_1(u)\right| + }{6} \frac{\left|\nabla_{ij}I_1(u)\right| + \left|\Delta_{ij}F_1(u)\right| + \left|\nabla_{ij}F_1(u)\right|}{6}$$

$d_{BPNSS}^{E}(\mathcal{B}_2, \mathcal{B}_3) = $

$$\sum_{j=1}^{n}\sum_{i=1}^{m} \frac{\left|\Delta_{ij}T_2(u)\right| + \left|\nabla_{ij}T_2(u)\right| + \left|\Delta_{ij}I_2(u)\right| + }{6} \frac{\left|\nabla_{ij}I_2(u)\right| + \left|\Delta_{ij}F_2(u)\right| + \left|\nabla_{ij}F_2(u)\right|}{6}$$

Then, $d_{BPNSS}^{E}(\mathcal{B}_1, \mathcal{B}_2) + d_{BPNSS}^{E}(\mathcal{B}_2, \mathcal{B}_3)$

$$(\sum_{j=1}^{n}\sum_{i=1}^{m} \frac{\begin{array}{l}\left|T^{2^+}{}_{\mathcal{B}_1(e_j)}(u_i) - T^{2^+}{}_{\mathcal{B}_2(e_j)}(u_i)\right| + \\ \left|T^{2^-}{}_{\mathcal{B}_1(e_j)}(u_i) - T^{2^-}{}_{\mathcal{B}_2(e_j)}(u_i)\right| + \\ \left|I^{2^+}{}_{\mathcal{B}_1(e_j)}(u_i) - I^{2^+}{}_{\mathcal{B}_2(e_j)}(u_i)\right| + \\ \left|I^{2^-}{}_{\mathcal{B}_1(e_j)}(u_i) - I^{2^-}{}_{\mathcal{B}_2(e_j)}(u_i)\right| + \\ \left|F^{2^+}{}_{\mathcal{B}_1(e_j)}(u_i) - F^{2^+}{}_{\mathcal{B}_2(e_j)}(u_i)\right| + \\ \left|F^{2^-}{}_{\mathcal{B}_1(e_j)}(u_i) - F^{2^-}{}_{\mathcal{B}_2(e_j)}(u_i)\right| + \\ \left|T^{2^+}{}_{\mathcal{B}_2(e_j)}(u_i) - T^{2^+}{}_{\mathcal{B}_3(e_j)}(u_i)\right| + \\ \left|T^{2^-}{}_{\mathcal{B}_2(e_j)}(u_i) - T^{2^-}{}_{\mathcal{B}_3(e_j)}(u_i)\right| + \\ \left|I^{2^+}{}_{\mathcal{B}_2(e_j)}(u_i) - I^{2^+}{}_{\mathcal{B}_3(e_j)}(u_i)\right| + \\ \left|I^{2^-}{}_{\mathcal{B}_2(e_j)}(u_i) - I^{2^-}{}_{\mathcal{B}_3(e_j)}(u_i)\right| + \\ \left|F^{2^+}{}_{\mathcal{B}_2(e_j)}(u_i) - F^{2^+}{}_{\mathcal{B}_3(e_j)}(u_i)\right| + \\ \left|F^{2^-}{}_{\mathcal{B}_2(e_j)}(u_i) - F^{2^-}{}_{\mathcal{B}_3(e_j)}(u_i)\right|\end{array}}{6})$$

$$\geq \sum_{j=1}^{n}\sum_{i=1}^{m} \frac{\begin{array}{l}\left|T^{2^+}{}_{\mathcal{B}_1(e_j)}(u_i) - T^{2^+}{}_{\mathcal{B}_3(e_j)}(u_i)\right| + \\ \left|T^{2^-}{}_{\mathcal{B}_1(e_j)}(u_i) - T^{2^-}{}_{\mathcal{B}_3(e_j)}(u_i)\right| + \\ \left|I^{2^+}{}_{\mathcal{B}_1(e_j)}(u_i) - I^{2^+}{}_{3_2(e_j)}(u_i)\right| + \\ \left|I^{2^-}{}_{\mathcal{B}_1(e_j)}(u_i) - I^{2^-}{}_{\mathcal{B}_3(e_j)}(u_i)\right| + \\ \left|F^{2^+}{}_{\mathcal{B}_1(e_j)}(u_i) - F^{2^+}{}_{\mathcal{B}_3(e_j)}(u_i)\right| + \\ \left|F^{2^-}{}_{\mathcal{B}_1(e_j)}(u_i) - F^{2^-}{}_{\mathcal{B}_3(e_j)}(u_i)\right|\end{array}}{6}$$

This implies that

$d_{BPNSS}^{H}(\mathcal{B}_1, \mathcal{B}_2) + d_{BPNSS}^{H}(\mathcal{B}_2, \mathcal{B}_3) \geq d_{BPNSS}^{H}(\mathcal{B}_1, \mathcal{B}_3)$.

Definition 4 The normalized hamming distance between any two BPNSS B_1 and B_2 is defined as

$d_{BPNSS}^{nH}(\mathcal{B}_1, \mathcal{B}_2) = $

$$\sum_{j=1}^{n}\sum_{i=1}^{m} \frac{\dfrac{\left(\begin{array}{l}\left|\Delta_{ij}T(u)\right| + \left|\nabla_{ij}T(u)\right| + \left|\Delta_{ij}I(u)\right| + \\ \left|\nabla_{ij}I(u)\right| + \left|\Delta_{ij}F(u)\right| + \left|\nabla_{ij}F(u)\right|\end{array}\right)}{6}}{mn}$$

where

$\Delta_{ij}T(u) = T^{2^+}{}_{\mathcal{B}_1(e_j)}(u_i) - T^{2^+}{}_{\mathcal{B}_2(e_j)}(u_i),$

$\nabla_{ij}T(u) = T^{2^-}{}_{\mathcal{B}_1(e_j)}(u_i) - T^{2^-}{}_{\mathcal{B}_2(e_j)}(u_i)$

Theorem 5 $d_{BPNSS}^{nH}(\mathcal{B}_1,\mathcal{B}_2) = \dfrac{d_{BPNSS}^{H}(\mathcal{B}_1,\mathcal{B}_2)}{mn}$

Proof: Since $d_{BPNSS}^{H}(\mathcal{B}_1,\mathcal{B}_2)$ satisfies the definition 1, for any positive m,n.

Therefore $d_{BPNSS}^{nH}(\mathcal{B}_1,\mathcal{B}_2) = \dfrac{d_{BPNSS}^{H}(\mathcal{B}_1,\mathcal{B}_2)}{mn}$ also satisfies 1.

Definition 6 Let U be the universe and E be the parameter then Euclidean distance is defined as

$d_{BPNSS}^{E}(\mathcal{B}_1,\mathcal{B}_2) =$

$$\sqrt{\sum_{j=1}^{n}\sum_{i=1}^{m}\frac{\left(\Delta_{ij}T(u)\right)^2+\left(\nabla_{ij}T(u)\right)^2+\left(\Delta_{ij}I(u)\right)^2+\left(\nabla_{ij}I(u)\right)^2+\left(\Delta_{ij}F(u)\right)^2+\left(\nabla_{ij}F(u)\right)^2}{6}}$$

where

$\Delta_{ij}T(u) = T^{2^{+}}{}_{B_1(e_j)}(u_i) - T^{2^{+}}{}_{B_2(e_j)}(u_i),$

$\nabla_{ij}T(u) = T^{2^{-}}{}_{B_1(e_j)}(u_i) - T^{2^{-}}{}_{B_2(e_j)}(u_i)$

Theorem 7 Let B_1, B_2 and B_3 be three BPNSS defined in the universe of discourse U. The introduced Euclidean distance satisfies the following properties:

i. $d_{BPNSS}^{E}(\mathcal{B}_1,\mathcal{B}_2) \geq 0$

ii. $d_{BPNSS}^{E}(\mathcal{B}_1,\mathcal{B}_2) = d_{BPNSS}^{E}(\mathcal{B}_2,\mathcal{B}_1)$

iii. $d_{BPNSS}^{E}(\mathcal{B}_1,\mathcal{B}_2) = 0$ if $B_1 = B_2$

iv. $d_{BPNSS}^{E}(\mathcal{B}_1,\mathcal{B}_2) + d_{BPNSS}^{E}(\mathcal{B}_2,\mathcal{B}_3) \geq d_{BPNSS}^{E}(\mathcal{B}_1,\mathcal{B}_3)$
     (for any B_2)

Proof

i.   Since
     $\left(\Delta_{ij}T(u)\right)^2, \left(T(u)\right)^2, \left(\Delta_{ij}I(u)\right)^2, \left(\nabla_{ij}I(u)\right)^2,$
     $\left(\Delta_{ij}F(u)\right)^2, \left(\nabla_{ij}F(u)\right)^2$ are all positive,
     $d_{BPNSS}^{E}(\mathcal{B}_1,\mathcal{B}_2) \geq 0$.

ii. $\left(\Delta_{ij}T(u)\right)^2$ is same for both $d_{BPNSS}^{E}(\mathcal{B}_1,\mathcal{B}_2)$
     and $d_{BPNSS}^{E}(\mathcal{B}_2,\mathcal{B}_1)$ since
     $\left(T^{2^{+}}{}_{B_1(e_j)}(u_i) - T^{2^{+}}{}_{B_2(e_j)}(u_i)\right)^2 =$
     $\left(T^{2^{+}}{}_{B_2(e_j)}(u_i) - T^{2^{+}}{}_{B_1(e_j)}(u_i)\right)^2$ and this is
     true for all membership degrees. Therefore,
     $d_{BPNSS}^{E}(\mathcal{B}_1,\mathcal{B}_2) = d_{BPNSS}^{E}(\mathcal{B}_2,\mathcal{B}_1)$.

iii. Since $\Delta_{ij}T(u) = T^{2^{+}}{}_{B_1(e_j)}(u_i) - T^{2^{+}}{}_{B_2(e_j)}(u_i) = 0$
     and $\nabla_{ij}T(u) = T^{2^{-}}{}_{B_1(e_j)}(u_i) - T^{2^{-}}{}_{B_2(e_j)}(u_i) = 0$
     for $B_1 = B_2$, then $d_{BPNSS}^{E}(\mathcal{B}_1,\mathcal{B}_2) = 0$ if $B_1 = B_2$.

iv. Let $d_{BPNSS}^{E}(\mathcal{B}_1,\mathcal{B}_2) =$

$$\sqrt{\sum_{j=1}^{n}\sum_{i=1}^{m}\frac{\left(\Delta_{ij}T_1(u)\right)^2+\left(\nabla_{ij}T_1(u)\right)^2+\left(\Delta_{ij}I_1(u)\right)^2+\left(\nabla_{ij}I_1(u)\right)^2+\left(\Delta_{ij}F_1(u)\right)^2+\left(\nabla_{ij}F_1(u)\right)^2}{6}}$$

$d_{BPNSS}^{E}(\mathcal{B}_1,\mathcal{B}_2) =$

$$\sqrt{\sum_{j=1}^{n}\sum_{i=1}^{m}\frac{\left(\Delta_{ij}T_2(u)\right)^2+\left(\nabla_{ij}T_2(u)\right)^2+\left(\Delta_{ij}I_2(u)\right)^2+\left(\nabla_{ij}I_2(u)\right)^2+\left(\Delta_{ij}F_2(u)\right)^2+\left(\nabla_{ij}F_2(u)\right)^2}{6}}$$

By the Euclidean norm definition, we take
$d_{BPNSS}^{E}(\mathcal{B}_1,\mathcal{B}_2) = \left\|\mathcal{B}_1 - \mathcal{B}_2\right\|_2$

$d_{BPNSS}^{E}(\mathcal{B}_2,\mathcal{B}_3) = \left\|\mathcal{B}_2 - \mathcal{B}_3\right\|_2$

Then, $\left\|\mathcal{B}_1 - \mathcal{B}_3\right\|_2 = \left\|\mathcal{B}_1 - \mathcal{B}_2 + \mathcal{B}_2 - \mathcal{B}_3\right\|_2$

By triangle inequality, $\left\|\mathcal{B}_1 - \mathcal{B}_3\right\|_2 \leq \left\|\mathcal{B}_1 - \mathcal{B}_2\right\|_2 + \left\|\mathcal{B}_2 - \mathcal{B}_3\right\|_2$

Hence
$d_{BPNSS}^{E}(\mathcal{B}_1,\mathcal{B}_2) + d_{BPNSS}^{E}(\mathcal{B}_2,\mathcal{B}_3) \geq d_{BPNSS}^{E}(\mathcal{B}_1,\mathcal{B}_3)$

Definition 8 The normalized Euclidean distance between any two BPNSS B_1 and B_2 is defined as

$d_{BPNSS}^{nE}(\mathcal{B}_1,\mathcal{B}_2) =$

$$\sqrt{\sum_{j=1}^{n}\sum_{i=1}^{m}\frac{(\left(\Delta_{ij}T(u)\right)^2+\left(\nabla_{ij}T(u)\right)^2+\left(\Delta_{ij}I(u)\right)^2+\left(\nabla_{ij}I(u)\right)^2+\left(\Delta_{ij}F(u)\right)^2+\left(\nabla_{ij}F(u)\right)^2)/6}{mn}}$$

Theorem 9 $d_{BPNSS}^{nE}(\mathcal{B}_1,\mathcal{B}_2) = \dfrac{d_{BPNSS}^{E}(\mathcal{B}_1,\mathcal{B}_2)}{\sqrt{mn}}$

Proof: Since $d_{BPNSS}^{E}(\mathcal{B}_1,\mathcal{B}_2)$ satisfies the definition 1. Therefore $d_{BPNSS}^{nE}(\mathcal{B}_1,\mathcal{B}_2) = \dfrac{d_{BPNSS}^{E}(\mathcal{B}_1,\mathcal{B}_2)}{\sqrt{mn}}$ also satisfies 1 for all m,n.

Numerical verification and Discussion of the Proposed Distance Measure

Example 1 Let $\mathcal{B}_1, \mathcal{B}_2$ and $\mathcal{B}_3 \in BPNSS(U)$ for

$U = \{u_1, u_2\}, E = \{e_1, e_2\}.$

Suppose

$$\mathcal{B}_1 = \left\{ \left(e_1, \left\{ \begin{matrix} (u_1, 0.5, 0.6, 0.9, -0.5, -0.9, -0.2), \\ (u_2, 0.6, 0.8, 0.9, -0.6, -0.7, -0.2) \end{matrix} \right\} \right) \right.$$

$$\left. \left(e_2, \left\{ \begin{matrix} (u_1, 0.7, 0.7, 0.8, -0.4, -0.7, -0.2), \\ (u_2, 0.3, 0.8, 0.8, -0.5, -0.7, -0.2) \end{matrix} \right\} \right) \right\}$$

$$\mathcal{B}_2 = \left\{ \left(e_1, \left\{ \begin{matrix} (u_1, 0.6, 0.5, 0.4, -0.6, -0.7, -0.1), \\ (u_2, 0.9, 0.5, 0.5, -0.5, -0.7, -0.2) \end{matrix} \right\} \right) \right.,$$

$$\left. \left(e_2, \left\{ \begin{matrix} (u_1, 0.9, 0.7, 0.6, -0.9, -0.7, -0.2), \\ (u_2, 0.9, 0.7, 0.6, -0.8, -0.7, -0.1) \end{matrix} \right\} \right) \right\}$$

and

$$\mathcal{B}_3 = \left\{ \left(e_1, \left\{ \begin{matrix} (u_1, 0.3, 0.6, 0.9, -0.7, -0.9, -0.2), \\ (u_2, 0.4, 0.4, 0.5, -0.6, -0.7, -0.5) \end{matrix} \right\} \right) \right.,$$

$$\left. \left(e_2, \left\{ \begin{matrix} (u_1, 0.9, 0.7, 0.8, -0.4, -0.7, -0.2), \\ (u_2, 0.5, 0.8, 0.8, -0.5, -0.7, -0.2) \end{matrix} \right\} \right) \right\}$$

Let us calculate the distance using the proposed distance measures.

$d^H_{BPNSS}(\mathcal{B}_1, \mathcal{B}_2) = \frac{1}{6} \sum_{j=1}^{2} \sum_{i=1}^{2} \{|0.5 - 0.6| + |0.6 - 0.5| + |0.9 - 0.4| + |-0.5 - (-0.6)| + |-0.9 - (-0.7)| + |-0.2 - (-0.1)| + |0.6 - 0.9| + |0.8 - 0.5| + |0.9 - 0.5| + |-0.6 - (-0.5)| + |-0.7 - (-0.7)| + |-0.2 - (-0.2)| + |0.7 - 0.9| + |0.7 - 0.7| + |0.8 - 0.6| + |-0.4 - (0.9)| + |-0.7 - (-0.7)| + |-0.2 - (-0.2)| + |0.3 - 0.9| + |0.8 - 0.7| + |0.8 - 0.6| + |-0.5 - (-0.8)| + |-0.7 - (-0.7)| + |-0.2 - (-0.1)|\} = 0.7333$

$d^E_{BPNSS}(\mathcal{B}_1, \mathcal{B}_2) = \frac{1}{6} \sum_{j=1}^{2} \sum_{i=1}^{2} \{ (0.5 - 0.6)^2 + (0.6 - 0.5)^2 + (0.9 - 0.4)^2 + (-0.5 - (-0.6))^2 + (-0.9 - (-0.7))^2 + (-0.2 - (-0.1))^2 + (0.6 - 0.9)^2 + (0.8 - 0.5)^2 + (0.9 - 0.5)^2 + (-0.6 - (-0.5))^2 + (-0.7 - (-0.7))^2 + (-0.2 - (-0.2))^2 + (0.7 - 0.9)^2 + (0.7 - 0.7)^2 + (0.8 - 0.6)^2 + (-0.4 - (0.9))^2 + (-0.7 - (-0.7))^2 + (-0.2 - (-0.2))^2 + (0.3 - 0.9)^2 + (0.8 - 0.7)^2 + (0.8 - 0.6)^2 + (-0.5 - (-0.8))^2 + (-0.7 - (-0.7))^2 + (-0.2 - (-0.1))^2 \}^{1/2} = 0.5032$

$d^{nH}_{BPNSS}(\mathcal{B}_1, \mathcal{B}_2) = \frac{1}{24} \sum_{j=1}^{2} \sum_{i=1}^{2} \{|0.5 - 0.6| + |0.6 - 0.5| + |0.9 - 0.4| + |-0.5 - (-0.6)| + |-0.9 - (-0.7)| + |-0.2 - (-0.1)| + |0.6 - 0.9| + |0.8 - 0.5| + |0.9 - 0.5| + |-0.6 - (-0.5)| + |-0.7 - (-0.7)| + |-0.2 - (-0.2)| + |0.7 - 0.9| + |0.7 - 0.7| + |0.8 - 0.6| + |-0.4 - (0.9)| + |-0.7 - (-0.7)| + |-0.2 - (-0.2)| + |0.3 - 0.9| + |0.8 - 0.7| + |0.8 - 0.6| + |-0.5 - (-0.8)| + |-0.7 - (-0.7)| + |-0.2 - (-0.1)|\} = 0.0305$

$$d^{nE}_{BPNSS}(\mathcal{B}_1, \mathcal{B}_2) = \frac{1}{24} \sum_{j=1}^{2} \sum_{i=1}^{2} \{ (0.5 - 0.6)^2$$
$$+ (0.6 - 0.5)^2 + (0.9 - 0.4)^2$$
$$+ (-0.5 - (-0.6))^2$$
$$+ (-0.9 - (-0.7))^2$$
$$+ (-0.2 - (-0.1))^2$$
$$+ (0.6 - 0.9)^2 + (0.8 - 0.5)^2$$
$$+ (0.9 - 0.5)^2$$
$$+ (-0.6 - (-0.5))^2$$
$$+ (-0.7 - (-0.7))^2$$
$$+ (-0.2 - (-0.2))^2$$
$$+ (0.7 - 0.9)^2 + (0.7 - 0.7)^2$$
$$+ (0.8 - 0.6)^2 + (-0.4 - (0.9))^2$$
$$+ (-0.7 - (-0.7))^2$$
$$+ (-0.2 - (-0.2))^2$$
$$+ (0.3 - 0.9)^2 + (0.8 - 0.7)^2$$
$$+ (0.8 - 0.6)^2$$
$$+ (-0.5 - (-0.8))^2$$
$$+ (-0.7 - (-0.7))^2$$
$$+ (-0.2 - (-0.1))^2 \}^{1/2} = 0.0209$$

That is,

$d^H_{BPNSS}(\mathcal{B}_1, \mathcal{B}_2) = 0.7333$, $d^E_{BPNSS}(\mathcal{B}_1, \mathcal{B}_2) = 0.5032$,

$d^{nH}_{BPNSS}(\mathcal{B}_1, \mathcal{B}_2) = 0.0305$, $d^{nE}_{BPNSS}(\mathcal{B}_1, \mathcal{B}_2) = 0.0209$.

Similarly, we get

$d^H_{BPNSS}(\mathcal{B}_2, \mathcal{B}_3) = 0.8000$, $d^E_{BPNSS}(\mathcal{B}_2, \mathcal{B}_3) = 0.5291$,

$d^{nH}_{BPNSS}(\mathcal{B}_2, \mathcal{B}_3) = 0.3333$, $d^{nE}_{BPNSS}(\mathcal{B}_2, \mathcal{B}_3) = 0.0220$.

That is,

$d^H_{BPNSS}(\mathcal{B}_1, \mathcal{B}_3) = 0.3500$, $d^E_{BPNSS}(\mathcal{B}_1, \mathcal{B}_3) = 0.2613$,

$d^{nH}_{BPNSS}(\mathcal{B}_1, \mathcal{B}_3) = 0.0145$, $d^{nE}_{BPNSS}(\mathcal{B}_1, \mathcal{B}_3) = 0.0108$.

Discussion

It shows that from definition 4.1,

i. Condition (i), that is $d(\mathcal{B}_1, \mathcal{B}_2) \geq 0$ holds since $d(\mathcal{B}_1, \mathcal{B}_2), d(\mathcal{B}_2, \mathcal{B}_3), d(\mathcal{B}_1, \mathcal{B}_3) \in [0,1]$.

ii. Condition (ii) $d(\mathcal{B}_1, \mathcal{B}_2) = d(\mathcal{B}_2, \mathcal{B}_1)$ is satisfied because of the use of square and absolute values.

iii. Condition (iii) that is $d(\mathcal{B}_1, \mathcal{B}_2) = 0$ if $\mathcal{B}_1 = \mathcal{B}_2$ is straightforward.

iv. Clearly, condition (iv) that is $d(\mathcal{B}_1, \mathcal{B}_2) + d(\mathcal{B}_2, \mathcal{B}_3) \geq d(\mathcal{B}_1, \mathcal{B}_3)$ holds for each proposed distance measure.

Finally, we observe that $d_{BPNSS}^{nE}(\mathcal{B}_1, \mathcal{B}_2) < d_{BPNSS}^{nH}(\mathcal{B}_1, \mathcal{B}_2) < d_{BPNSS}^{E}(\mathcal{B}_1, \mathcal{B}_2) < d_{BPNSS}^{H}(\mathcal{B}_1, \mathcal{B}_2)$ which implies that $d_{BPNSS}^{nE}(\mathcal{B}_1, \mathcal{B}_2)$ is the accurate measure.

Conclusion and Future Work

We defined a new entropy and distance measure. Finally, we have observed the accurate measures among Hamming distance, Euclidean and their normalized versions. The four proposed methods are the distance measures of BPNSS when the elements of the BPNSS are equal. Our future work is to propose distance measures when the elements of the sets are unequal.

References

Anitha, S., Francina Shalini, A. (2022). Bipolar Pythagorean Neutrosphic Soft Set, Int. Conf. on Recent Strategies in Mathematics and Statistics, 81.

Aslam, M., Abdullah, S., and Ullah, K. (2013). Bipolar Fuzzy Soft Sets and Its Applications in Decision Making Problem, arXiv:1303.6932v1 [cs. AI] 23.

Ali, M., Son, Le Hoang, Deli, I., and Tien, Nguyen Dang (2017). Bipolar netrosophic soft sets and applications in decision making, Journal of Intelligent and Fuzzy System, 33(2017), 4077–4087.

Broumi, S., Deli, I., and Smarandache, F. (2014). Relations on interval valued neutrosophic soft sets, Journal of New Results in Science 5, 1–20.

Deli, I., and Broumi, S. (2015). Neutrosophic soft relations and some properties, Annals of Fuzzy Mathematics and Informatics 9(1), 169–182.

Deli, I., and Broumi, S. (2015). Neutrosophic soft matrices and NSM decision making, Journal of Intelligent and Fuzzy Systems 28(5), 2233–2241.

Deli, I., Ali, M. and Smarandache, F. (2015). Bipolar Neutrosophic Sets and Their Application Based on Multi-Criteria Decision Making Problems, Proceedings of the 2015 Inter-national Conference on Advanced Mechatronic Systems, Beijing, China, 2015.

Grzegorzewski, P. (2004). Distances between intuitionistic fuzzy sets and/or interval-valued fuzzy sets based on the Hausdorff metric, Fuzzy Sets and Systems, 148(2), pp. 319–328 [Online].

Jhansi, R., Mohana, K., and Smarandache, Florentin, Correlation measure for pythagorean Neutrosophic sets with T and F as dependent neutrosophic components

Lee, K. M. (2000). Bipolar-valued fuzzy sets and their operations, Proc Int Conf on Intelligent Technologies, Bangkok, Thailand, pp. 307–312.

Smarandache, F. (2002). Neutrosophy and Neutrosophic Logic, First International Conference on Neutrosophy, Neutrosophic Logic, Set, Probability, and Statistics University of New Mexico, Gallup, NM 87301, USA.

Smarandache, F. (2010).Neutrosophic set- a generalization of Intuitionistic fuzzy set, Jour. of Defense Resources Management, 107–116.

Szmidt, E. and Kacprzyk, J. (2000). Distances between intuitionistic fuzzy sets, Fuzzy Sets and Systems, 114(3), 505–518.

Peng, Xindong, Yang, Yong (2015). Some results for pythagorean fuzzy sets, International Journal of Intelligent systems, 30(2015), 1133–1160.

Yager, R. R. (2013). Pythagorean fuzzy subsets, In:Proc JoBPFint IFSA World Congress and NAFIPS Annual Meeting, Edmonton,Canada, 2013, 57–61.

Zadeh, L. A., Fuzzy Sets, Inform and Control 8(1965) 338 – 353.

Mastitis Disease Diagnosis in Dairy Cattle Using Artificial Neural Networks

Nivetha Martin,[a,] G. Hannah Grace,[b] N. Ramila Gandhi,[c] and P. Pandiammal[d]*

[a]Assistant Professor, Department of Mathematics, Arul Anandar College (Autonomous), Karumathur, Tamil Nadu, India, Pin Code: 625514
[b]Assistant Professor, Department of Mathematics, VIT University, Chennai, Tamil Nadu, India, Pin Code: 600127
[c]Associate Professor, Department of Mathematics, PSNA College of Engineering & Technology, Dindigul, Tamil Nadu, India, Pin Code: 624622
[d]Assistant Professor, Department of Mathematics, GTN Arts College (Autonomous), Dindigul, Tamil Nadu, India, Pin Code: 624005
E-mail: *nivetha.martin710@gmail.com

Abstract

Mastitis is one of the infectious diseases that cause inflammation in the mammary glands of dairy cattle. As this disease results in abnormalities of udder, the secretion of milk is highly affected and hence it is a threat to the dairy industry. This may be caused due to physical, chemical or heat damage to the udder tissue. Early diagnosis of this disease is required to mitigate the fatality of cattle and loss of sales. In this article, multi-layer artificial neural networks (ANN), a deep learning tool is used in the diagnosis of Sub-clinical mastitis disease affecting dairy cattle. The data set of 100 dairy cattle under nine attributes from the region of Chellampatti block of Madurai district is collected and used in this research article. Different learning algorithms are also applied to determine the efficacy of the predicting system in the R programming environment. The outcome of this research is to determine the most optimal algorithm that makes the diagnosis of mastitis disease more effective and feasible for further prediction.

Keywords: Mastitis, disease diagnosis, ANN, R programming.

Introduction

Dairy farming is a subclass of agriculture that involves rearing, breeding, and growing of dairy animals for producing milk and milk products. Generally, dairy cattle are a special concern of every dairy industry and dairy farming is exercised more naturally in rural areas in comparison to the urban regions. The dairy industries are watchful in providing best feed and care for the dairy cattle as the quality of the milk products highly depends on the quality of milk secreted, but inspite of that, the occurrences of diseases in dairy cattle remain a huge threat to these industries.

Mastitis is one of the diseases that affect milk secretion and sometimes causes fatality. The contagious nature of this disease increases its severity within a short time span. Microorganisms such as bacteria, virus, fungi, yeast and algae cause infection of the udder tissue and result in inflammation of mammary glands. Based on the degree of inflammation this disease is classified into clinical, sub-clinical and chronic mastitis. (Das, Guha, Biswas, Jana, Chatterjee, and Samanta 2017) have stated that sub-clinical mastitis do not exhibit characteristic symptoms and so the fatality rate is high among dairy cattle in the latter stages of this disease. This disease is termed as an expensive disease as it causes huge loss to the dairy industry and also there is a possibility of transmission of this zoonotic disease to human beings henceforth early diagnosis of this disease is required. Researchers have applied different techniques to diagnose the presence of mastitis of various kinds such as screening, bacteriology test, use of serum iron as indicators, protein-based enzymes, polymerase-based chain reaction, tools of molecular biology, meta taxonomic approach, spectroscopy, latent class analysis, putative biomarkers, Draminski mastitis detector, interpretation criteria, biomarkers, Zagreb mastitis test, counts of somatic cells, esterase tests, Molecular diagnosis, infrared thermography, biosensing techniques. In addition to the above biological diagnosis methods, the researchers have also applied machine learning algorithms in the diagnosis of mastitis and also integrated them with other advanced diagnostic tools. (Golzarian, Soltanali, Doosti Irani, and Ebrahimi 2017) used thermal image processing,(Sharifi, Pakdel, Ebrahimi, Reecy, Fazeli Farsani, and Ebrahimie 2018), (Hyde, Down, Bradley, Breen, Hudson, Leach, and Green 2020), (Feng, Niu, Wang, Ivey, Wu, Qi, and Cao 2021) integrated machine learning algorithms of Random forest supervised learning in the diagnosis of mastitis and have also applied artificial neural networks. The researchers have applied machine learning algorithms to diagnose clinical mastitis and this has motivated us to develop an ANN-based diagnostic model in this research work to identify sub-clinical mastitis present in dairy cattle. The efficacy of the diagnostic model under different learning algorithms of ANN is determined using the R programming language.

The rest of the work is organized as follows. Section 2 contains a brief description of ANNs. Section 3 describes materials and methods. Section 4 discusses the results, and the final section concludes work on future extensions of this study.

ANN and Its Training Algorithms

Artificial Neural Networks (ANN) is the simulation of the human-brain process. The framework of ANN is influenced by the biological nervous system. Just like a brain comprising of neurons that are connected, an ANN model consists of neurons interlinked with weighted links. The ANN model is a composition of three layers: an input layer, a hidden layer, and an output layer. Each layer shall comprise of many nodes and generally the output layer consists of a single node. The nodes of the input layer are all connected to all the nodes of the hidden layer and the nodes of the hidden layer are all connected to the output layer with different weights. The training algorithms update the initial weights to derive the optimal weights. The two types of ANN are feedforward and feedback. The former type has invariable input and output with unidirectional flow of information and without feedback, but the latter type has feedback. ANN has the capacity to learn on the training subject to the learning strategies of supervised learning, unsupervised learning, and reinforcement learning. ANN with supervised learning is predominantly applied in pattern recognition and is used in clustering under unsupervised learning.

The training algorithms used in ANN to update weights and they are grouped into five categories namely Gradient Descent, Conjugate Gradient, Quasi-Newton, Resilient Backpropagation (BP), and Levenberg-Marquardt Algorithms. The description of the most commonly used algorithms are presented in Table 1 as follows.

Table 1. ANN Algorithms

Algorithm	Characterization	Output Function	Description
Gradient Descent Algorithms	standard BP algorithm	$x_{k+1} = x_k - \alpha_k g_k$	x_k - current vector; α_k - learning rate; g_k - current gradient of the error; x_{k+1} - new weight vector; K - number of iterations
Resilient Backpropagation	Heuristic learning algorithm	$\Delta x_k = -sign\left(\dfrac{\Delta E_k}{\Delta x_k}\right)\Delta k$	Δx_k - Changes in the current weight vectors; ΔE_k - Error function
Conjugate Gradient Algorithms	Optimization method with low memory requirement	$p_k = -g_k + \beta_k p_{k-1}$	p_k - search direction; β_k - constant
Quasi-Newton Algorithms	Basic local method that uses second order information	$x_{k+1} = x_k - H_k^{-1} g_k$	H_k - Hessian matrix; x_{k+1} - new weight vector; x_k - current vector
Levenberg-Marquardt Algorithm	Standard technique to solve non-linear least square problems	$x_{k+1} = x_k - [J^T J + \mu I]^{-1} J^T e$	x_{k+1} - new weight vector; x_k - current vector; J - Jacobian matrix; e - vector of network errors

Materials and Methods

This section presents the problem definition, objectives, and methodology of this research work used in developing a diagnosis model.

Definition of the Problem

The livelihood of the rural people is highly dependent on livestock rearing and dairy cattle is one of their primary sources of income. The survival rate of the rural-based dairy industries are also dependent on these dairy cattle and these industries are burdened with the tasks of maintaining the well-being of these mammary animals. As the risk of infectious diseases is quite common in a rural setup, these dairy cattle are vulnerable to Mastitis disease. Though these industries have a good treatment facilities, the hidden characteristic symptoms of sub-clinical mastitis cause fatality. To avoid such risks, the industries are advised to maintain the clinical records of the cattle, especially an information system with a complete database. As environmental factors play a vital role in causing such diseases if such records are maintained it will certainly be helpful for doctors to make clinical decisions. Presently many industries embrace decision support systems at low costs with few required features, but the rural-based industries are hesitant in adopting such artificial intelligence-based decision systems due to huge investment and expensive maintenance charges. But still such AI integrated information system is required to make clinical decisions with hidden symptoms to avoid medical errors and mistreatment. In Spite of profound knowledge and experience of the medical practitioners, the clinical decisions are made based on their intuitions, but if such decisions are integrated with computer-based records, the diagnostic errors will be minimized and the quality of clinical decisions will be enhanced.

Objective

The underlying objective of this research work is to develop a diagnosis system for sub-clinical mastitis disease in dairy cattle using the technique of data mining. This system enables in unveiling and deriving of the hidden knowledge associated with sub-clinical mastitis disease from historical data. It helps medical practitioners to make smart clinical decisions which is not possible with the conventional decision support system. In this proposed model, the system can classify dairy cattle based on risk level by applying the technique of data mining into two classes as healthy and diseased. Hence this diagnosis will help medical practitioners to make correct and optimal diagnostic decisions.

Data Collection

The data for this research work is collected from a veterinary hospital located in Chellampatti block of Madurai district. The database has several attributes and out of which nine core attributes are considered and presented in Table 1

Table 1. Description of Attributes

Attribute Name	Type	Attribute Domain Values
Age	Continuous	Age in years
Temp	Discrete	Temperature
Milk yield	Discrete	0-Normal and 1- low
Udder redness	Discrete	0 - No and 1- Yes
Udder swelling	Discrete	0 - No and 1- Yes
Udder hardness	Discrete	0 - No and 1- Yes
Appearance of the milk	Discrete	0-Normal, 1 – watery, 2- flakes,3-clots
Sunken eyes	Discrete	0 - No and 1- Yes
Diarrhea	Discrete	0 - No and 1- Yes

Data Pre-processing

The collected data is subjected to data cleaning. The duplications, redundancy are removed, missing values are filled and the data are coded in accordance to the values of the domain attribute provided by practising veterinarians.

Steps Involved in Diagnosing Mastitis Disease Using ANN

The sequential steps involved in the diagnosis of Mastitis disease are presented in the below fig. 1 as follows.

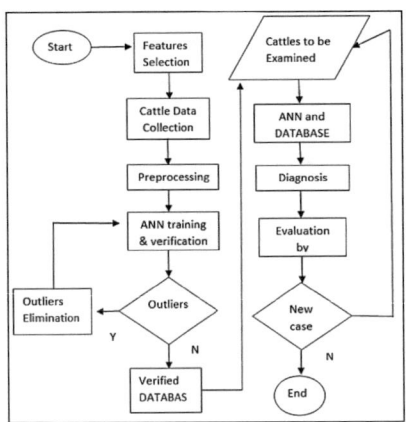

Fig. 1. ANN Flowchart

Experimental Results

The preprocessed data is subjected to ANN in the R programming environment and the following results are obtained to make further analysis. The confusion matrix is very useful to evaluate the classifiers. It is a contingency table with the values pertaining to actual and predicted and it is presented in Table 1.

Table 1. Actual vs Predicted

		Actual	
Predicted		True Positive (TP)	False Positive (FP)
		False Negative (FN)	True Negative (TN)

The efficacy of the ANN is estimated using the parameters of Sensitivity (SE), Specificity (SP), and Accuracy (AC)

$$SE = \frac{TP}{TP+FN} * 100$$
$$SP = \frac{TN}{TN+FN} * 100$$
$$AC = \frac{TP+TN}{TP+TN+FN+FP} * 100$$

Among 100 observations, 70 percentage is subjected to training and a remaining of 30 percentage is subjected to testing. The algorithms such as back propagation (backprop), resilient back propagation with (rprop+) and without weight backtracking (rprop-) and globally convergent algorithms (sag,slr) are used. The obtained ANN model is presented in Figure. The confusion matrix and the performance parameters are obtained using R programming is found to be optimal with Backpropagation algorithm and it is presented below as follows in fig. 2.

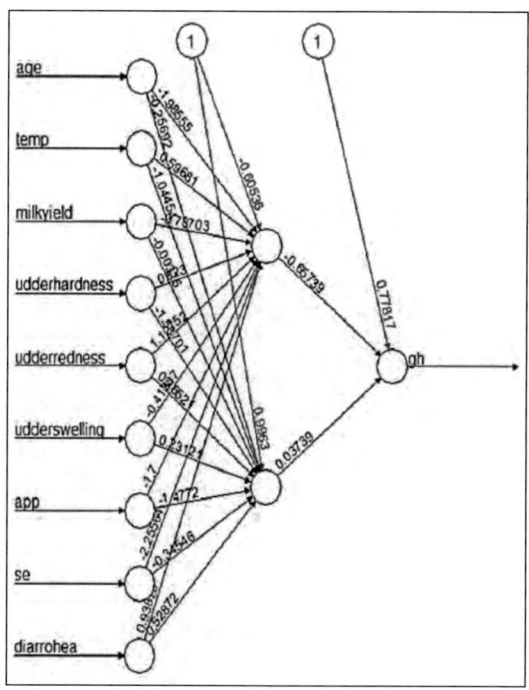

Fig. 2. ANN Architecture

The values corresponding to TP, TN, FP, and FN are presented in Table 2 and the quality parametric values are presented in Table 3

Table 2. Actual vs Predicted values

	Actual	
Predicted	31	7
	5	57

Table 3. ANN Performance Parametric values

Algorithms	Sensitivity	Specificity	Accuracy
Back propagation	86.11%	91.93%	88%

Among 100 data samples, 31 observations are TP, 57 observations are TN, 7 are FP and 5 are FN. The accuracy rate using the Backpropagation is high in comparison with other training algorithms.

Conclusion

In this research paper, a multi-layer artificial neural network is developed to make diagnosis of Sub-clinical Mastitis disease that affects dairy cattle. The collected data is subjected to training by different algorithms in R programming and the Backpropagation algorithm is found to yield optimal results. The ANN model developed has nine input layers, two hidden layers, and a single output layer. The results show that the ANN makes an efficient diagnosis of the Sub-clinical Mastitis disease.

References

Das, C. Guha, Biswas, U., Jana, P. S., Chatterjee, A., and Samanta, I.. (2017). Detection of emerging antibiotic resistance in bacteria isolated from subclinical mastitis in cattle in West Bengal, Veterinary World, 10(5), 517–520.

Ebrahimie, E., Mohammadi-Dehcheshmeh, M., Laven, R., and Petrovski, K. R. (2021). Rule discovery in milk content towards mastitis diagnosis: Dealing with farm heterogeneity over multiple years through classification based on associations. Animals, 11(6), 1638.

Feng, Y., Niu, H., Wang, F., Ivey, S. J., Wu, J. J., Qi, H.,., and Cao, Q. (2021). Social Cattle: IoT-based mastitis detection and control through social cattle behavior sensing in smart farms. IEEE Internet of Things Journal, 9(12), 10130–10138.

Golzarian, M. R., Soltanali, H., Doost"i Irani, O., and Ebrahimi, S. H. (2017). Possibility of early detection of bovine mastitis in dairy cows using thermal images processing. Iranian Journal of Applied Animal Science, 7(4), 549–557.

Hyde, R. M., Down, P. M., Bradley, A. J., Breen, J. E., Hudson, C., Leach, K. A., and Green, M. J. (2020). Automated prediction of mastitis infection patterns in dairy herds using machine learning. Scientific reports, 10(1), 1–8.

Upper Vertex Square Free Detour Number of a Graph

K. Christy Rani,[a] and G. Priscilla Pacifica[b]

[a]Research Scholar, Register No. 20122212092002, Department of Mathematics,
St. Mary's College (Autonomous), Thoothukudi (Affiliated to Manonmaniam Sundaranar
University, Abishekapatti, Tirunelveli-627012), Tamilnadu, India
[b]Assistant Professor, Department of Mathematics, St. Mary's College (Autonomous),
Thoothukudi, Tamilnadu, India
E-mail: christy.agnes@gmail.com, (priscillamelwyn@gmail.com)

Abstract

In this paper, we introduce the concepts of minimal vertex square free detour sets. A vertex square free detour set S_u of G is called a minimal vertex square free detour set if no proper subset of S_u is a vertex square free detour set of G, for some vertex u in G. For every vertex $u \in G$, the upper vertex square free detour number $dn_{\square f_u}(G)$ is the maximum order of its minimal vertex square free detour set of G. The upper vertex square free detour number of certain classes of graphs is determined. It is shown that for every pair of positive integers α and β with $1 \leq \alpha \leq \beta$, there exists a connected graph G in which $dn_{\square f_u}(G) = \alpha - 1$ or $dn_{\square f_u}(G) = \alpha$ and $dn^+_{\square f_u}(G) = \beta - 1$ or $dn^+_{\square f_u} = \beta$ for some vertex u in G.

Keywords: Minimal detour set, Upper detour number, Minimal vertex square free detour set, Upper vertex square free detour number.

Introduction

The shortest and the longest distance in a simple, finite, connected graph $G=(V,E)$, $d(u,v)$ and $D(u,v)$, where $u,v \in G$ are known as the geodesic distance and the detour distance. The concept of geodetic number and detour number was studied by Chartrand et al., (1993, 2003). The detour concept was developed into various parameters in recent years. This detour concept was extended to the connected and triangle-free concepts by S. Athisayanathan et al. in Ramalingam et al. (2016, 2017) and Santhakumaran and Athisayanathan (2009). Inspired by these great minds, we define the upper vertex square free detour number of a graph and determine the upper vertex square free detour number for some standard graphs. Furthermore, we characterize the graph by which the realization result is proved. For the basic terminologies refer to Chartrand (2003).

Preliminaries

The following definition and theorems are used in the sequel.

Definition 1: In a connected graph G, let u be any vertex. A u-square free detour set S_u is said to be a minimal u-square free detour set if no proper subset of S_u is an u-square free detour set. The upper u-is defined as square free detour number denoted by $dn^+_{\square f_u}(G)$ is defined as the maximum order of a minimal u-square free detour set ($dn^+_{\square f_u}$-set) of G.

Theorem 2: In a connected graph G, let u be any vertex.

i. Every end-vertex of G other than the vertex u (whether it is an end–vertex or not) belongs to every u-square free detour set.

ii. No cut-vertex of G belongs to any $dn_{\square f_u}$-set.

Theorem 3: Let G be a connected graph.

i. If G is a complete graph K_n, then $dn^+_{\square f_u}(G)$ is 1 for every vertex u in K_n.

ii. If G is a complete bipartite graph

$$K_{m,n}(2 \le m \le n),\ \text{with partitions } X \text{ and } Y,$$

then $dn_{\square f_u}(G) = \begin{cases} m-1 \ if\ u \in X. \\ n-1 \ if\ u \in Y. \end{cases}$

iii. If G is a cycle C_n, then $dn_{\square f_u}(G) = 1$ for every vertex u in K_n.

iv. If G is a wheel $W_n = K_1 + C_{n-1}$, then

$$dn_{\square f_u}(G) = \begin{cases} \dfrac{n-1}{3} & if\ u \in K_1, n \ge 4 \\ 2 & if\ u \in C_{n-1},\ n \ge 6 \\ 1 if\ u \in C_{n-1}, 4 \le n \le 5 \end{cases}$$

Upper Vertex Square Free Detour Number of a Graph

In this section, we investigate the upper vertex square free detour number of some standard graphs.

Theorem 1: Let G be the complete graph K_n of order $n(n \ge 3)$. For any vertex u in G, a set $S_u \subseteq V$ is a S_u of G iff S_u consists of exactly one vertex of G other than u.

Proof. Consider a vertex u in $G = K_n$ and $S = \{x_i, x_j : 1 \le i \ne j \le n\}$ be a set in G. Let $u = x_i \in V$ for some $i(1 \le i \le n)$. By Theorem 3 (i), every vertex x of V lies on x_i-x_j square free detour of length 2 for $j(1 \le i \ne j \le n)$ and so S is the square free detour set of G. Now suppose that S is a $dn^+_{\square f_u}$-set of G. If S' is a subset S with $|S'| = 1$, then S' is not a square-free detour set of G. Hence S is the $dn^+_{\square f_u}$-set of G. Also by Theorem 2, $u = x_i \notin S_u$ and so $S_u = \{x_j\}$. Thus S_u is a $dn^+_{\square f_u}$-set of G with $|S_u| = 1$.

Conversely, consider that S_u is the $dn^+_{\square f_u}$-set of G. Assume that $|S_u| = 1$. Then by Theorem 3(i), S_u

consists of exactly one vertex of G other than u. If $|S_u| > 1$, then by the same Theorem, any subset S_u of S_u is a $dn^+_{\square f_u}$-set of G. Hence S_u is not a $dn^+_{\square f_u}$-set of G, which contradicts our assumption. Therefore, S_u consists of exactly one vertex of G other than u.

Corollary 2: Let G be a complete graph K_4. For any vertex u in G, the set $S_u \subseteq V$ is the $dn^+_{\square f_u}$-set of G iff S_u consists of the antipodal vertex of u.

Theorem 3: Let $G = (V,E)$ be a complete bipartite graph $K_{m,n}(2 \le m \le n)$ with partitions X and Y such that $|X| = m$ and $|Y| = n$. Then a set $S_u \subseteq V$ is a $dn^+_{\square f_u}$-set of G iff

i. S_u consists of $m-1$ vertices of X if u is a vertex in X.

ii. S_u consists of $n-1$ vertices of Y if u is a vertex in Y.

Proof. This follows from Theorem 3(ii).

Theorem 4: Let $G = (V,E)$ be a C_n of order $n(3 \le n \le 5)$. Then a set $S_u \subseteq V$ is a $dn^+_{\square f_u}$-set of G iff S_u contains

i. only one vertex adjacent to u when n is odd.

ii. exactly one vertex antipodal to u when n is even.

Proof. Let $G = C_n : x_1 x_2 x_3 \ldots x_n x_1$ be a cycle of order $n(3 \le n \le 5)$. Let $u = x_i$.

(i) Let n be odd. Let $S_u = \{x_j : 1 \le j \le n, i \ne j\}$ be a set of exactly one adjacent vertex of u in G. We find that all the vertices of V lie on x_i-x_j square free detour in G so that S_u is a square free detour set of G. To prove that S_u is minimal, consider any subset S_u of S_u. Then by Theorem 3(iii), S_u is not a square-free detour set of G and so S_u is a $dn^+_{\square f_u}$-set of G.

Now, suppose S_u is a minimal $dn^+_{\square f_u}$-set of G. Let S_u be any set consisting of exactly one vertex v adjacent to u. Then as in the first part of case (i) of this theorem, S_u is a $dn^+_{\square f_u}$-set of G. Hence $|S_u| = |S^*_u| = 1$. Let $S_u = \{v\}$ in G. If u and v are not adjacent, then there are internal vertices of the u–v geodesic that do not lie on any u–v square free detour in G so that S_u is not a dn-set of G, which is a contradiction. Thus S_u consists of a vertex adjacent to u in G.

Theorem 5: Let G be a C_n ($n \geq 6$). A set $S_u \subseteq V$ is a minimal $dn^+_{\Box f_u}$-set of G, for any $u \in G$ iff S_u consists of two independent vertices which are neither adjacent nor antipodal to u in G.

Proof. Let $G = C_n$: $x_1 x_2 x_3 \ldots x_n x_1$ be a cycle of order $n \geq 6$. Let $u = x_i$ and $S_u = \{x_j, x_k : 1 \leq j < k \leq n\}$ be an independent set of two non-adjacent and non-antipodal vertices of length $\dfrac{n-1}{3}$ in G. Then the vertices of G lie on the x_i–x_j square free detour or on x_i–x_k square free detour and so S_u is a square-free detour set of G. To prove that S_u is minimal, consider any subset S_u of S_u with only one vertex. Then S_u is not a square-free detour set of G and so S_u is a $dn^+_{\Box f_u}$-set of G.

Conversely, consider that S_u is a $dn^+_{\Box f_u}$-set of G. If $|S_u| = 2$. Then by the first part of this theorem, S_u consists of exactly one vertex of G other than u. Assume $|S_u| \geq 3$. Then S_u must be an independent set of non-adjacent and non-antipodal vertices of G and so S_u is a minimal square-free detour set of G. Hence S_u is not a $dn^+_{\Box f_u}$-set of G, contradicting our assumption. Thus S_u consists of two independent vertices which are neither adjacent nor antipodal to u in G.

Theorem 6: Let $G = (V, E)$ be a wheel $W_n = K_1 + C_{n-1} (n \geq 4)$. For the hub of the wheel, the set $S_u \subseteq V$ is a $dn^+_{\Box f_u}$-set of G iff S_u consists of 2 vertices of C_{n-1}.

Proof. Let $W_n = K_1 + C_{n-1} (n \geq 4)$ be the wheel where $V(K_1) = \{x_0\}$ and $V(C_{n-1}) = \{x_1, x_2, \ldots, x_{n-1}\}$. Let $u = x_0$ be the hub of the wheel W_n. Let $S_u = \{x_1, x_3, \ldots, x_l\}$ be a set of $\dfrac{n-1}{2}$ vertices of G where $x_l = x_{n-1}$ if n is odd, $x_l = x_n$ if n is even.

Then all the vertices of G lie on square free detour $x_0 - x_{2i+1}$ of length 2 for some $i \left(0 \leq i \leq \dfrac{n-3}{2} \right)$.

Therefore $S^+_{x_0}$ is a square-free detour set of G. We now prove that $S^+_{x_0}$ is minimal. Let us consider that $|S^*_{x_0}| \leq \dfrac{n-3}{2}$ is any subset of $S^+_{x_0}$. Then $S^+_{x_0}$ is not a square-free detour in G. Thus $S^+_{u=x_0}$ is a $dn^+_{\Box f_u}$-set in G.

Conversely, assume that $S_{u=x_0}$ is a $dn^+_{\Box f_{x_0}}$-set of G. Let S_{x_0} be any set of $\dfrac{n-1}{2}$ vertices. Then as in the first part of this Theorem, S_{x_0} is a $dn^+_{\Box f_{x_0}}$-set. Let $\left| S_{x_0} \right| > \dfrac{n-1}{2}$, then by Theorem 3(iv), any subset $S^*_{x_0}$ of S_{x_0} is a $dn^+_{\Box f_u}$-set of G. Hence S_{x_0} is not a $dn^+_{\Box f_u}$-set of G, which contradicts our assumption. Hence S_u is a $dn^+_{\Box f_u}$-set with $\dfrac{n-1}{2}$ vertices of the rim when u is the hub of the wheel.

Theorem 7: Let G be a wheel W_n of order $n \geq 10$. For any vertex u on C_{n-1}, a set $S_u \subseteq V$ is a minimal $dn^+_{\Box f_u}$-set of G iff S_u consists of the hub of the wheel and any two vertices of C_{n-1} at distance $\dfrac{n-1}{3}$ from u.

Proof. Let G be $W_n = K_1 + C_{n-1}$ of order $n \geq ü$ For any vertex $u = x_i (1 \leq i \leq n-1)$ on C_{n-1}. Let $S = \{x_0, x_i, x_j, x_k : 1 \leq i < j < k \leq n-1\}$ where x_0 is the hub of the wheel and x_i, x_j, x_k are the vertices on the rim of the wheel with $D_{\Box f}(x_i, x_j) = D_{\Box f}(x_i, x_k) = \dfrac{n-1}{3}$. Then every vertex y of G lies on x_i–x_j or x_i–x_k square free detour in G. Then $S_{u=x_i} = \{x_0, x_j, x_k : 1 \leq j < k \leq n-1\}$. Thus $|S_u| = 3$ and so S_u is a $dn_{\Box f_u}$-set of G. To show that S_u is minimal, let us consider that $|S_{x_i}| = 3$ be any subset of S_{x_i}. By Theorem 3(iv), S_{x_i} is not a square free detour set in G. Therefore, S_u is a $dn^+_{\Box f_u}$-set in G.

Conversely, let S_u be a $dn^+_{\square f_u}$-set of G. Assume S^*_u is any set containing two vertices at distance $\dfrac{n-1}{3}$ from u and a hub of the wheel. Then S_u is a $dn^+_{\square f_u}$-set in G. Let $|S_u| \geq 4$, then by Theorem 3(iv), any subset S'_u of S_u is a $dn^+_{\square f_u}$-set of G. Hence S_u is not a minimal $dn^+_{\square f_u}$-set of G that leads to a contradiction. Therefore, S_u contains the hub of the wheel and any two vertices of C_{n-1} at a distance $\dfrac{n-1}{3}$ from u.

Theorem 8: Let G be a wheel W_n of order n $(6 \leq n \leq 9)$. For any vertex u on C_{n-1}, a set $S_u \subseteq V$ is a minimal $dn^+_{\square f_u}$-set of G iff S_u consists of the hub of the wheel and exactly one vertex adjacent or antipodal to u in the rim of the wheel.

Proof. Let G be $W_n = K_1 + C_{n-1}$ of order $n(6 \leq n \leq 9)$. For any vertex $u = x_i$ $(1 \leq i \leq n-1)$ on C_{n-1}. Let $S = \{x_0, x_i, x_j : 1 \leq i < j \leq n-1\}$ where x_0 is the hub of the wheel. Then we have two cases.

Case 1: Let n be odd. Then x_j is an adjacent vertex or antipodal to $u=x_i$. Then every vertex y of G lies on x_i-x_j square free detour in G. Then by Theorem 2, $S_{u=x_i} = \{x_0, x_j : 1 \leq j \leq n-1\}$. Thus $|S_u| = 2$, S_u is a $dn_{\square f_u}$-set of G. To show that S_u is minimal, let us consider that $|S'_{u=x_i}| = 1$ be any subset of S_{x_i}. By Theorem 3(iv), S'_{x_i} is not a square free detour in G. S_u is a minimal square-free detour set in G.

Case 2: Let n be even. Then x_j is an adjacent to $u=x_i$. Then every vertex y of G lies on x_i-x_j square free detour in G. Then by Theorem 2, $S_{u=x_i} = \{x_0, x_j : 1 \leq j \leq n-1\}$. Thus $|S_u| = 2$, S_u is a $dn_{\square f_u}$-set of G. To show that S_u is minimal, let us consider that $|S_{x_i}| = 1$ be any subset of S_{x_i}. By Theorem 3(iv), S_{x_i} is not a square free detour in G. Thus S_u is a $dn^+_{\square f_u}$-set in G.

Conversely, let S_u be a $dn^+_{\square f_u}$-set of G. Let S^*_u be any set consisting of a vertex of the rim and a hub of the wheel. Then S_u is a $dn^+_{\square f_u}$-set in G. Hence $|S_u| = |S^*_u| = 2$. Thus S_u consists of any adjacent vertex or antipodal vertex of u in the rim of the wheel and the hub of the wheel.

Theorem 9: Let G be a wheel $W_n = K_1 + C_{n-1}$ of order n=4 or 5. For any vertex u on C_{n-1}, a set $S_u \subseteq V$ is a $dn^+_{\square f_u}$-set of G iff S_u consists of a vertex adjacent to u or a vertex antipodal to u in the rim of the wheel.

Proof. This follows from Theorems 3(iv) and 4.

Theorem 10: Let G be a connected graph.

a. If G is the tree with k end-vertices,
$$dn^+_{\square f_u}(G) = \begin{cases} k-1 & if\ u\ is\ an\ end-vertex. \\ k & if\ u\ is\ a\ cut-vertex. \end{cases}$$

b. For any vertex u in the complete graph K_n, $dn^+_{\square f_u}(K_n) = 1$.

c. If G is the complete bipartite graph $K_{m,n}(2 \leq m \leq n)$, with partitions X and Y, then $dn^+_{\square f_u}(G) = \begin{cases} m-1 & if\ u \in X. \\ n-1 & if\ u \in Y. \end{cases}$

d. If G is the cycle C_n of order n,
$$dn^+_{\square f_u}(C_n) = \begin{cases} 1, & if\ 3 \leq n \leq 5. \\ 2, & if\ n \geq 6. \end{cases}$$

e. If G is the wheel $W_n = K_1 + C_{n-1}$
$$dn^+_{\square f_u}(W_n) = \begin{cases} 1 & if\ u \in C_{n-1}, n = 4,5 \\ 2 & if\ u \in C_{n-1}, n \geq 6. \\ \dfrac{n-1}{2} & if\ u \in K_1, n \geq 4. \end{cases}$$

Proof. (a) This is similar to that of Theorem 2.

(b) This is similar to that of Theorem 1 and Corollary 2.

(c) This follows from Theorem 3.

(d) This follows from Theorems 4 and 5.

(e) This follows from Theorems 6, 7, 8, and 9.

Theorem 11 For any pair of integers α and β with $1 \le \alpha \le \beta$, there exists a connected graph G in which

$$dn_{f_u}(G) = \alpha - 1 \text{ or }$$

$$dn_{f_u}(G) = \alpha \text{ and }$$

$$dn^+_{f_u}(G) = \beta - 1 \text{ or } dn^+_{\square f_u} = \beta \text{ for some vertex}$$

u in G.

Proof. When $\alpha = \beta = 1$, $K_n(n \ge 2)$ possesses the desired properties. When $\alpha = \beta$ and $\beta \ge 2$, any tree T with order $n \ge 3$ and β end-vertices is sufficient. Then by Theorem 2(i), $dn_{\square f_u}(G) = dn^+_{\square f_u}(G) = \beta - 1$ for any end-vertex and $dn_{\square f_u}(G) = dn^+_{\square f_u}(G) = \beta$ for any cut-vertex in a tree·

When $1 \le \alpha < \beta$. Let F_1 be a bistar $B_{3,\alpha-1}$ obtained by joining the central vertices r_0 and s_0 of $K_{1,\alpha-1}$ and $K_{1,3}$ respectively where $V(K_{1,3}) = \{S_0, S_1, S_2, S_3\}$ and $V(K_{1,\alpha-1}) = \{r_0, r_1, r_2, ..., r_{\alpha-1}\}$. Let $P_3^{(i)} : t_i u_i v_i (1 \le i \le \beta - \alpha + 1)$ be the i copies of the path P_3. Consider F_2 is the graph derived from F_1 by joining $t_i (1 \le i \le \beta - \alpha + 1)$ to s_3 of bistar. Let F_3 be the required graph G shown in Figure 1, derived from F_2 by joining s_2 and v_1 with a chord.

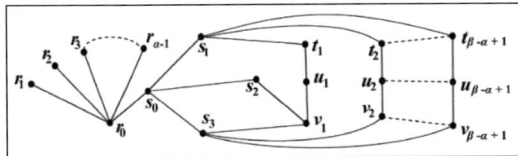

Figure 1. G

Suppose a set $S^e = \{r_1, r_2, ..., r_{\alpha-1}\}$ containing the end vertices of G and a set $S^c = \{r_0, s_0\}$ containing the cut vertices of G. Let $S^* = \{r_i, s_1 : 1 \le i \le \alpha - 1\}$ where $r_i \in S^e$ and $|S^*| = \alpha$. Now it is obvious that $I_{D_{\square f}}[S^*] = V(G)$. Let S_u be a u-square free detour set in G. By Theorem 2(i), vertex u does not belong to $dn_{\square f_u}$-set. Thus

$|S_u| = |S^*| - 1 = \alpha - 1$ for any vertex $u \in S^e$ in G. Suppose now $S^{**} = S^* \cup S^c$. We find that $I_{D_{\square f}}[S^{**}] = V(G)$. Then by Theorem 2(i), $|S_u| = |S^{**}| - |S^c| = \alpha + 2 - 2 = \alpha$. Hence $dn_{\square f_u}(G) = \alpha - 1$ for every end-vertex of G or $dn_{\square f_u}(G) = \alpha$ for every cut-vertex of G.

Next, we prove that $dn^+_{\square f_u}(G) = \beta - 1$ or $dn^+_{\square f_u}(G) = \beta$. Suppose $S^+ = S^e \cup \{s_1\} \cup \{u_1, u_2, ..., u_{\beta-\alpha+1}\}$ with $|S^+| = (\alpha - 1) + 1 + (\beta - \alpha + 1) = \beta + 1$. Let S_u be a u-square free detour set for some u in G. Then Theorem 5(i), $|S_u| = |S^+| - 1 - 1 = (\beta + 1) - 2 = \beta - 1$ when $u \in S^e$ and by theorem 2(ii), $|S_u| = |S^+| - 1 = (\beta + 1) - 1 = \beta$ when $u \in S^c$. Obviously, S_u is a u-square free detour set for some vertex u in either S^e or in S^c of G. We now claim that S_u is minimal u-square free detour set. To prove by contradiction, assume that there is a subset M_u such that $M_u \subset S_u$. Let $u_2 \in M_u$ which is not in S_u. Then u_2 must be in $\{u_1, u_2, ..., u_{\beta-\alpha+1}\} \subset V - \{S^+ \cup S^e\}$. We note that u_2 does not belong to any u-$u_j (1 \le j \le \beta - \alpha + 1)$ square free detour set where $u \in S^+$. This leads to a contradiction that M_u is not a u-square free detour set. Thus S_u is a $dn^+_{\square f_u}$-set in G. Hence $dn^+_{\square f_u}(G) = \beta - 1$ when $u \in S^e$ and $dn^+_{\square f_u}(G) = \beta$ when $u \in S^c$.

Conclusion

In this paper, we determined the upper vertex square free detour number for some standard graphs. Furthermore, we proved the realization result. Derivation of similar results in the context of some other classes of graphs is an open area of research.

References

Chartrand, G., Johns, G. L., and Tian, S. (1993), Detour distance in graphs. In Annals of discrete mathematics, Elsevier. 55, pp: 127–136.

Chartrand, G., Johns, G. L., and Zhang, P. (2003). The detour number of a graph. Utilitas Mathematica, 64.

Ramalingam, S. S., Asir, I. K., and Athisayanathan, S. (2017). Upper Vertex Triangle Free Detour Number of a Graph. Mapana Journal of Sciences, 16(3), 27–40.

Ramalingam, S. Sethu, I. Keerthi Asir, and S. Athisayanathan (2016). Vertex Triangle Free Detour Number of a Graph. Mapana Journal of Sciences, 15(3), 9–24.

Santhakumaran, A. P., and S. Athisayanathan. On the connected detour number of a graph. Journal of Prime Research in Mathematics, 5 (2009), 149–170.

AI-Driven Proctoring System

Subramani, R.,ᵃ and P. Arulpandy²

ᵃ,ᵇDepartment of Mathematics, CHRIST(Deemed to be University), Bengaluru-560029, India
E-mail: subramani.r@christuniveristy.in, arulpandy.p@christuniversity.in

Abstract

Remote learning has become increased drastically during pandemic, especially in last two years. Nevertheless, there has yet to be a satisfactory answer to academic tests. Some colleges have gathered assignments that students can copy directly from the internet, on the other hand, some have established remote proctoring, in which a manual proctor monitors student action. One should find a solution to this until our lives come back to normal mode. In this research, we present a method for developing an AI-based integrated system that can aid in the prevention of exam cheating. We have used Computer Vision techniques for building that system. We have used YOLOv3 algorithm and built the proctoring system for Multiple Person detection and Mobile Phone detection.

Keywords: Artificial Intelligence, Proctoring System, Computer Vision, YOLOv3

Introduction

COVID-19 has marked the beginning of a new era in online education. Educational Testing Service (ETS), which administers the TOEFL and GRE, is allowing the registered ones to take the examinations sitting right in their home place and to the whole duration of the exam will be observed in a person. This kind of process requires many people and it may not be that convenient to do. Massive open online courses (MOOCs) and many related types of long learning are growing in popularity and reach. The capability to successfully control and invigilate the tests which are conducted online is a big limitation for education to grow. Proctoring manually, which entails either compelling test takers to visit an examination center or watching them visually and vocally throughout exams through a webcam, is currently the most common technique of assessment. Such approaches, on the other hand, are time-consuming and costly.

Instead of taking classes in a traditional way of class teaching in the institution, students can now take classes utilizing a computer from any place across the world, where the people who teach provide knowledge through different ways and kinds of platforms. It also contains remote methods for validating the person who takes the exam as the individual who is supposed to be the person registered to the exam. In addition, any automated processes that aid secure a test administration event are included in the definition of online proctoring.

The phrase "online proctoring" is more informative than "remote proctoring" and is preferred. It underlines the importance of using the Internet and automated processes to create a secure system for testing taker monitoring. Remote proctoring, on the other hand, is a phrase that can be used to describe any type of proctoring that takes place outside of a traditional testing location. For several decades, the popular "find your own proctor" paradigm, which is also known as remote proctoring, has been a less-than-ideal, non-technology-based alternative for supervising exam administration for distance education courses.

Online proctoring consists of two main components. The first component consists of turning on the webcam in order to observe what is going on around the student. So this can be seen and observed by a person throughout the entire exam. The student can be get caught while doing any malicious activities during the examination such as any lip movement, found talking to others, using a mobile phone, copying answers from a book or another lappy and any other suspicious activity like looking away from the screen. The second option is a lockdown, which prevents students from accessing any other computer software, including the Internet browser and not allowing any means of copying and cheating. In our model we implemented to proctoring the students without the supervision of the proctor by using AI-driven tools. As a result, we developed an Artificial Intelligence-based Python program that can proctor the students with the help of the webcam on the lappy and microphone, which will help the invigilators to monitor many pupils at the same time.

Object detection – detecting and identifying an object is the process of passing this intelligence to computers. Object detection has many applications in various different domains that include surveillance of video's, and retrieving images from videos, including video surveillance, image retrieval systems, and autonomous vehicle driving systems etc., These tasks can be achieved using various algorithms and techniques, but in our paper, we will focus on YoloV3 algorithm.

The paper's organization is as follows: Section two provides the literature survey of the paper. Section three deals with YOLOv3 architecture. In section four depicts the implementation of the model with an algorithm and results analysis. Conclusions and future scope are provided in section five.

Literature Survey

The need for online learning has grown dramatically over the years. Researchers have offered a number of approaches for proctoring exams in online mode in an effective manner while maintaining academic integrity. The authors of (Y. L Prasanna, et al, 2021; S S Teja, et. al 2021) claim that by proposing 8 control techniques that allow professors to enhance the difficulty and hence lower the possibility of cheating, they may encourage academic honesty. The authors of (H. S. G. Asep et. al 2019; N. Malhotra, et. al 2022) propose a secure web-based test system as well as a network design that should prevent cheating.

Researchers have also developed various comprehensive monitoring solutions, such as (A. J S, H. S. Kumaran, et. al 2021), in which they employ snips to lower the cost of bandwidth for broadcasting enormous data in video format. In the paper of (N. Soman, et. al 2017) there developed a robot with a 360-degree camera and sensors that detect motion to attempt semi-automated machine proctoring. If any suspicious motion or footage is obtained, this robot sends the video to a monitoring center. Authors used MOOP system in (M Tanuj, et. al 2021), in that they included many ways like which incorporate both robust and collective ways to identify cheating in online tests. They have three pieces of hardware: 2 webcams and one tracking the gaze.

More than the applications related to education, previous work on audio-visual-based behavior recognition has been done in the multimedia community. The authors of A. J S, H. S. Kumaran, et al. (2021); S. S. Teja Gontumukkala, et al. (2021) evaluate conversational and emotional behavioral aspects of people who are interacting using both voice and video recordings of moving of the person's faces in the interaction. N. V. Sai Prakash Nagulapati et al. (2022) uses nonverbal clues derived from audio-visual data to predict the hire ability of applicants and interviewers in real job interviews. They use audio-video cues to significantly predict the ups and downs in degrees of combined cohesion in N. V. Sai Prakash Nagulapati et al., (2022).

YOLOv3

Version 3 of YOLO which is YOLOv3 mainly detects objects in videos, films, and photos etc., it is basically used for real-time object detection. It uses deep CNN's for learning features. Compared to the other two version of YOLO. It can be worked using the deep learning libraries of Keras and OpenCV. In order to identify different items in a particular class, various classification libraries are used by deep neural network algorithms. Things in photographs will be divided into certain clusters inside that certain objects with the same type of attributes will be clustered together, and the remaining will be taken care of whenever it is required. The information learned by the convolutional layers is handed on to a classifier, which makes the detection prediction, as is customary for object detectors. In the algorithm, the prediction will be dependent on that convolutional layer with 1x1 convolutions.

YOLO is very efficient that it maintains the highest accuracy compared to other methods. Unlike other algorithms, if an input image is given at the testing phase, it will take a quick and complete scan of the image so that we can draw conclusions based on the whole information of the image. Many other algorithms can give the regions a score value similar to the YOLO by some comparison techniques. Some regions have a high score, in those regions only we mainly observe positive detections.

Architecture

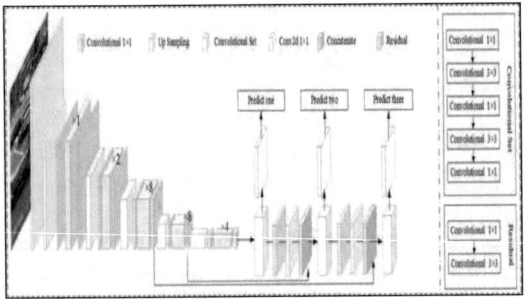

Fig. 1. YOLOv3 Architecture

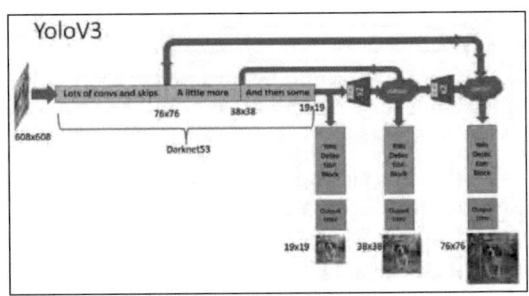

Fig. 2. YOLOv3 Architecture

A grid structure is followed by the algorithm for dividing the input image. If we consider the highest score regions of current grid boundaries and already defined ones, they will contribute to certain placements and predictions. So this anchor box will recognize only one item. Hence the prediction accuracy will also be dependent on the confidence score of the anchor boxes which are also called boundary boxes. The ground truth boxes' dimensions are clustered to identify similar shapes and sizes, and also anchor boxes are designed by finding the most common shapes and sizes from the standard data.

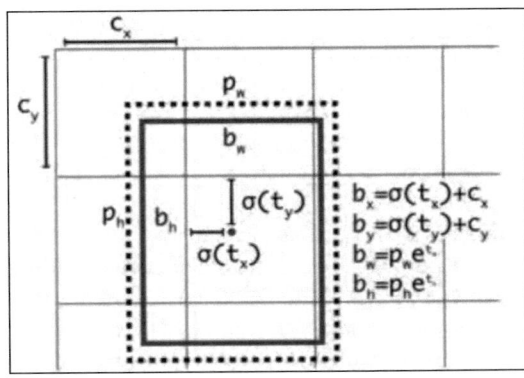

Fig. 3. Bounding Box Representation

R-CNN (Region-based Convolutional Neural Networks), Fast R-CNN, and Mask R-CNN will be different comparable algorithms that are designed for the same purpose. And YOLO algorithm compared to the above three is designed to do a regression of the anchor boxes and classification both at the exact instant.

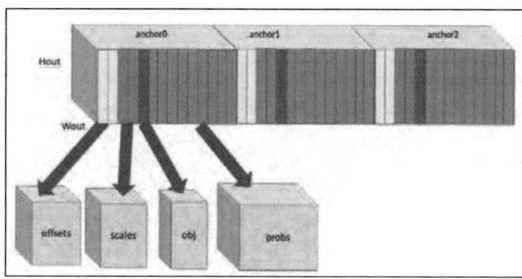

Fig. 4. YOLOv3 Output Layer Scheme

Accuracy and Specificity

Precision accuracy, speed of the algorithm and specificity measure are the main metrics considered for differentiating YOLOv3 from the other algorithms.

	Type	Filters	Size	Output
	Convolutional	32	3 × 3	256 × 256
	Convolutional	64	3 × 3 / 2	128 × 128
1×	Convolutional	32	1 × 1	
	Convolutional	64	3 × 3	
	Residual			128 × 128
	Convolutional	128	3 × 3 / 2	64 × 64
2×	Convolutional	64	1 × 1	
	Convolutional	128	3 × 3	
	Residual			64 × 64
	Convolutional	256	3 × 3 / 2	32 × 32
8×	Convolutional	128	1 × 1	
	Convolutional	256	3 × 3	
	Residual			32 × 32
	Convolutional	512	3 × 3 / 2	16 × 16
8×	Convolutional	256	1 × 1	
	Convolutional	512	3 × 3	
	Residual			16 × 16
	Convolutional	1024	3 × 3 / 2	8 × 8
4×	Convolutional	512	1 × 1	
	Convolutional	1024	3 × 3	
	Residual			8 × 8
	Avgpool		Global	
	Connected		1000	
	Softmax			

Fig. 5. Darknet-53

In terms of accuracy, speed, and architecture, YOLO version 2 and version 3 are poles apart. And on 2016, YOLOv2 was released, two years of release of YOLO v3.

Darknet-19 is involved mainly in extracting the features in YOLO version 2, whereas Darknet-53 is now the backbone feature extractor for YOLOv3. The YOLO founders also developed Darknet-53, a standalone. Darknet-53 is made up of 53 convolutional layers and 19 have 19 layers. This increase in number of layers makes a great advantage for 53. Quickness and Precision of YOLO version 3 van be observed in mean average precision (MAP) and values of IOU.

Usage of YOLOv3

One can download weights and cfg files from the original designer of YOLOv3's website. One can also be able to access the pre-trained weights by writing the model as equal to YOLOv3() equation. When you use the weights that are already trained on COCO, the YOLO can only be used for detecting the objects with one of the 80 pre-trained classes included in the dataset.

```
'person', 'bicycle', 'car', 'motorcycle', 'airplane', 'bus', 'train', 'truck',
'boat', 'traffic light', 'fire hydrant', 'stop sign', 'parking meter', 'bench',
'bird', 'cat', 'dog', 'horse', 'sheep', 'cow', 'elephant', 'bear', 'zebra',
'giraffe', 'backpack', 'umbrella', 'handbag', 'tie', 'suitcase', 'frisbee',
'skis','snowboard', 'sports ball', 'kite', 'baseball bat', 'baseball glove',
'skateboard', 'surfboard', 'tennis racket', 'bottle', 'wine glass', 'cup', 'fork',
'knife', 'spoon', 'bowl', 'banana', 'apple', 'sandwich', 'orange', 'broccoli',
'carrot', 'hot dog', 'pizza', 'donut', 'cake', 'chair', 'couch', 'potted plant',
'bed', 'dining table', 'toilet', 'tv', 'laptop', 'mouse', 'remote', 'keyboard',
'cell phone', 'microwave', 'oven', 'toaster', 'sink', 'refrigerator', 'book',
'clock', 'vase', 'scissors', 'teddy bear', 'hair drier', 'toothbrush'
```

Fig. 6. Classes in COCO dataset

Implementation

In our paper, we have utilized phone detection and person counting using YOLOv3 algorithm. This YOLOv3 algorithm is trained on COCO dataset to detect people and mobile phones in video streams. We can get an alert if the total count is not the same and the index of the mobile phone in the dataset is around 67, so we should be able to identify the similar indexes of the class and then should report that it is a mobile phone.

With TensorFlow and OpenCV in Python, we used instance segmentation (YOLOv3) to identify the no of people using its pre-trained weights. We have three different algorithms that is YOLOv3 Faster CNN and SSD. Among these YOLOv3 has fastest speed. The YOLOv3 pre-trained model can categorize 80 objects and is nearly as quick and accurate as SSD. It has 53 convolutional layers, each with a batch normalization layer and a leaky RELU activation layer between them. Instead of employing pooling, they employed a stride of 2 in convolutional layers to downsample. The input image must be in RGB have a type of float32, and have dimensions of 320x320, 416x416, or 608x608. Anchor boxes aid in the model's specialization. IOU is used in non-maximal suppression or NMS. The ratio between intersection and union of two boxes is called IOU. The box has the highest chance of being detected is chosen.

Algorithm

1. Import the essential packages.
2. Assign the initial weights to the layers in the model which are called darknet weights. Also, create a function for construction of the convolutional layers and check the batch normalization is required or not.
3. Construct the functions for residual layers and block layers which will in turn work with convolutional layers which we have created earlier and overall a functional block should be built to create the whole DarkNet model.
4. Create the convolutional layers for YOLOv3 model and also the outputs of the model.
5. Create YOLOv3 model (Define the way to extract the outputs. Additionally define anchor boxes before creating a function that integrates everything to construct our model).
6. Convert webcam images to RGB, resize them to 320x320, 416x416, or 608x608, change their datatype to float32, extend them to four dimensions, and divide them by 255.

Results

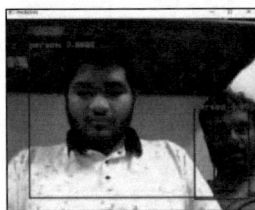

Fig. 8. Mobile Phone, Multiple person & No person Detection

Conclusion

Taking online-proctored tests would necessitate additional assistance for students with various forms of disability. Furthermore, there will be a positive opinion in the use of this kind of learning in many open online courses and education around the world. In this paper, we have implemented the Artificial Intelligence-based proctoring tool for the genuineness of the candidate during their examination especially capturing a live video of the examinee and identifying whether he is using a phone or not, whether there is another person with him in front of a camera or not.

This has been achieved by constructing the functions for residual layers and block layers by creating the convolutional layers for YOLOv3 model. Moreover, we defined anchor boxes for the construction of our model. The future dimension of the research can be extended by incorporating more variables such as eyeball rotation and other small objects.

References

Prasanna, Y. L., Tarakaram, Y., Mounika, Y. and Subramani, R. (2021), Comparison of Different Lossy Image Compression Techniques, 2021 International Conference on Innovative Computing, Intelligent Communication and Smart Electrical Systems (ICSES), pp. 1–7.

Asep, H. S. G., and Bandung, Y. (2019). A Design of Continuous User Verification for Online Exam Proctoring on M-Learning, 2019 International Conference on Electrical Engineering and Informatics, pp. 284–289, doi: 10.1109/ICEEI47359.2019.8988786.

Malhotra, N., Suri, R., Verma, P., and Kumar, R. (2022). Smart Artificial Intelligence Based Online Proctoring System, 2022 IEEE Delhi Section Conference (DELCON), pp. 1–5, doi: 10.1109/DELCON54057.2022.97.

Chandana Sarvani, A. H. N. S., Sai Bharath, B., Vijaya Bharathi Reddy, P. R. and S. R, (2022), AI-Driven Medical Imaging Analysis for COVID-19 Detection, 2022 International Conference on Electronics and Renewable Systems (ICEARS), pp. 1799–1804, doi: 10.1109/ICEARS53579.2022.9751986.

Teja Gontumukkala, S. S., Varun Godavarthi, Y. S., Ravi Teja Gonugunta, B. R., Subramani, R. and Murali, K. (2021), Analysis of Image Classification using SVM, 2021 12th International Conference on Computing Communication and Networking Technologies (ICCCNT), pp. 1–6, doi: 10.1109/ICCCNT51525.2021.9579803.

A. J S, H. S. Kumaran, S. U, K. P. B. V. Rajesh, and L. R, (2021), Deep Learning based Approach for Facilitating Online Proctoring using Transfer Learning, 2021 5th International Conference on Computer, Communication and Signal Processing, pp. 306-312.

Tanuj, M., Virigineni, A., Mani, A., and Subramani, R. (2021), Comparative Study of Gradient Domain Based Image Blending Approaches, 2021 International Conference on Innovative Computing, Intelligent Communication and Smart Electrical Systems (ICSES), pp. 1–5, doi: 10.1109/ICSES52305.2021.9633858.

Soman, N., Devi, M. N. R. and Srinivasa, G., (2017). Detection of anomalous behavior in an examination hall towards automated proctoring, 2017 Second International Conference on Electrical, Computer and Communication Technologies (ICECCT), pp. 1–6, doi: 10.1109/ICECCT.2017.8117908.

Sai Prakash Nagulapati, N. V., Venati, S. R., Chandran, V., and S. R. (2022), "Pedestrian Detection and Tracking Through Kalman Filtering," 2022 International Conference on Emerging Smart Computing and Informatics (ESCI), pp. 1–6, doi: 10.1109/ESCI53509.2022.9758215.

Raj, R. S. V., Narayanan, S. A. and Bijlani, K., (2015). Heuristic-Based Automatic Online Proctoring System, 2015 IEEE 15th International Conference on Advanced Learning Technologies, pp. 458–459, doi: 10.1109/ICALT.2015.127

Preliminary Research for Adhesive Peeling Device

Khushali Pokar,[*] *Mugdha Tambe,*[b] *Pratibha Sharma,*[c] *Sagarika Pawar,*[d] *Samruddhi Patil,*[e] *Sharvari Damle,*[f] *Zahra Bhuraniwala,*[g] *and K. K. Singh*[e*]

Karnavati University, Gandhinagar, India
E-mail: [*]kk@karnavatiuniversity.edu.in

Abstract

After some studies, it was found that adhesive stickers are required but the residual adhesions left on the face after peeling are problematic. In "Discovery Avery," Removing stickers is considered a game for them. According to research, they have accepted the fact that adhesive labels will not come off easily and it is okay to remove them with various methods such as alcohol use, citrus cleaner. Placing sticky labels on compressed toys makes the hair stick and pull it out while removing. Avery., October 23, 2020 found that there are separate updates to the label labels. Product labels, tutorial labels, warning labels, and important price labels; many of them are legally required. Every sticky label leaves behind a sticky residue. People Associated Age group with review Young teens find the label very annoying, hating sticky labels as found in everything around us.

Introduction

This paper provides a basic treatment for appropriate sticker removal for people annoyed with it. This review of the literature aims to understand the problems associated with labels/adherence stickers and the problems associated with existing solutions. Many research papers are thoroughly researched to study the causes of various agricultural problems and risks and what can be done to reduce them. Glem Hudak, Paul, and Kiln say, the label is a way, the need to identify, to distinguish, to differentiate - in other words, to create order in the human world is very much possible, a feature of the general human being. Labeling is one way to achieve this concept of order. It allows us to build common boundaries so that we can see our surroundings and find shared foundations for our judgment. Labels value and hold the wealth of descriptive information. Attachment labels are found to be very annoying when removing them. We all know how important it is to be a product to show education, pricing, methods, uses and much more but should they be too sticky to bother someone to remove? Kasman says his annoyance with the label labels: - "The problem arises when it is attached to an item we are buying and we want to remove the label before using the item. This is especially annoying when something is intended as a gift. I always try to remove any entry labels before giving a gift but it is not always easy. Obtain relevant response information from 3 different years and their existing solutions that can help us find a permanent solution for adhesive labels and also get the right details of the adhesive stickers' response to different types of products.

Background of Current Research (Present Scenario of Current Research)

Based on the above problem people tend to find some cheaper and some expensive solutions to solve the problems. Also, use of thinners and alcohol is widely used for removing stickers from hard and non-reactive surfaces. Stickers from the food items, especially from the fruits are removed by 21 soaking fruits in the water for a few minutes or by scrapping the piece of fruit which has a sticker on it. The stickers and labels are put on the product to track the required information of the respective product also it contains basic data such as serial number, manufacturing date, price of the product, nowadays each and everything we buy has label and other informative stickers on it for convenience and guidance of users.

Literature Review

A literature review to overview the previously published works on a particular topic. Search and review publications available on your given topic or in the area of your chosen topic. Writes a state of the art in relation to the topic or topic you are writing about. It examines books, academic articles, and other sources relating to a particular subject, research field, or theory, and in so doing, provides a comprehensive, concise, and critical assessment of these activities in relation to the research problem under investigation. Basically, a book review is a summary of all the reviews you have read or read and put them in one text, under which we review the ways or means by which we can learn the existing solution, various methods etc. We have lessons of some types of stickers and labels that seem to reach the surface of the material they are attached to, and trying to remove these complex adhesives and adhesive residues they leave behind can be really stressful.

According to research, it has been found that 10-30 yrs old is very annoying, sticky labels as they are found in everything in the environment. Because of their limited strength and endurance, they are annoyed at this age group. For elementary school children, it is very annoying as it sticks to their hands. People from the 30-60 yrs age group are fine if glue residues are left behind. As they try other available solutions such as alcohol-based cleansers, vinegar, orange cleaner, etc. And people from the age group of 60-90 yrs are satisfied with the labels of the

goods as they are quiet and knowledgeable about removing it.

Adhesive labels are divided into 3 categories such as:

- Permanent adhesion: most of the labels we use are permanent adhesive because they work for a wide range of labels, and applications and are found to be very friendly. These labels form a strong bond with the face thus removing it can cause further damage.

  Removable adhesive: this type of label can be easily removed after a certain period of operation without damaging the surface. These labels/stickers are ready to temporarily tag items, store coupons, and limited-time promotions.

  Reusable attachment: this is designed for temporary removal in cases where the label needs to be repositioned or reused. It saves your time and resources over time when the labels need to be removed to be relocated without losing much of the stickiness.

  The factors on which adhesive labels depend are:
- Strength of adhesion.
- The longer the time, the label adheres to it.
- The temperature label is open.
- The type of place is attached.

Solutions available: - If you remove the clipboard - at the back of the face you may be irritated by the clipper left on the back. If you are tempted to scratch the remains of a sticker with a knife or any other sharp tool you can scratch the surface of the gunk.

Other stickers to remove sticky residues using homemade products for washing soda, peanut butter, hair dryers, alcohol detergents, baking soda and coconut oil, mayonnaise, and vinegar.

Problem Statement

Certain types of stickers and labels appear to blend in on the surface of the object, and to try to remove these problematic adhesives, the residual residue left behind is truly disturbing. Removing remnants takes effort and time, and leaves residues of glue.

Aim

To develop an adhesive peeling device.

Objectives

Study the available methods and techniques for stripping stickers - read.

- Studying the problems people face as a result of pasting stickers in the workplace/ home - a visual lesson.
- Building a tangible object of peeling.

Methodology

The survey questionnaire also had such questions, two changes, and asked for opinions, results, and benefits. So we got a lot of information by observing the stickers on different items. There are different types of stickers on different items. We got a lot of information from this survey. And it got a lot of information. Feels attractive, sometimes having stickers on products like electronic gadgets it doesn't bother but in few items like fruits does bother. This survey result presents the conclusion of how to remove sticker residue from the surface. The questionnaire included objective questions and brief questions, Steel utensils such as dishes, and glass have stickers exactly in the middle of usable parts, and unnecessary and harmful snickering should be avoided.: Methodology: We used Google Forms as a tool to conduct the survey. And stickers stick to the material of those objects. How do people of different ages react to it? This helped us to know the opinions of the people. What is your opinion on products having stickers on them? This includes how people react to the problem. It consists of interviews, and questionnaire. There was also a good response from the people It should have but maybe something easily coming off Should have stickers that do not leave any adhesive after removal Similarly, we interviewed people of different ages to find out the difficulties of these stickers from children to the elderly. And they were hard to remove.

Result

Also, 20.6% of people try to remove stickers by DIY method. The number of people removing stickers by flexible method is high. Also, 32.3% of people rarely succeed in removing stickers. 75.4% of people remove stickers from products and 7.7% of people do not remove stickers from products. This graph shows the results of a survey in which 61.9 people remove stickers using a flexible method. 7.9% of people use items available in the market, while 9.5% use other items to remove stickers. This graph shows the results of how many people removed the sticker from the product and keep it as it is. Considering the graph of how often you succeed in removing stickers quickly, it is clear that 43.1% of people are very successful in removing stickers. Also, 49.2% of people think that one or two items have stickers on them. The graph results show that if the stickers are not peeled off, will it make a big difference in your life? 57.8% of people think that if the stickers are not peeled off, it will make a big difference in their lives. A lot of research documentation was looked at in detail to find the right data. This graph shows that most labels leave behind sticky residues with 51.6% of people believing that labels leave behind sticky residues and 8.1% of people believing that labels do not leave sticky residues behind. Also, 1.5% of people think that not every item has stickers. The graph shows the result of when you damaged the product while removing the label 35.5% of people damage the product while removing the label and 37.1% of people do not damage the product while removing the label.

Conclusion

As for the person having OCD, for them it can do problems. Their attention would be stuck there, and they can not do some other work, without removing the sticker. The result enhanced understanding of how people reply towards the glutinous tags and what are their views on it. The result enhanced understanding of how outsides of a product reply towards the glutinous tags and how the overall product is affected. This fact is also noticed that the adhesive is the problem, not the tags. So concluding the entire exploration Its understood the problems faced worldwide on residue of the adhesive is really frustrating, and also deleterious if it comes to toddlers. We concluded the study that the main problem lies in the size which is used to stick stickers and not the entire paper. The study includes the result of how worldwide humans respond to suchlike problem areas, and also how humans always find side results to overcome suchlike problems. Also the veneers get damaged no matter what we use to remove them. We need to find a result that's soft, and white for the product and for yourself too. In Observation, It seems a small problem but it can yield serious issues in the future, to the natural body and product too. The results that subsist in requests are working but they've some problems that need to be worked on. The age group which is generally affected by this problem is 16–42. The possible result is to come with a result that can help to remove this without using any hazardous substance and are skin friendly for all age group. The investigation includes the result of how worldwide humans respond to matching problem areas, also humans always find side results to overcome matching problems. Also, the shells get damaged no matter what we use to remove them. The result meliorated understanding of how people respond towards the sticky tags and what are their views on it. This fact is also noticed that the glue is the problem, not the tags. As for the person having OCD, for them, it can produce problems. Their attention would be stuck there, and they can not do some other work, without removing the sticker. The result perfected understanding of how people respond towards the tenacious tickets and what are their views on it. The influence perfected understanding of how skins of products respond towards the tenacious tickets and how an overall product is affected. This fact is also noticed that the size is the problem not the ticket.

References

Hudak, Glam, Kiln, Paul (2001). Labeling: Pedagogy and Politics. A study on labels and what they refer to.

Mary Wickison (1997). How to Remove Sticky-Gooey-Icky Adhesive Labels. Clinical Chemistry, 43(3), 1 March 1997, p.548, Adhesive Labels Cause High Thyroxine Results Carolina Peixe, Conceição Casanova, Joana Lia Ferreira, Inês Coutinho. (2019) Glass Crystal Models Quanqing Yu, Rui Xiong, Chuan Li, Michael G. Pecht. Water-Resistant Smartphone Technologies. IEEE Access 2019, 7, 42757–42773. Https://doi.org/10.1109/ACCESS.2019.2904654

Zhang, Zedong, Lan, Dongming, Zhou, Pengfei, Li, Jun, Yang, Bo, Wang, Yonghua (2017). Control of sticky deposits in wastepaper recycling with thermophilic esterase. Cellulose, 24, 311–321. Https://doi.org/10.1007/s10570-016-1104-x

Karwoski, A. C., and Plaut, R. H. (July 2005) https://onlinelibrary.wiley.com/action/dosearch? Contribauthorst ored=Karwoski%2 C+A+C

Jerry, C. (2018) https://whyevolutionistrue. com/2018/09/30/i-hate-fruit-stickers/

Bhatti, A., Ash, J. Gokani, Singh, S., https://spiral. imperial.ac.uk/handle/10044/1/67488

Karwoski, A. C., Plaut, R. H. (July 2005). https:// onlinelibrary.wiley.com/action/dosearch? Contribautho rstored=Karwoski%2 C+A+C

Jerry, C. (2018) https://whyevolutionistrue.com/2018/ 09/30/i-hate-fruit-stickers

Bhatti, A, Ash, J Gokani, S. Singh, https://spiral. imperial.ac.uk/handle/10044/1/67488 Survey

https://www.reddit.com/r/doesanybodyelse/ comments/qjwnf/d ae_have_an_irration al_ fear_of_stickers/?Utm_source=amp&utm_ medium=&utm_c ontent=post_body

Mercedes Benz (Aug 2018) 25 https://www.drivemb. com/blogs/1653/uncategorized/hatebum per-stickersfamilystickers-cars/ Bill Kasman (July 2, 2018)

Guidelines for Material Selection for Brackets to Treat Orthodontic Patients

Sharad Shetty,[a,] Varshin Vala,[b,*] Aliasgar Chunawala,[c] Chakravarthy,[d] Ashima Banker,[e] and Abhijoy Banerjee[f]*

Karnavati University, Gandhinagar, India
E-mail: *varshin@karnavatiuniversity.edu.in

Abstract

The study was conducted to check how material plays an important role in creating brackets for orthodontics as well as their chemical nature and how it interacts with human muscles. In this study, even emphasis was given to how treating a material with another coat helps to overcome corrosion problems. Orthodontic Brackets are used as a temporary attachment to the teeth during the period of Orthodontic treatment. In this method, wires are used to reshape the tooth structure by application of force.

Keywords: Orthodontic Brackets, Stainless Steel

History

Orthodontic brackets first came into the picture when the human desire to align crooked teeth arose. This resulted in the origin of orthodontics. As per the history, the initial inception of Orthodontic appliances to correct misaligned teeth was found in Etruscan in the Egyptian artifact (Kamak, 2015). It was found that the metal wire was wrapped around individual teeth to realign its shape. During the era of 1825–1923, many improvements were made in the field of orthodontics by Calvin S Case (1847–1923) and Norman W Kingsley (1825–1896) they came up with a solution named extraction for orthodontic purposes. This philosophy greatly influenced the basic design of orthodontics braces. Edward Hartley Angle (1855–1930) (Craig, 2002) was named as the "Father of Modern Orthodontics" because he played a crucial role in recognizing orthodontics as a separate (Kohl, 1964) science from general dentistry.

Material for Orthodontic

Material selection is an important criterion for Orthodontic brackets as it has to be feasible as well as it shouldn't react to a person who is using the brackets on top of that it shouldn't react with the food that person consumes. Initially, materials like Gold of either 14 karats or 18 karts were used to create Orthodontic brackets. The first person to use Stainless Steel was Rudolf Schwarz (Hotz, 1973). During this period many different materials were experimented with but people mostly use this material as listed below.

Stainless Steel

When it comes to Orthodontics Stainless Steel is one of the most widely used materials it is composed of Iron and carbon alloy which contains other materials like Nickel

Steel is an alloy composed of iron and carbon that contains Nickel, Chromium, and other materials that impart the property of resisting Corrosion (Yoo et al., 2008). This is again further divided into three main groups which consist of martensitic, ferritic, and Austenitic. The materials that are used in orthodontics are usually made from an Austenitic type that has 18% chromium, 8% nickel, 0.2% carbon (Kohl, 1964). To impart corrosion resistance in pure iron and around 11% chromium is added along with the addition of carbon which results in the formation of the alloy to prevent resistance.

Different Types of Stainless Steel

When it comes to stainless steel consists of basically 5 types which depend on their microstructure and chemical composition: duplex martensitic, ferritic, precipitation-hardening, and gaustenitic (Yoo, 2008).

Martensitic Stainless steel

As per the Study conducted by American Iron and Steel Institute (AISI), the concerned alloys comes under the family called 400 Series (Anusavice, 2004). To Increase its strength and improve hardness this alloy is treated with heat. But this material has lower corrosion resistance and after heat treatment, this can be further reduced [6]. This is used to manufacture blades and cutlery which belong to the family of AISI 440 and 420 types.

Ferritic stainless steel

This also shares 400 Series. The Chromium content in the alloy is usually in the range of 12% and 29% of and it also contains a very low amount of nickel, usually below 2%. They have strong corrosion-resistant, ductile, and magnetic and cannot be heat-treated (Anusavice, 2004). As this material will be difficult to weld and work so it's not preferred in dentistry.

Austenitic Stainless Steel

This Stainless Steel is one of the most commonly used materials for Preparing orthodontic brackets (Kocijan and Conradi, 2010). and wires (Anusavice, 2004, Izquierdo et al., 2010) Austenitic Steel has good corrosion resistance and good formability, weldability (Padilla, 1999), ductility and wear resistance (Kocijan and Conradi, 2010). Austenitic alloys tend to

have stress-corrosion cracking and intergranular corrosion (Cobb, 2010).

Duplex stainless Steel

Duplex stainless steel is made up of delta-ferritic phases and is authentic (Cardarelli, 2008); Platt, 1997). This alloy contains chromium around 18% to 26% and nickel around 4%–7%. This material has a higher tensile and yield strength along with that it has high weldability. It has more strength than aesthetic or ferritic stainless steel, high toughness (Cardarelli, 2008), and great resistance to uniform corrosion when compared with austenitic types (Cobb, 2010).

Super Stainless Steel

Stainless steel is mostly used in Orthodontic applications but there is also a study that says about the allergic reactions caused by nickel. Therefore, the use of sr-50A which is known as super Austenitic steel is widely used because of its localized corrosion resistance similar to that of titanium alloy (Oh et al., 2004, 2005).

Material Reaction

Saliva

Saliva can cause corrosion of orthodontic stainless steel because saliva contains viruses, bacteria, yeast, and fungi in its product (Anandkumar and Maruthamuthu, 2008). It has been evaluated by several authors that stainless steel has a good property to resist corrosion which can be used in Orthodontics (Costa, 2007; Lin, 2006). Following corrosion type is summarized.

Galvanic Corrosion: This type of corrosion occurs when two different metals are placed together in an electrolyte such as brackets and wires that are made of different alloys in the oral cavity this corrosion occurs (Iijima et al., 2006).

Corrosion fatigue: This type of corrosion occurs if metals are under some cyclic stress so when archwires are left in the oral cavity for a longer period and load stresses the corrosion is triggered.

Microbiologically-influenced corrosion: This corrosion occurs when microorganisms leave their by-products which affect the metals and cause corrosion which gradually increases with an increase in the amount of waste.

Fluoride Solutions

Fluoride solutions are very important for orthodontic alloys including stainless steel (Walker et al., 2007). Because the hydrogen-absorbing stainless steel was slightly affected compared to nickel-titanium and beta-titanium (Kao et al., 2007).

Heat Treatment

Heat treatment can improve the elastic strength of the material (Backofen and Gales, 1951). But this treatment cannot be performed at a temperature over 650C (Brantley and Eliades, 2001). Interlevel recommends maintaining the temperature of 18/8 stainless steel belove 400c to reduce corrosion resistance (Ingerslev, 1966).

Coating of Stainless Steel

A few of the coatings that can be used are Ceramic Coating (Pelaez-Vargas, 2005), Silver-platinum Coating (Ryu et al., 2010), Coating of Orthodontic brackets with polytetrafluoroethylene (Demling et al., 2010). This coating helps in reducing the friction between wires and brackets which in turn improves anchorage control (Mugurumam, 2011).

Other techniques to coat orthodontic appliances are called thermal and chemical procedures. Thermal procedures include Vapour Deposition (VD), Thermal Phase Separation (TPS), and Chemical procedures consisting of electrode position, electrophoresis, and sol-gel (Pelaez-Vargas, 2005)

Conclusion

Austenitic Stainless Steel is the preferred alloy for the manufacturing of wires, and brackets because of its good corrosion resistance and notable mechanical properties. However, a few metals have their own drawbacks considering the fact such as allergies, heat treatment, and even fluoride solutions. Therefore, better knowledge and guidelines for choosing the materials is an important step for a better result to treat orthodontics.

References

Hotz RP. The changing pattern of European orthodontics. Br J Orthod 1973, 1, 4–8.

Kamak, Hasan (2021). Orthodontic brackets selection, placement and debonding. Journal of Orthodontic Research, 3(3), 2015, p. 208. Gale Academic OneFile. Accessed 26 May 2021.

Craig, R., Powers, J. (2002). Restorative Dental Materials. 11 ed. St. Louis: Mosby.

Kohl, R. (1964). Metallurgy in Orthodontics. Angle Orthod, 34(1), 37–52.

Yoo, Y. R., Jang, S. G., Oh, K. T., Kim, J. G., Kim, Y. S. (2008). Influences of Passivating Elements on the Corrosion and Biocompatibility of Super Stainless Steels. J Biomed Mater Res B Appl Biomater. 86(2), 310–20.

Anusavice, K. Phillips Ciencia de los materiales dentales. 11a ed. Spain: Elsevier; 2004.

Kocijan, A., and Conradi, M. (2010). The Corrosion Behaviour of Austenitic and Duplex Stainless Steels in Artificial Body Fluids. Mater technol. 44(1), 21–4.

Izquierdo, P. P., De Biasi, R. S., Elias, C. N., Nojima, L. I. (2010). Martensitic Transformation of Austenitic Stainless Steel Orthodontic Wires during Intraoral Exposure. Am J Orthod Dentofacial Orthop. 138(6), 714 e1–5.

Padilla, D. (1999). Aplicaciones de los aceros inoxidables. Rev Inst investig Fac minas metal cienc geogr. 2(3), 11–21.

Cobb, H. M. (2010). The History of Stainless Steel. Ohio: asm International.

Cardarelli, F. (2008). Materials Handbook: A Concise Desktop Reference. 2nd ed. London: Springer.

Platt, J. A., Guzman, A., Zuccari, A., Thornburg, D. W., Rhodes, B. F., Oshida, Y., et al. (1997). Corrosion Behavior of 2205 Duplex Stainless Steel. Am J Orthod Dentofacial Orthop. 112(1), 69–79.

Oh, K. T., Choo, S. U., Kim, K. M., Kim, K. N. (2005). A Stainless Steel Bracket for Orthodontic Application. Eur J Orthod. 27(3), 237–44.

Oh, K. T., Kim, Y. S., Park, Y. S., Kim, K. N. (2004). Properties of Super Stainless Steels for Orthodontic Applications. J Biomed Mater Res B Appl Biomater. 69(2), 183–94.

Brantley, W., and Eliades, T. (2001). Orthodontic Materials. Scientific and Clinical Aspects. Stuttgart: Thieme.

Park, J., and Kim, Y. (1995). Metallic Biomaterials. The Biomedical Engineering Handbook. Boca Ratón, FL: CRC Press.

Damon, Q., Orange, C. A. (2011). Ormco Corporation. [Internet] [Cited 2011 Sept 11]. Available from: http:// www.ormco.com/index/damon-products-damonq-featuresbenefits-2.

Anandkumar, B., Maruthamuthu, S. (2008).Molecular Identification and Corrosion Behaviour of Manganeses Oxidizers on Orthodontic Wires. Curr Sci. 2008, 94(7), 891–6.

Costa, M. T., Lenza, M. A., Gosch, C. S., Costa, I. (2007). Ribeiro-Dias F. In Vitro Evaluation of Corrosion and Cytotoxicity of Orthodontic Brackets. J Dent Res. 86(5), 441–5.

Lin, M. C., Lin, S. C., Lee, T. H., Huang, H. H. (2006). Surface Analysis and Corrosion Resistance of Different Stainless Steel Orthodontic Brackets in Artificial Saliva. Angle Orthod. 76(2), 322–9.

Iijima, M., Endo, K., Yuasa, T., Ohno, H., Hayashi, K., Kakizaki, M., et al. (2006). Galvanic Corrosion Behavior of Orthodontic Archwire Alloys Coupled to Bracket Alloys. Angle Orthod. 76(4), 705–11.

Walker, M. P., Ries, D., Kula, K., Ellis, M., Fricke, B. (2007). Mechanical Properties and Surface Characterization of beta titanium and Stainless Steel Orthodontic Wire following Topical Fluoride Treatment. Angle Orthod. 77(2), 342–8.

Kao, C. T., Ding, S. J., He, H., Chou, M. Y., Huang, T. H. (2007). Cytotoxicity of Orthodontic Wire Corroded in Fluoride Solution in Vitro. Angle Orthod. 2007, 77(2), 349–54.

Backofen, W. A., and Gales, G. F. (1951). The Low-Temperature Heat-Treatment of Stainless Steel for Orthodontics. Angle Orthod. 21(2), 117–24.

Brantley, W., and Eliades, T. (2001). Orthodontic Materials. Scientific and Clinical Aspects. Stuttgart: Thieme.

Ingerslev, C. H. (1966). Influence of Heat Treatment on the Physical Properties of Bent Orthodontic Wire. Angle Orthod. 36(3), 236–47.

Pelaez-Vargas, A. (2005). Evaluación de la toxicidad in vitro, la adherencia y la nanotopografía de recubrimientos aplicados por sol-gel para implantes dentales. Medellín: Universidad Nacional de Colombia.

Ryu, H. S., Bae, I. H., Lee, K. G., Hwang, H. S., Lee, K. H., Koh, J. T., et al. (2010). Antibacterial Effect of silver-Platinum Coating for Orthodontic Appliances. Angle Orthod. 82(1), 151–7.

Demling, A., Elter, C., Heidenblut, T., Bach, F. W., Hahn, A., Schwestka-Polly, R., et al. (2010). Reduction of Biofilm on Orthodontic Brackets with the Use of a Polytetrafluoroethylene Coating. Eur J Orthod. 32(4), 414–8.

Muguruma, T., Iijima, M., Brantley, W. A., Mizoguchi, I. (2011). Effects of a Diamond-Like Carbon Coating on the Frictional Properties of Orthodontic Wires. Angle Orthod. 81(1), 141–48.

An Efficient Prediction of Kidney Disease Using Machine Learning Algorithms

Ajmeera Kiran,,a C. Vinothini,b C. Sathya,c M. Jayanthi,d K. Sai Prasad,e and P. Chinnasamyf*

a,e,fDepartment of Computer Science and Engineering, MLR Institute of Technology, Hyderabad, India
b,cDepartment of Computer Science and Engineering, Dayananda Sagar College of Engineering, Bangalore, India
dDepartment of Computer Science and Engineering, PSNA College of Engineering and Technology, Dindigul, India
E-mail: *kiranphd.jntuh@gmail.com, chinnasamyponnusamy@gmail.com

Abstract

Everyone is aiming to be health-conscious in today's environment, even though, owing to work and a tight schedule, one only pays attention to one's health when symptoms appear. Chronic Kidney Disease, on the other hand, is difficult to forecast, diagnose, and prevent since it has no symptoms or, in some situations, no disease-specific signs. It can cause lasting health harm. Machine learning, on the other hand, may be a solution to this issue because it excels at prediction and analysis. Using different machine learning techniques, we will assess datasets from patients with chronic kidney disease with many characteristics as well as 400 records. To develop a model that accurately detects whether a person has chronic kidney illness.

Keywords: Machine Learning, Logistic Regression, Classification, Decision Tree, Random Forest

Introduction

Chronic kidney disease (CKD) has gained a great deal of attention due to its high death rate. The World Health Organization claims that chronic diseases are becoming a significant threat to developing nations. Cardiovascular illness is brought on by the build-up of industrial wastes in the arteries, which also results in other health issues like high and low pressures, hypertension, sciatica, and musculoskeletal issues. Renal failure is another name for chronic kidney disease. It's a serious kidney condition that causes kidney function to deteriorate over time. The patient may have the following symptoms if CKD is not detected and treated early: High blood pressure causes anemia, weak bones, poor nutritional health, and nerve damage. As a result, it is vital to detect CKD at an early stage, although this is challenging due to the disease's gradual progression and lack of distinct symptoms. Machine learning can be beneficial in predicting whether or not a patient has CKD because some patients have no symptoms at all. The prediction model is trained with historical CKD patient data using different algorithms.

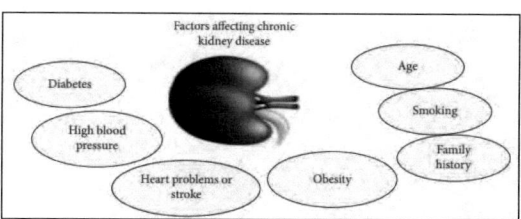

Fig. 1. Example Factors Scenario for kidney diseases

Literature Survey

Many scientists are working on predicting CKD using a variety of different classification techniques and those researchers get the model's expected result.

The Multiclass Decision Forest methodology outperforms earlier methods, according to research by Prasad et al. (2020) and Senen et al. (2021), with a reliability of about 99% for a small number of features of 14 attributes. Through the use of various machine learning-based classification approaches, Peera et al. (2017), and Wibwa et al. (2018), seem to have been able to speed up diagnosis and improve diagnosis precision. The intended research examines the severity-based classification of various CKD stages. utilizing techniques like RF, RBF, and the Basic Dissemination Multilayer Perceptron. With an efficiency of 85.3%, the analysis's findings demonstrate that the RBF algorithm performs more effectively than the other predictors. Ayesha et al. (2016) employed J48 and the Systematic Negligible Optimization (SMO) technique to evaluate and quantify CKD using the Weka tool. They suggested contrasting the Decision Tree and Support Vector Machine (SVM) approaches to diagnose CKD. The decision tree outperformed the SVM recognized incorrect classification in the classification step. Asif Salekin and John Stankovic employ a machine learning method to identify CKD in a unique way. A collection of 400 documents and 25 characteristics that indicate whether or not individuals have CKD is used to gather the researchers' observations by Devika et al. (2018) and Chittora et al. (2021).

Methodology

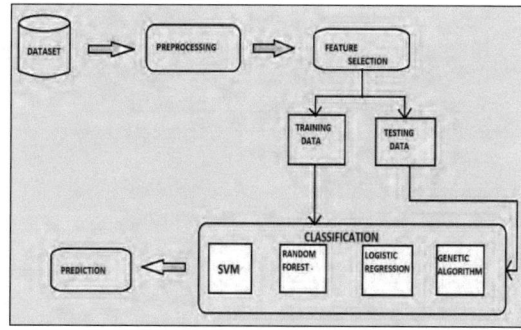

Fig. 2. Proposed Method data flow diagram

Dataset: A dataset of chronic renal disease prediction using a machine learning algorithm is accessible from the UCI repository (2015). The dataset comprises 400 patient records in total. They have 25 properties as well, but we only require 8 for model building. Specific gravity, hypertension, albumin, hemoglobin, red blood cell count, diabetes mellitus, hunger, and Pus cell are some of the characteristics.

Attribute Name	Attribute Type	Attribute Values	Attribute Code
Age	numeric	years	age
Blood Pressure	numeric	mm/Hg	bp
Spesific Gravity	numeric	1.005, 1.010, 1.015, 1.020, 1.025	sg
Albumin	numeric	0, 1, 2, 3, 4, 5	al
Sugar	numeric	0, 1, 2, 3, 4, 5	su
Red Blood Cells	nominal	normal, abnormal	rbc
Plus Cell	nominal	normal, abnormal	pc
Plus Cell Clumps	nominal	present, notpresent	pcc
Bacteria	nominal	present, notpresent	ba
Blood Glucosa Random	numeric	mgs/dl	bgr
Blood Urea	numeric	mgs/dl	bu
Serum Creatine	numeric	mgs/dl	sc
Sodium	numeric	mEq/l	sod
Potassium	numeric	mEq/l	pot
Hemoglobin	numeric	gms	hemo
Packed Cell Volume	numeric	-	pcv
White Blood Cell Count	numeric	cells/cumm	wbcc
Red Blood Cell Count	numeric	millions/cumm	rbcc
Hypertension	numeric	yes, no	htn
Diabetes Mellitus	numeric	yes, no	dm
Coronary Artery Disease	nominal	yes, no	cad
Appetite	nominal	good, poor	appet
Pedal Edema	nominal	yes, no	pe
Anemia	nominal	yes, no	ane
Class	nominal	ckd, notckd	class

Fig. 3. Chronic renal disease prediction

id	age	bp	sg	al	su	rbc	pc	pcc	ba	bgr	bu	sc	sod	pot	hemo	pcv	wc	rc	htn	dm	cad	appet	pe	ane	classification
0	48	80	1.02	1	0		normal	notpreser	notpreser	121	36	1.2			15.4	44	7800	5.2	yes	yes	no	good	no	no	ckd
1	7	50	1.02	4	0		normal	notpreser	notpresent		18	0.8			11.3	38	6000		no	no	no	good	no	no	ckd
2	62	80	1.01	2	3	normal	normal	notpreser	notpreser	423	53	1.8			9.6	31	7500		no	yes	no	poor	no	yes	ckd
3	48	70	1.005	4	0	normal	abnormal	present	notpreser	117	56	3.8	111	2.5	11.2	32	6700	3.9	yes	no	no	poor	yes	yes	ckd
4	51	80	1.01	2	0	normal	normal	notpreser	notpreser	106	26	1.4			11.6	35	7300	4.6	no	no	no	good	no	no	ckd
5	60	90	1.015	3	0			notpreser	notpreser	74	25	1.1	142	3.2	12.2	39	7800	4.4	yes	yes	no	good	yes	no	ckd
6	68	70	1.01	0	0		normal	notpreser	notpreser	100	54	24	104	4	12.4	36			no	no	no	good	no	no	ckd
7	24		1.015	2	4	normal	abnormal	notpreser	notpreser	410	31	1.1			12.4	44	6900	5	no	yes	no	good	yes	no	ckd
8	52	100	1.015	3	0	normal	abnormal	present	notpreser	138	60	1.9			10.8	33	9600	4	yes	yes	no	good	no	yes	ckd
9	53	90	1.02	2	0	abnormal	abnormal	present	notpreser	70	107	7.2	114	3.7	9.5	29	12100	3.7	yes	yes	no	poor	no	yes	ckd
10	50	60	1.01	2	4		abnormal	present	notpreser	490	55	4			9.4	28			yes	yes	no	good	no	yes	ckd
11	63	70	1.01	3	0	abnormal	abnormal	present	notpreser	380	60	2.7	131	4.2	10.8	32	4500	3.8	yes	yes	no	poor	yes	no	ckd
12	68	70	1.015	3	1		normal	present	notpreser	208	72	2.1	138	5.8	9.7	28	12200	3.4	yes	yes	yes	poor	yes	no	ckd
13	68	70						notpreser	notpreser	98	86	4.6	135	3.4	9.8				yes	yes	yes	poor	yes	no	ckd
14	68	80	1.01	3	2	normal	abnormal	present	present	157	90	4.1	130	6.4	5.6	16	11000	2.6	yes	yes	no	poor	yes	no	ckd
15	40	80	1.015	3	0		normal	notpreser	notpreser	76	162	9.6	141	4.9	7.6	24	3800	2.8	yes	no	no	good	no	yes	ckd
16	47	70	1.015	2	0		normal	notpreser	notpreser	99	46	2.2	138	4.1	12.6				no	no	no	good	no	no	ckd
17	47	80						notpreser	notpreser	114	87	5.2	139	3.7	12.1				yes	no	no	poor	no	no	ckd
18	60	100	1.025	0	3		normal	notpreser	notpreser	263	27	1.3	135	4.3	12.7	37	11400	4.3	yes	yes	yes	good	no	no	ckd
19	62	60	1.015	1	0		normal	notpreser	notpreser	100	31	1.6			10.3	30	5300	3.7	yes	no	yes	good	no	no	ckd
20	61	80	1.015	2	0	abnormal	abnormal	notpreser	notpreser	173	148	3.9	135	5.2	7.7	24	9200	3.2	yes	yes	yes	poor	yes	yes	ckd
21	60	90						notpreser	present		180	76	4.5			10.9	32	6200	3.6	yes	yes	no	good	no	no	ckd
22	48	80	1.025	4	0	normal	normal	notpreser	notpreser	95	163	7.7	136	3.8	9.8	32	6900	3.4	yes	no	no	good	no	yes	ckd
23	21	70	1.01	0	0		normal	notpreser	notpresent										no	no	no	good	no	yes	ckd
24	42	100	1.015	4	0	normal	abnormal	notpreser	present		50	1.4	129	4	11.1	39	8300	4.6	yes	no	no	poor	no	no	ckd
25	61	60	1.025	0	0		normal	notpreser	notpreser	108	75	1.9	141	5.2	9.9	29	8400	3.7	yes	yes	no	good	no	yes	ckd
26	75	80	1.015	0	0		normal	notpreser	notpreser	156	45	2.4	140	3.4	11.6	35	10300	4	yes	yes	no	poor	no	no	ckd
27	69	70	1.01	3	4	normal	abnormal	notpreser	notpreser	264	87	2.7	130	4	12.5	37	9600	4.1	yes	yes	yes	good	yes	no	ckd
28	75	70			1	3		notpreser	notpreser	123	31	1.4							no	yes	no	good	no	no	ckd

Fig. 4. Chronic renal disease attribute classification

Data Preprocessing

Organizing the data by formatting, cleaning, and sampling. **Formatting:** Selected data will not be in a suitable format or pattern. This data will be in a relational database and now it is feasible to have it in a flat file or proprietary file layout.

Cleaning: It is the deletion or fixing of missing data. We are no longer sure whether the records present are used for our mannequin advent or not being used and there are instances the place requires field's information is missing. These anomalies may need to be approached. Some of the attributes have sensitive information that may not be needed so we removed them.

Feature Selection: Following the computation of the missing values, it is necessary to determine the relevant features that have a strong and positive connection with aspects that are significant for disease diagnosis.

Classification: SVM, logistic regression, genetic algorithms, and random forest are 4 main machine learning algorithms that produce the greatest diagnostic findings. Machine learning algorithms are used to create predictive models.

Logistic Regression: It is a supervised learning method that examines absolute dependent values using a variable from among the needed blocks of independent values. Logistic regression of the absolute dependent values can be used to calculate the output values. The answers can thus be represented in either absolute or differential terms. Numerical or binary variables include yes or no, 0 or 1, true or false, and so on. However, this model reflects the range of potential values between 0 and 1.

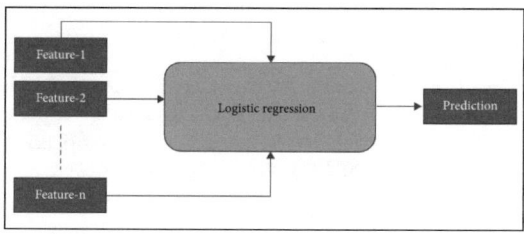

Fig. 5. Working of Logistic regression

Random Forest: Ensemble learning theory underpins random forest. It's a method that employs a range of classifiers to overcome the technology's compounded faults. The random forest, which consists of numerous decision trees that rely on different subsets of the provided information, uses the average to improve determining the speed of the dataset. The final result is the most common outcome for each observation. Each classification model receives a majority vote once a new observation is introduced into all of the trees.

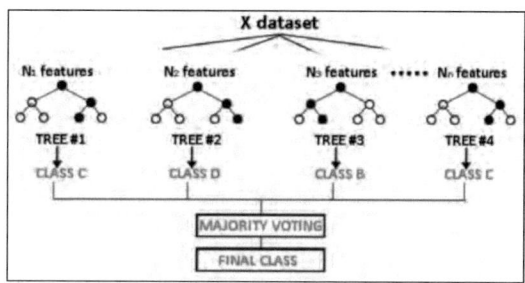

Fig. 6. Dataset Attribute classification

Support Vector Machine The fundamental goal of this model is to provide an accurate linear or deterministic partition that divides n-proportional space into groups, allowing for easy access to merging freshly created data into their relevant 755 modules for future reference. The hyperplane is a term used to describe a method of sorting data in an accurate linear manner. The process is known as the Support vector algorithm because the outermost vector points that contribute to the building of the hyperplane are referred to as support points.

Genetic Algorithm/TPOT: TPOT (Tree-based Pipeline Optimization Utility) is an AutomL tool that allows you to build optimized pipelines quickly using genetic programming. For data transformation, feature decomposition, feature selection, and model selection TPOT is an open-source library that uses sci-kit-learn components.

Prediction: A flask interface, which is fundamentally a Python architecture that links the web service and the models, transmits the input from the keyboard to the prototype. This flask would then present the outcome for person's perceptions on a website.

Performance Evaluation

```
             precision   recall  f1-score   support

          0      0.92      0.93      0.92        58
          1      0.93      0.92      0.93        62

   accuracy                          0.93       120
  macro avg      0.92      0.93      0.92       120
weighted avg     0.93      0.93      0.93       120
```

Fig. 7. Performance comparison of accuracy using Logistic Regression

```
Accuracy: 0.975
[[55  3]
 [ 0 62]]
             precision   recall  f1-score   support

          0      1.00      0.95      0.97        58
          1      0.95      1.00      0.98        62

   accuracy                          0.97       120
  macro avg      0.98      0.97      0.97       120
weighted avg     0.98      0.97      0.97       120
```

Fig. 8. Performance comparison of accuracy using Radom Forest

```
print(model2)

0.9416666666666667


print(metrics.classification_report(y_test,predict2))

             precision   recall  f1-score   support

          0      0.93      0.95      0.94        58
          1      0.95      0.94      0.94        62

   accuracy                          0.94       120
  macro avg      0.94      0.94      0.94       120
weighted avg     0.94      0.94      0.94       120
```

Fig. 9. Performance comparison of accuracy using Support Vector Machine

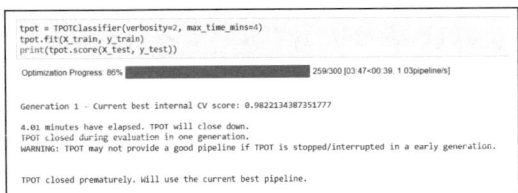

Fig. 10. The results comparison of the Proposed Method

Table 1. Comparison between Existing Algorithms

Classifier Techniques	Accuracy
LR	92.5
Random Forest	97.5
SVM	94.16
Gaussian Algorithm	100

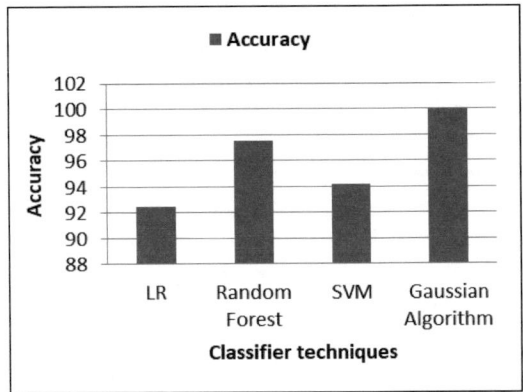

Fig. 10. Comparison between existing algorithms

Conclusion

To forecast accuracy, we looked at 8 different CKD patient factors and employed machine learning techniques like Regression Analysis, SVM, Random Forest, and Evolutionary Algorithms. According to the analysis of the data, the effectiveness of the above-mentioned is 92.5%, 94.1%, 97.5%, and 100%, respectively. The Genetic Algorithm provides the maximum accuracy of 100% in the shortest amount of time. Predicting CKD with an evolutionary algorithm will help patients maintain a healthy.

References

Prasad, K. S., Reddy, N. C. S., and Puneeth, B. N. (2020). A Framework for Diagnosing Kidney Disease in Diabetes Patients Using Classification Algorithms. SN COMPUT. SCI. 1, 101.

Ebrahime Mohammed Senan, Mosleh Hmoud Al-Adhaileh, Fawaz Waselallah Alsaade, Theyazn H. H. Aldhyani, Ahmed Abdullah Alqarni, Nizar Alsharif, M. Irfan Uddin, Ahmed H. Alahmadi, Mukti E Jadhav, Mohammed Y. Alzahrani (2021).

Diagnosis of Chronic Kidney Disease Using Effective Classification Algorithms and Recursive Feature Elimination Techniques, Journal of Healthcare Engineering, vol. 2021, Article ID 1004767, 10 pages.

Wickramasinghe, , Perera, , and Kahandawaarachchi, (2017). Dietary prediction for patients with Chronic Kidney Disease (CKD) by considering blood potassium level using Machine-learning algorithms, 2017 IEEE Life Sciences Conference (LSC), Sydney, NSW, pp. 300–303.Wibawa, , Malik, I., and Bahtiar, N. (2018). Evaluation of Kernel-Based Extreme Learning Machine Performance for Prediction of Chronic Kidney Disease, 2018 2nd International Conference on Informatics and Computational Sciences (ICICoS), Semarang, Indonesia, 2018, pp. 1–4.Dulhare, ,and Ayesha, M. (2016). Extraction of action rules for chronic kidney disease using Naïve Bayes classifier, 2016 IEEE International Conference on Computational Intelligence and Computing Research (ICCIC), Chennai, pp. 1–5.

Devika, , Avilala, , and Subramaniyaswamy, V. (2019). Comparative Study of Classifier for Chronic Kidney Disease prediction using Naive Bayes, KNN and Random Forest, 2019 3rd International Conference on Computing Methodologies and Communication (ICCMC), Erode, India, pp. 679–684

Chittora, , Chaurasia, S., Chakrabarti, P., et al. (2021). Prediction of chronic kidney disease-a machine learning perspective, IEEE Access, 9, pp. 17312–17334.

UCI Machine Learning Repository: Chronic_Kidney_ Disease DataSet, Archive.ics.uci.edu, 2015.

Prediction of Liver Diseases at Earliest Using Machine Learning Algorithms

P. Chinnasamy,[a] P. Siva Padmini,[b] S. Geetha,[c] N. Kumareshan,[d]
K. Karunambiga,[e] and Arulananth T S[f]

[a]Department of Computer Science and Engineering, MLR Institute of Technology, Hyderabad, India.
[b]Department of CSD, Marri Laxman Reddy Institute of Technology, Dundigal, Hyderabad, India
[c]Department of Computer Science and Engineering, BNM Institute of Technology,
Bangalore, Karnataka, India
[d]Department of Electronics and Communication Engineering, Sri Eshwar College of Engineering
Coimbatore, India
[e]Professor, Department of CSE, Karpagam Institute of Technology, Affiliated to Anna University,
Coimbatore 641105
[f]Department of ECE, MLR Institute of Technology, Hyderabad, India,
E-mail: *chinnasamyponnusamy@gmail.com, arulananthece@mlrinstitutions.ac.in

Abstract

Medical practitioners are finding it difficult to predict the illness in its earliest stages due to the vague symptoms. Often, abnormalities don't show up until it's too severe. Our study will employ algorithms to enhance liver disease diagnoses to solve this issue. The current study's main purpose is to differentiate between people with liver disease and without liver disease using algorithms for classification. The research uses performance metrics to compare the categorization method. To help the medical community with liver disease diagnosis in patients, a GUI will be designed in python. The Graphical User Interface is easy to use for medical professionals to screen for liver diseases.

Keywords: Classification algorithms, Machine Learning, Liver Disease Patients

Introduction

Liver, a significant organ presents solely in vertebrates that conducts a variety of vital biological tasks, including detoxification and the creation of proteins as well as bio-chemicals required for metabolism and survival. Diseases relating to the liver are considered fatal as they show very minimal symptoms which makes it harder for anyone to conclude the disease at its initial stages. Although there are numerous types of liver disorders, the liver will function normally though it is partially damaged. When it comes to the diagnosis and treatment of liver disorders and infections, time is important. Recognizing the disease at the preliminary phases and taking the appropriate medications will drastically improve the rate of survival of the patient and minimize the chances of disease, making the patient's medical condition critical. As a result, detecting them at an initial stage is challenging. It is important to detect it and take the appropriate procedures to treat it. Detection of liver-related issues at the preliminary phases will help to increase the patient's chances of surviving. A patient medical condition whose liver disease has been diagnosed at the initial stages has a better chance of successful treatment than one who is diagnosed at later stages. From several recent pieces of research, it has been found that alcohol consumption is increasing at an alarming rate and this is a serious problem because alcohol is one of the leading causes of liver damage. Indians are more susceptible to liver failure. As a result, inventing the machine that aids in disease diagnosis will be extremely beneficial for the medical industry. The technologies help doctors in taking proper patient diagnoses, and by using tools such as the automatic classification of illnesses of the liver, anyone can lessen the queue of patients at liver specialists Rahman et al. (2019).

The major goal of the study is to employ algorithms for classification to distinguish between people with liver disease and without liver disease. The performance of 4 classification techniques was compared using liver patient data: ANN, Logistic Regression, KNN, and SVM. Furthermore, the most accurate model is developed into GUI that is easy to use using Python. It has 585 patient records taken from the state of Andhra Pradesh.

Literature Survey

Several neural networks models, like a diagnosis support system and expert system, have been constructed in recent research works to aid clinicians in the identification of liver disorders in the medical area by Kalaiselvi et al. (2021). Also, to add, a system proposed by Christopher for identifying illnesses takes six benchmarks into account: liver disease, heart disease, etc. The authors worked on two WSO and C4.5-based systems with 70.1 percent of accuracy when utilizing nineteen rules from the liver disease dataset and 61.7 percent when using number forty-three rules from C4.5 and WSO, respectively. Priya et al. (2018) conducted a critical investigation on the diagnosis of liver illnesses by assessing several classification algorithms, including the nave Bayes classifier, C4.5, BPNN, support vector, and KNN. The accuracy of the Nave Bayes classifier was 51.59 percent, the C4.5 method was 55.94 percent, the BPNN was 66.66 percent, the KNN was 62.6 percent, and the support vector machine was 62.6 percent.

The liver disorder dataset's low performance in training and testing was due to a lack of data in the dataset. As a result, Pasha et al. (2022) developed a strategy to successfully correct data shortage depending upon the extra sampling used within minor classes. For the investigation, the author evaluated two decision tree algorithms. The algorithms include C4.5 and CART, and BUPA liver disease information was also used in the research. These previously designed approaches were adequate, but further effort is needed to increase the detection rate of liver disease diagnosis. In this circumstance, this will make liver disease diagnosis more successful and efficient by preventing liver condition misdiagnosis. Developing a system that performs better than earlier efforts would aid in preventing disease misdiagnosis and offering the finest and necessary treatment to the person with liver disease.

The Proposed Method

Dataset

The dataset consisting of liver disease patients of India included 583 patients with 10 different features. This information was gathered from an online source, Kaggle. On the basis of liver disease, the patients were classified as 1 or 2. Table contains a thorough explanation of the dataset. The table contains information about the attribute and its kind. All of the features, except gender, are numerical values, as seen in the table. Gender is turned into an integral value during the pre-processing step of the data (0 and 1).

Data-Preprocessing

Pre-processing data is a key part of any machine-learning task. Most datasets utilized in Machine Learning challenges must be processed, cleaned, or modified before being fed into a Machine Learning algorithm. Missing value imputation, encoding categorical variables, scaling, and other pre-processing techniques are among the most often utilized. These approaches are simple to grasp. However, when dealing with data, things might become clumsy.

Classification Techniques

SVM

The goal of the Support Vector Machine is to determine the best hyperplane for separating data into distinct classes. Support Vector Machine is implemented in Python using the sci-kit-learn library. The data is split into two sets: 25 percent as test data and 75 percent as training data. The SVM is used to build a hyper plane. It is considered that when the margin is higher the generalization error of the classifier would be lower. The hyper plane with the most distance to the very near training data point produces a fair distinguishing.

Logistic Regression

Logistic regression can be considered among the most basic categorization models. Due to its parametric nature, it may be partially interpreted by staring at the parameters, so it can be used for experimenters looking for correlations between variables.

K-Nearest Neighbor

The full training dataset is the model for KNN. When an unknown data instance necessitates a forecast, the K-Nearest Neighbor method will look for the k number of most comparable instances in the training dataset. The most comparable examples' attributes of prediction are combined and reappeared as the prediction for the unknown instances.

The type of data determines the similarity metric. The Euclidean distance can be employed with real-valued data. Other sorts of information, like category or binary data, can be employed with hamming distance. The KNN algorithm is part of the instance-based and lazy learning algorithms families.

ANN

A neural network with backpropagation is created. At the input layer of this network, there were ten input neurons. The dataset's total number of attributes is represented by a number of inputs.

To achieve the required rate of recognition, which can diagnose a patient's liver condition. To achieve the desired optimum result, in neural network models, several parameters must be adjusted. The 3 parameters are used. They include hidden neurons, momentum rate, and learning rate. Back propagation neural networks have all of these parameters. The learning rate determines the system's learning capability, whereas the momentum rate dictates the system's learning pace. To achieve the best result, the network's number of hidden neurons must be modified.

The hidden layer's no. of neurons is being experimented with in order to find the optimal neurons capable of accurately representing the features in the input dataset and produce the desired result output. By changing neurons, the hidden layer's required number of neurons was determined. Because of its ability to flip smoothly and the derivatives' simplicity, the sigmoid function was utilized in the output layer. The artificial neural network was created with the Keras module in Python, which uses the tensor-flow backend.

In the table below, the back propagation neural network is described.

Table 1. ANN Description

No. of Inputs	10
No. of hidden Layers	2
No. of neurons in 1st hidden layer	400
No. of neurons in 2nd hidden layer	400
No. of Output	1
Learning Rate	0.26
Epoch	100

Results and Evaluation

The major goal of this research was to apply machine learning to diagnose liver disease. Support Vector Machine, Logistic Regression, K-Nearest Neighbor, and Artificial Neural Networks are utilized to make predictions. They were all more accurate in their predictions. We observed accuracy and precision with each algorithm, which can be summed up as below:

Accuracy: Fraction of the sum of genuine positives and genuine negatives to the sum of all predictions.

$$Accuracy = \frac{no.\,of\,TP + no.\,of\,TN}{no.\,of\,TP + FP + FN + TN}$$

Precision: Fraction of genuine positives to all positive outcomes is defined as precision.

$$Precision = \frac{no.\,of\,TP}{no.\,of\,TP + FP}$$

In the table below, the results of each categorization algorithm are given.

Table 3. Results of classification algorithms

Classification Algorithm	Accuracy	Precision
Logistic Regression	72.72	75.39
K-Nearest Neighbors	68.53	73.77
Support Vector Machine	72.02	72.34
Artificial Neural Network	93.88	76.85

Artificial Neural Networks produced the best outcomes, as seen in the table. As a result, the ANN model is chosen to predict liver problems from the given dataset.

Development of GUI

The artificial neural network was the model that was most accurate with the test data. As a result, the GUI is created using an artificial neural network. Python's Tkinter library was used to develop the user interface. One for forecasting and the other for training fresh data are created. All of the dataset's attributes have input areas in the GUI. Based on the learned model, if the patient has liver disease, the machine will detect it. The GUI will assist medical staff in detecting liver illness in individuals early on. The developed GUI is depicted in the image below.

Test Cases

To explain the technique, the following test cases are used.

Test Case	Input	Output
1	Age: 65 Gender: Female Total Bilirubin: 0.7 Direct Bilirubin: 0.1 Alkaline Phosphatase: 187 Alamino Aminotransferase: 16 Aspartate Aminotransferase: 18 Total Proteins: 6.8 Albumin: 3.3 Albumin and Globulin Ratio: 0.9	1 (Yes)
2	Age: 17 Gender: Male Total Bilirubin: 0.9 Direct Bilirubin: 0.3 Alkaline Phosphatase: 202 Alamino Aminotransferase: 22 Aspartate Aminotransferase: 19 Total Proteins: 7.4 Albumin: 4.1 Albumin and Globulin Ratio: 1.2	2 (No)

Test Case 1

Chances of having Liver Disease is more, please consult a Doctor.

Symptoms

Classic symptoms of liver disease include:

- nausea
- vomiting
- right upper quadrant abdominal pain, and
- jaundice (a yellow discoloration of the skin due to elevated bilirubin concentrations in the bloodstream).

Conclusion

In this paper, we suggested methodologies for detecting patients with liver illness using techniques of ML. Support Vector Machine (SVM), Logistic Regression, K-Nearest Neighbors (KNN), and ANN were used. All models were used to create the system, and their performance was evaluated. Certain performance metrics were used to evaluate performance. With a 98 percent accuracy, ANN was the model that produced the best results. When comparing this research to earlier studies, it was determined that ANN was quite effective. ANN was used to create a GUI that may be utilized as a tool to predict liver disease.

References

Kalaiselvi, , Meena, K., and Vanitha, V., Liver Disease Prediction Using Machine Learning Algorithms, 2021 International Conference on Advancements in Electrical, Electronics, Communication, Computing and Automation (ICAECA), pp. 1–6, doi: 10.1109/ICAECA52838.2021.9675756.

Rahman, A. S., Shamrat, F. J. M., Tasnim, Z., Roy, J., and Hossain, S. A. (2019). A comparative study on liver disease prediction using supervised machine learning algorithms. International Journal of Scientific & Technology Research, 8(11), 419–422.

Priya, M. B., Juliet, P. L., and Tamilselvi, P. R. (2018). Performance Analysis of Liver Disease Prediction Using Machine Learning Algorithms.

Pasha, Syed Nawaz, Ramesh, Dadi, Mohmmad, Sallauddin, Navya, P., Anil Kishan, P., and Sandeep, (2022). Liver disease prediction using ML techniques, AIP Conference Proceedings, 2418, 020010. https://doi.org/10.1063/5.0081787.

Detection of Lung Cancer Using Deep Learning Techniques

D. Roja Ramani,[a] K. Karunambiga,[b] P. Chinnasamy,*[,b] K. Manjula,[c] S. Dhanasekaran,[d] and D. Divya Priya[e]

[a]Department of CSE, New Horizon College of Engineering, Bangalore, Karnataka, India
[b]Professor, Department of CSE, Karpagam Institute of Technology, Affiliated to Anna University, Coimbatore 64110
[b*,e]Department of Computer Science and Engineering, MLR Institute of Technology, Hyderabad, India
[c]Department of CSE, Marri Laxman Reddy Institute of Technology and Management Dundigal, Hyderabad, India
[d]Department of IT, Kalasalingam Academy of Research and Education (Deemed to be University), Srivilliputtur, India
E-mail: *chinnasamyponnusamy@gmail.com

Abstract

Lung cancer is a deadly form of cancer that is difficult to detect. As a result, it is more important for care to evaluate nodules swiftly and appropriately for both men and women. As a result of this research, CT was found. Images are better for getting accurate results. As a result, the majority of CT Images from scans are used to diagnose cancer. Therefore, a neural network helps in great way to detect cancer cells. CNN, image recognition, and Artificial intelligence together yield accurate values to detect cancer. Just by looking at a person's age, body weight, eating and drinking habits, and other diseases prediction of lung cancer can be done. Then using deep learning algorithms detection of the lung cancer can be performed if the predictions are positive. Deep learning performs beyond the typical machine learning approaches.

Keywords: Deep Learning, CT scan, Lung Cancer, Health Prediction

Introduction

Lung cancer arises when cells multiply uncontrollably. A variety of methods for detecting lung cancer in its early stages have been developed. In 2015, lung cancer was in the second place for the cause of death, according to a WHO poll, and it is now the fifth top cause of death in 2017 Smokers are the most affected, It accounts for 85% of all instances by (Bharathi and Arulananth (2017)). In recent years, various computer-aided diagnostic (CAD) systems have been developed. Lung cancer must be detected early to avoid deaths and improve survival chances. Detection of lung cancer can be done by Artificial Intelligence i.e. deep learning. Therefore, neural networks help in a great way to detect cancer cells. CNN, image recognition, and machine learning together yield accurate values.

Literature Survey

Lung cancer incidence is directly inversely correlated with the frequency of heavy smokers.

Using different classifiers including Naive Bayes, SVM, Decision Trees, and Logistic Regression, the melanoma prediction was examined by Radhika et al. (2019). Additionally, it examines the type of data—benchmark or independently gathered data—used to diagnose the diseases in question. Furthermore, taking into account the various available approaches, directions for future research have been determined and illustrated by Pradharan et al. (2020). Although CT scan reports are more accurate than mammography, individual CT scan images are segmented into normal and bad categories. Segmenting the aberrant photos allows the tumor region to be the main focus. assessment performed using image-extracted features by Rahane et al. (2018). Srinivasan et al. (2020) identify carcinogenic pulmonary nodules from either the provided input pulmonary image and also categories lung cancer according to its aggressiveness. This methodology utilizes cutting-edge learning to locate malignant pulmonary nodules.

Proposed Method

The proposed system has three important phases; prediction, detection, and visualization as shown in Fig. 1.

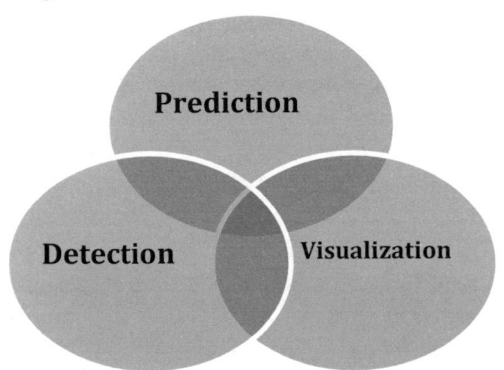

Fig. 1. The Phases of the Proposed Method

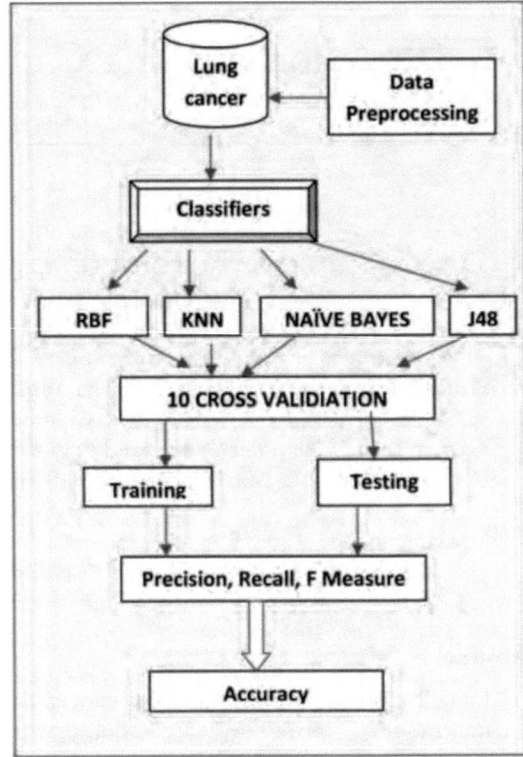

Fig. 2. The Process of the Proposed Method

A. Prediction

These limitations brought on by the enormous dimensions of the data can be solved using a machine-learning approach. To estimate the likelihood of developing cancer, we are employing techniques for machine learning in this study. The different algorithms used as shown in Fig. 2.

B. Detection

In order to reduce the number of components and optimize the network's design for image processing, convolutional neural networks are constructed. It consists of layers that are organised according to their characteristics and purposes.

Rather than transforming existing data, we will acquire new information. Data collection is a key element in deep learning since we need large amounts of data and it's not always possible to gather hundreds or even millions of photos, so data augmentation is necessary for the scene. The proposed model is depicted in Fig. 3.

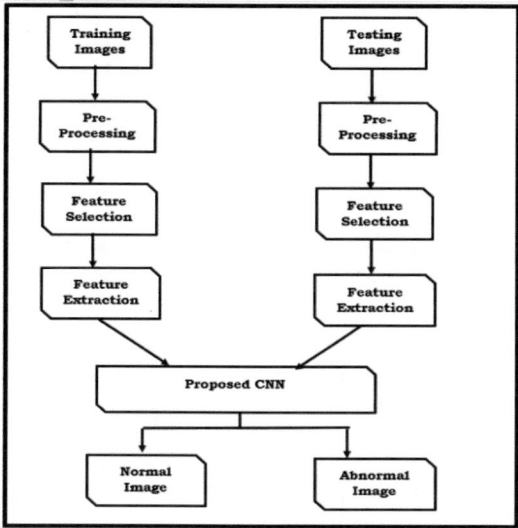

Fig. 3. The Proposed Model

There are two parts to the job: training and testing. There are a total of 201 shots, with 171 functioning as training images and the remaining 30 as testing images. Images are first pre- processed in the testing section, where they are resized and blurred using histogram equalization.

Pre-Processing

These phases are input into CNN layers, which classify the images and the accuracy of the images was calculated using sample weights.

CNN Layer

The convolutional layer in a CNN is the most important layer. The characteristics of the layer contain a number of grains, which are classifiers. Predicting and detecting lung cancer by primitives like age, lifestyle & habits are done to screen patients and identify potential cancer patients quickly. The data is first entered through the form

then based upon the data it predicts whether the person has any chance of getting lung cancer. If there are chances then it automatically detects the u\lung cancer cells earlier than before models and then it displays the images in 3D format in a plane. It makes the patient and doctors visualize and analyze the cells as well.

Visualization

The final predicted images can be visualized using matplotlib, inline, and image to create 3d images in a plane. Mplot3d and art3d are imported from poly 3d collection to generate the 3d images. The proposed method execution is clearly shown in Figs. 4–8.

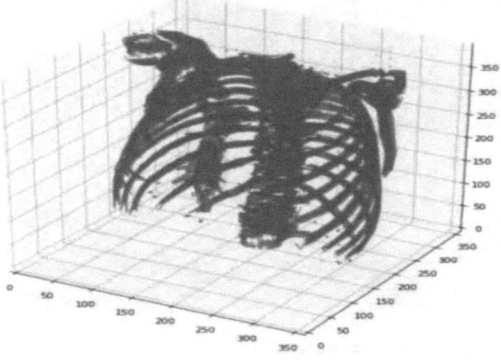

Fig. 4. A Sample input Image

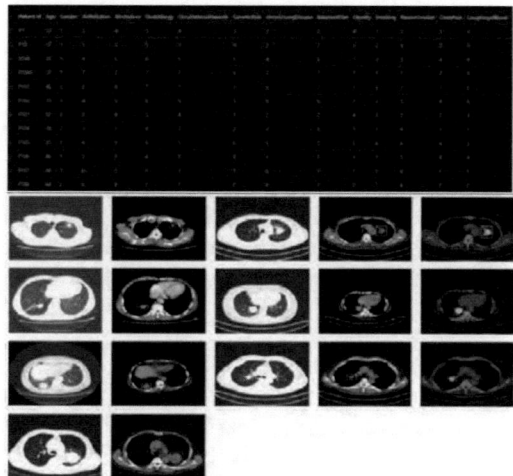

Fig. 5. A Sample Test Images and values

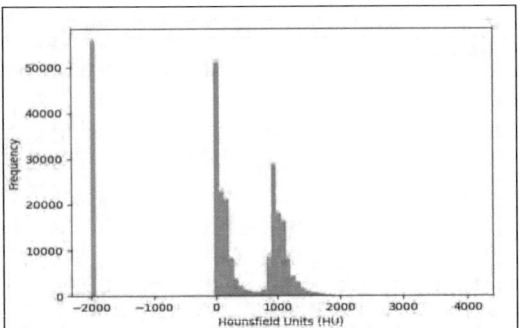

Fig. 6. The Detection of the Lung Cancer

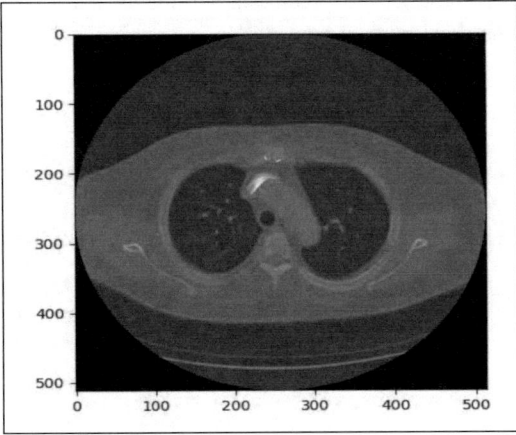

Fig. 7. The Output images

Conclusion

The proposed method was used to predict lung cancer earlier by analyzing the eating, drinking, smoking, body weight, body mass, etc. This helps to take early treatment for lung cancer. If the predictions are positive and the person's habits can lead to lung cancer then detection of lung cancer can be done using CT scans. The CNN algorithm gives accurate and efficient results by giving 3d visualized images which help to analyze the images more clearly. In the Future, Advanced algorithms can be employed to increase the model's efficiency and accuracy.

References

Bharathi, H., and Arulananth, T. S. (2017). A review of lung cancer prediction system using data mining techniques and self organizing map (SOM). International Journal of Applied Engineering Research, 12(10), 2190–2195

R. P. R., R. A. S. Nair and V. G. (2019). A Comparative Study of Lung Cancer Detection using Machine Learning Algorithms, 2019 IEEE International Conference on Electrical, Computer and Communication Technologies (ICECCT), 2019, pp. 1–4, doi: 10.1109/ICECCT.2019.8869001.

Pradhan, Kanchan, and Chawla, Priyanka (2020) Medical Internet of things using machine learning algorithms for lung cancer detection, Journal of Management Analytics, 7, 4, 591–623, doi: 10.1080/23270012.2020.1811789

Rahane, , Dalvi, H., Magar, Y., Kalane, A. and Jondhale, S. (2018). Lung Cancer Detection Using Image Processing and Machine Learning HealthCare, 2018 International Conference on Current Trends towards Converging Technologies (ICCTCT), 2018, pp. 1–5, doi: 10.1109/ICCTCT.2018.8551008.

Asuntha, A., Srinivasan, A. (2020). Deep learning for lung Cancer detection and classification. Multimed Tools Appl, 79, 7731–7762 (2020). https://doi.org/10.1007/s11042–019-08394-3.

Brain-Computer Interface Based Home Automation

Manoj Kumar[a,], Jayanthi D[b,*], Bittu kumar[c,*], Chinimilli Vaibhav Surya[d], and Karuturi Nitin Sai[e]*

[a,c-e]Department of Electronics and Communication Engineering, MLR Institute of Technology, Hyderabad, India
[b]Department of Electronics and Communication Engineering, R.M.D Engineering College Kavaraipettai, Chennai,India
E-mail: [*]manojiitism@gmail.com, slvjayanthi@gmail.com, bittu.mlrit@gmail.com

Abstract

Brain-Computer Interface (BCI) is a communication channel between the human brain and a computer. It is used to study brain waves and drastically help paralyzed or handicapped people by restoring their communication, movements, and control. The brain waves are obtained using electroencephalography (EEG), a method to capture electrical activity using an electrode placed on the scalp. Inside our brain, when there is an exchange of information between neurons, there is a minute electrical discharge. This electrical activity is the result of neural interaction. Different states of the brain will result in different patterns of neural interaction. Waves of various amplitudes and frequencies are obtained from these patterns. Our project uses this wave information as the core input to achieve BCI based Home Automation System. The signals generated from the brain will be received by Think Gear ASIC Module (TGAM). It will divide the information into packets and transmit the data in a wireless medium (Bluetooth). The raw data of the brain wave can be converted into the required format using Arduino or MATLAB GUI platform. Using this data, the platform will send instructions to the home appliances system to operate those modules. The attention values and blink data are used to ON/OFF the appliances.

Keywords: MATLAB, TGAM, Brain-Computer Interface (BCI), Electroencephalography, Home automation

Introduction

A Brain-Computer Interface (BCI) is a new communication channel to capture human thoughts. By using BCI, it has become possible to link brain activity to the operation of computers and devices, creating a direct communication channel between mind and machine. Disabled people can use BCI technology to improve their independence and maximize their capabilities at home. This project is designed to help disabled people to control the appliance with an increase in accuracy (Mtshali et al., 2019).

This paper presents a new architecture for home device control systems via thoughts. Such a system will be helpful for people suffering from paralysis or similar condition who are limited to movement. For disabled people, controlling a household device such as light, fan, air-conditioning, etc., will be difficult without any assistance. In the early days of BCI, the main applications were focused on motor rehabilitation. Nowadays, BCI is used beyond medical applications like entertainment, military, research, etc. Most of the BCI applications are motor-related and strongly correlate to why animals need brains. According to Daniel Wolpert, the need for a brain is related to movement.

Thanks to neuroscience, a lot of the brain was reverse-engineered. The 10/20 electrode placement for EEG offers satisfactory spatial coverage for most applications, but for sophisticated applications like silicon sensor placement for ECoG, fMRI play a key role. Today's BCI requirements are good temporal and spatial resolution, mobility, user-friendliness, ethics, discreteness, and robustness. For modern BCIs, EEG and ECoG offer optimal performance and mobility. ECoG is a highly invasive method and involves higher costs. Still, the silicone-based sensors offer a higher signal-to-noise ratio, better spatial resolution, and ease of use, and most of them provide discreteness.

Literature Survey

Intelligent home systems use mobile phone applications and operate the devices using the internet. Implementation of automated home appliances targeting support for physically disabled persons and operated using voice operating devices like Alexa, Apple Siri, google assistance etc. These devices will record the voice commands of the disabled persons and operate the home electrical devices to turn ON and OFF. This system though applicable for disabled, dumb people, cannot use this system to operate the machines. This stood as a limitation of the voice-operated system. In AbdWahab et al. (2016), authors proposed a design where home appliances are operated by using the Internet of Things. The household appliances are operated wirelessly using CytronBlue Bee Bluetooth module using a mobile phone. The device needs to operate through the mobile phone using an application installed. But, illiterate people cannot use mobile phones, and it is not easy to operate the application on mobile. In Kshirsagar et al. (2002) researchers proposed a new model where home appliances are operated using hand gestures. A new glove was developed, which automatically identifies the gestures and operates the home appliances. This design has a limitation of persons who cannot move their hands will not use the design (Murali Chandra Babu et al., 2019). Home automation can be helpful for those who cannot communicate with the surrounding people and cannot operate mobile

phones too. These people can operate the devices by using their brain signals. The brain signals can be collected by using the Brain-computer interface (BCI) technology (Gao and Xu et al., 2003). Arduino can be used as a controller and process the received data to operate the appliances (Zickler et al., 2009). The proposed system is cost-effective with ease in design. The system is designed for paralyzed and disabled people to help themselves operate the appliances according to the reported designs.

General BCI Architecture

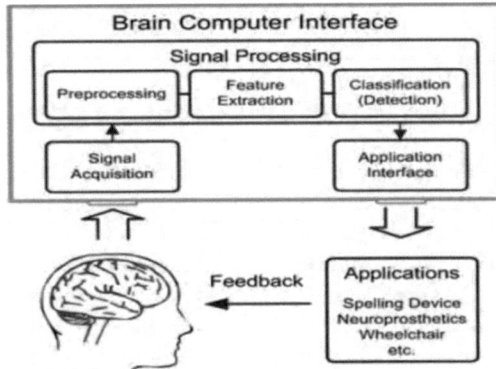

Fig. 3. BCI Architecture

BCI is a system where it is able to capture and identify the human brain's electrical movements with the help of an electrode and display it using some software. So this human-computer interface will capture the movements which are running in the brain without any physical movements. The principle in other terms portrays the need of human brain, an executor (Here it is a computer), and an interface between the brain and computer. To make the computer know what the brain wants to communicate and it could able to capture the brain activity. The next sub-sections are contained types of BCI Technology (S. P. Levine et al., 2000, Vidal, 1997).

The EEG electrode, which architecture is shown in Fig. 3, is placed on the human brain to extract the brain waves or signals. The signals which are acquired from the brain are not amplified or processed so it contains the noise in it (F. Miralles et al., 2015). So to amplify and process it we need a signal processing system. So this brain-computer interface (BCI) consists of a signal processing unit where the signal will first get acquisition and then it will be sent to the preprocessing so here in preprocessing the signal will be processed and filter out some noise in it. So after that the feature extraction takes place and after that, the classification takes place so in the classification the noise detection will happen and the noise will get removed. So this processed signal will go to the application interface. From here we can control the appliance. So here we call the "neutral state" when nothing changes or happens, the "active state" when the BCI will do something, the "neutral EEG set" as the composite of those EEG trials that elicit a neutral response, and the "active EEG set" the complement of the neutral EEG set. The normal BCI is a two-state machine where state changes happen at a rate produced by the BCI period and they are determined by a Boolean variable B1 which is an activation that will become true when the BCI detects a member of the active EEG, and wrong if it does not detect.

Proposed System

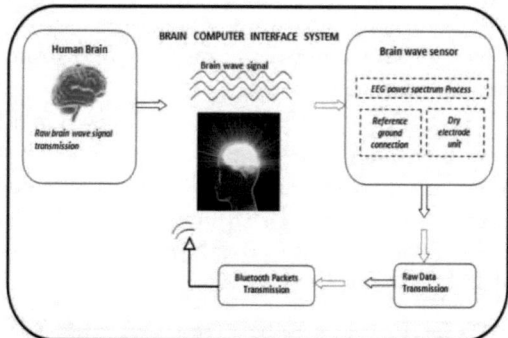

Fig. 4. Block diagram of the proposed system

The proposed approach which block diagram is shown in Fig. 4, helps people physically disabled to control home appliances using electroencephalogram signals (EEG). BCI technology will collect the signals from the brain through electrodes and operate the corresponding device. The brain signals can be recorded in EEG, and it measures the voltage variations within the neurons of the brain. These signals are calculated by using electrodes placed around the scalp. The signals are transferred to the microcontroller, which will process the information and turn ON/OFF the necessary device as required. The following modules or parts are shown in Figs. 5–8, is used to make this system in real-time.

Electrodes

We will be using dry non-invasive scalp sensors to measure the electrical activity of our brain Specific electrodes would be placed at the motor cortex area and the rest would be placed in many specific areas of the brain to measure brain activity.

Fig. 5. Electrodes

ASIC Module

Fig. 6. ThinkGear AM

The signals obtained from the scalp are very weak. They are in microvolts and contain noise. Therefore, we need to amplify and filter out the frequencies that are not useful to us. Here we will be using the ThinkGear ASIC module to do the job for us. The TGAM will process the necessary information and store it in the form of packets. These packets are used to transmit data.

Transmission of Data

Wireless transmission of data will take place using Bluetooth here. The HC-05 Bluetooth modules are used to transmit the data from the Think Gear ASIC Module (TGAM) to the Arduino. The HC-05 of Arduino is configured in the command mode to connect with the HC-05 of the TGAM. Once configured, they will connect automatically when switched on.

Fig. 7. Bluetooth Module

Micro Controller

The Arduino is used here and coded in a way to receive the packet information from TGAM. We focus on receiving the attention data from the module and using it to control the home appliances. A three-channel relay is used here to control mini-scale appliances as an example.

Fig. 8. Arduino

Results and Discussion

Wireless transmission of data will take place using Bluetooth here. The HC-05 Bluetooth modules are used to transmit the data from the Think Gear ASIC Module (TGAM) to the

Arduino. The HC-05 of Arduino is configured in the command mode to connect with the HC-05 of the TGAM. Once configured, they will connect automatically when switched on.

Making of EEG Headset

We were successfully able to assemble the brain head set components. The electrode we used is silver chloride (AgCl) Dry electrode. We chose this over wet electrodes because it doesn't require additional electrolytic gel and scalp preparations, making it quicker and easier to use.

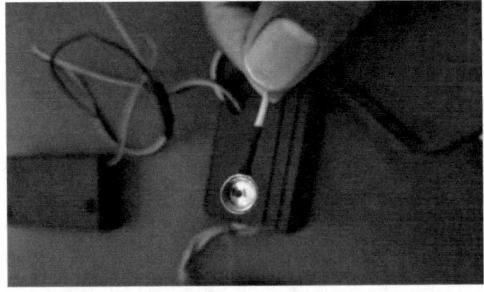

Fig. 9. AgCl dry electrode

The Think Gear Application-Specific Integrated Circuit Module (TG ASIC Module or TGAM) was unavailable at the time of project phase. We were able to acquire it by dissembling a third-party headset available.

The AgCl electrode (shown in Figs. 9 and 10 including the electrode clip) was connected to TGA module and an electrode clip is used as reference electrode here. The TGA Module (shown in Fig. 11) is interfaced with HC-05 Bluetooth module and battery power.

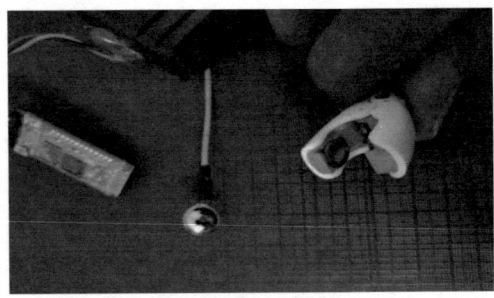

Fig. 10. The electrode clip

Interfacing the Brain Head Kit with Computer

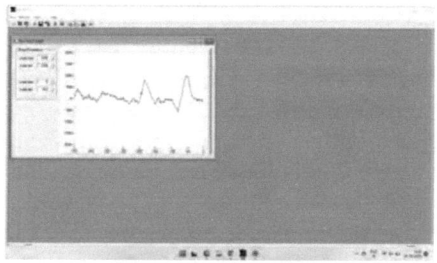

Fig. 11. Raw Wave Graph obtained from TGAM

After interfacing the HC-05 of brain sensor module with the computer, we were able to observe the raw data of our brain signals in visual form through NeuroView. We confirmed the working of our EEG head kit to be successful using the visualizers. We checked if the blinks are being detected in the waveform. The hard blinks appeared as a curve of sinusoidal waves but sharper.

Configuration of Bluetooth Module Using Command Mode

The configuration of HC-05 with TGAM's HC-05 is done by turning it into command mode. The first few attempts were a failure. After certain research, the HC-05 is configured into proper command mode using an Arduino code.

Fig. 12. X-CTU application and commands sent

X-CTU (shown in Fig. 12) is used to send commands to HC-05 and configure it to connect with TGAM's Bluetooth module.

The Bluetooth module after successfully being configured finally established connection and communication with the HC-05 of the brain sensor kit.

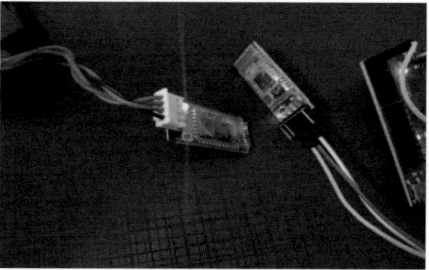

Fig. 13. Two Bluetooth modules connected with each other

Obtaining the Attention Values

The packet format of TGAM is referred to as code in Arduino IDE for obtaining the attention values, as shown in Fig. 14.

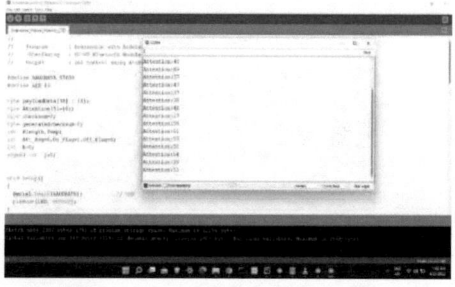

Fig. 14. Attention using MATLAB platform

A similar logic is implemented in MATLAB to obtain attention values and wave graphs, as shown in Fig. 14 of MATLAB plate form.

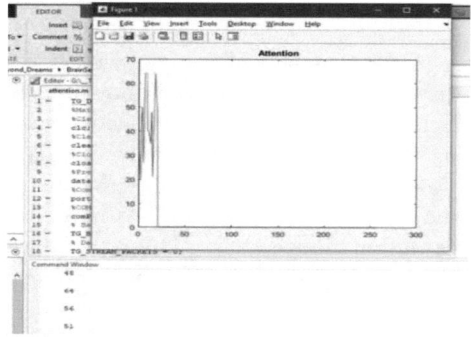

Fig. 15. Graph of Attention using MATLAB platform

Home Automation

The system, as shown in Fig. 16, is built using a three-channel relay that is interfaced with Arduino. Here we used a light bulb, a mini fan, and a motor as an example of home appliances (Prasad et al., 2017, 2021).

Fig. 16. Home Appliances interfaced with microcontroller

The attention value is used to switch ON/OFF the home appliances. The captured image is shown in Fig. 17.

Fig. 17. Triggering of light using attention values

Conclusion

Though the current technology of the EEG is effective up to an extent, it is nowhere near perfect. Disabled persons and elder people facing difficulty in operating home appliances can overcome the limitation with this BCI technology. The proposed Arduino-based home automation system supports physically handicapped persons with brain-computer interfaces. As mentioned in this work, the proposed BCI system that supports paralyzed and disabled persons to work independently and operate the appliances is easy to build with a cost-effective design. This design can be enhanced to any number of loads and can be extended to various applications like automobiles, Industries, Remote Controls and many more. Yet, attention and blink detection is still not 100% accurate, an issue that will most likely be solved as blink detection technology gets better and better. Algorithms that calculate necessary information from raw brainwave data can get even better, but until mankind learns how to manipulate and control individual brainwave frequencies, absolute and complete control of brainwaves will remain impossible. This project does hold promise though for the future of EEG and brain waves-related products.

BCI technology is a good innovation for paralyzed people. Not only for the paralyzed people but also for the latest technological innovations. The neural link technology which was proposed by Elon musk was similar to the concept of this BCI system. So by using this BCI technology we can create wonders and change the whole future. The controlling of computers through neural and electrical signals from the brain is really difficult to assess as the field of research is still in the learning process. If the advanced technology of 2022 matches the advances of the last few years, direct neural-brain communication between human beings and computers will finally be widely used. So one day the newly purchased computers and laptops will come up with an in-built signaling sensor and recognition software which is similar to the mouse and keyboard which we are using today.

References

Mtshali, , and Khubisa, F. (2019). A Smart Home Appliance Control System for Physically Disabled People, 2019 Conference on Information Communications Technology and Society (ICTAS), Durban, South Africa, pp. 1–5.

Abd Wahab, IoT-based home automation system for people with disabilities, 2016 5th International Conference on Reliability, Infocom Technologies and Optimization (Trends and Future Directions) (ICRITO), Noida, pp. 51–51, Noida.

Kshirsagar, , Sachdev, S., Singh, N., Tiwari, A., and Sahu, S. (2020). IoT Enabled Gesture-Controlled Home Automation for Disabled and Elderly, 2020 Fourth International Conference on Computing Methodologies and Communication (ICCMC), Erode, India, pp. 821–826, 2020.

Timler, ,and Lent, (2002). Power gain and dissipation in quantum-dot cellular automata, J. Appl. Phys., vol. 91, no. 2, pp. 823–831, Jan. 2002.

Murali Chandra Babu, K., Harsha Vardhini, P. A., Koteswaramma, N. Design and Implementation of Arduino based Riders Safe Guard 2.0, International Journal of Innovative Technology and Exploring Engineering (IJITEE), pp. 3078–3083, 9(1), November 2019.

Gao, D. Xu, Cheng, M., and Gao, S. (2003). A Bci-based Environmental Controller For The Motion-disabled, Ieee Transactions On Neural Systems And Rehabilitation Engineering, 11, pp. 137–140, June 2003.

Zickler, C., Di Donna, V., Kaiser, V., Al-Khodairy, A., Kleih, S., Kübler, A., Malavasi, M., Mattia, D.,

Mongardi, S., Neuper, C. and Rohm, M. (2009). BCI applications for people with disabilities: defining user needs and user requirements. Assistive technology from adapted equipment to inclusive environments, AAATE, 25, pp. 185–189.

Levine, , Huggins, , Bement, , Kushwaha, , Schuh, , Rohde, , Passaro, , Ross, , Elisevich, , and Smith, (2000). A Direct Brain Interface Based On Event-related Potentials, IEEE Transactions On Rehabilitation Engineering, 8, pp. 180–185, June 2000.

Prasad, S. V. S., Pittala, C. S., Vijay, V., Vallabhuni, R. R. (2021). Complex Filter Design for Bluetooth Receiver Application, Proceedings of the 6th International Conference on Communication and Electronics Systems, ICCES 2021, art. no. 9489020, pp. 442–446, 2021.

Prasad, S. V. S., Savithri, T. S., Krishna, I. V. M. (2017). Performance Evaluation of Svm Kernels On Multi spectral Liss III Data For Object Classification, International Journal on Smart Sensing and Intelligent Systems, 10(4), pp. 863–878.

Defect Exposure in Vegetables and Fruits Using Machine Learning Algorithms

P. Chinnasamy,[a,*] K. B. Sri Sathya,[b] Ramesh Kumar Ayyasamy,[c]
V. Praveena,[d] and Arulananth T S[e,*]

[a]Department of Computer Science and Engineering, MLR Institute of Technology, Hyderabad, India
[b]Department of CSE, KPR Institute of Engineering and Technology, Coimbatore, India
[c]Assistant Professor, Department of Information Systems, Faculty of Information and Communication Technology, Universiti Tunku Abdul Rahman (UTAR), Malaysia, rameshkumar@utar.edu.my
[d]Department of CSE, Dr. N.G.P. Institute of Technology, Coimbatore, India
[e]Department of ECE, MLR Institute of Technology, Hyderabad, India
E-mail: *chinnasamyponnusamy@gmail.com, arulananthece@mlrinstitutions.ac.in

Abstract

Defects to agricultural food items like fruits and vegetables, resulting in a decrease of the quality and quantity of agricultural consuming items. The regular or most used day-to-day approach for defect detection is continuous watching and observation of crops either by farmers or further by some professionals. We have used image processing and classification techniques in machine learning to identify and detect diseases on agricultural products like vegetables and fruits. The main goal of the system being proposed here is to monitor and detect diseases in fruits and vegetables in their early stages to farmers for a healthy yield. Our System can deliver the most accurate results in a very short time while consuming minimal computational resources.

Keywords: Agriculture, Disease Prediction, Machine Learning, Fruits diseases

Introduction

Farmers normally observe visual symptoms of the disease on fruits and vegetables. The cause for diseases in fruits and vegetables can be excess or lack of mineral uptake by the plants from the soil. There are many varieties of defects that affect these agricultural items which potentially lead to cause social and financial losses. Prior exposure on crop situations like it's health conditions would help in the control of disease extending or spreading further. The goal of this model is to find those defects or diseases on these agricultural items like vegetables and other eatable items in the starting stage itself. This model takes input as an image of a fruit or vegetable and tell us from which defect it is suffering and with mentioned accuracy percentage. According to the problem we had observed, giving the solution to it at most quickly and accurately would help a lot in solving this problem.

Literature Survey

According to the study and survey country India consists most of it parts as agricultural part, almost 1/3rd of our country's people stay in rural areas, where agriculture is their primary source of occupation and living. Prior identification of defects is very demanding content for research and work on. Almost there are many numbers of diseases caused by fungi, bacteria, nematodes, etc. Defects and infections in farming fields and crops will lead to a dynamic downfall in yield and standards of cultivated products. Early identification of defects and symptoms of defects employing our naked eye is a difficult task for any human. This leads to the diffusion of defects to the whole plant or crop. Identification of crop defects and their safety mainly in large farms is done by using digital image processing by Bharathi and Arulananth, T. S (2017).

Existing System

The native approach or mostly farmers are used to their day-to-day observation of their crop to identify whether any infection or defects started to affect the crop. After they come to know that, "Yes," they will get to know about that issue by the help of their native farmers or government centers provided in their nearby areas in Praveena et al. (2020).

Disadvantages:

a. Any rural cultivating people have to travel to faraway places or the offices of professionals or experts to contact them.
b. Due to the slow response of government procedures, they may consult any private officers, but they are too much expensive, and farmers cannot afford those expenses.
c. Maybe they can identify regular diseases but non-native defects which are increasing day by day due to environmental changes, it's difficult for them to identify them.
d. Cannot see the exact problem the farmer is facing
e. Sometimes wrong disease detection and diseases spread within short span
f. Lead in destruction of the crop.

Proposed Method

Our model takes information and collections of different images and stores them in databases for the identification of fruit and vegetable defects. One among it is for training the model and the other is for testing the model. In our model, we formed a lot of data sets of images of different fruits and vegetables by using resources like the Internet and domain websites for data sets like Kaggle. Images are bounded

into different forms of classes. Data sets hold the total images of the infected crop with their own disease names.

Advantages

a. Detect symptoms immediately
b. Avoid losses during the next period
c. Fast and accurate result
d. Controls through chemical applications and avoid an economic loss to the farmer.

Feasibility Study

We further moving forward with analysis. It is to cross-check the feasibility of the proposed trained model. All projects or models of applications are feasible by following their own unlimited resources and with no time bound. The main factors here are time and stock of the products. If a model is feasible then it should adhere to its bound of these two resources.

There are three types of feasibilities:

- Technical feasibility
- Operational feasibility
- Economic feasibility

Technical Feasibility

To evaluate any model to be feasible it should be easily available to the users in a user convenient way. i.e. it should have a good user interface and web design so that it satisfies this principle of feasibility. Web is the most widely used tech in every project that is designed worldwide. There should not be any technical errors or defects for the system at any time under any circumstances like overload, peak time, etc...

Operational Feasibility

If there is no operator or person required to explain the functionality and working of the product and is easily operable for every user then this feasibility is said to be satisfied. Here we check the awareness among the users to use the product. Every user should understand the technologies which are used.

Economic Feasibility

We should consider various metrics in order to conclude that a particular project is said to be economically feasible:

a. Cost-benefit analysis
b. Long-term returns
c. Maintenance costs

To access easily the mandatory requirement nowadays is information highway aka the internet. This should be available and affordable to every organization which has demanded of it. It should not become overhead and put on more burden.

Results and Evaluation

Supervised Learning

There are many types of learning and their implementation of algorithms, but the most famous one among all those and much feasible is supervised learning. As in this type of learning it has already the required data and expected output. In this type of learning, here the model would be trained first by using the provided data, where we can split the amount of data that is used for training and the other for testing purposes. It is the real-time example and most trusted learning and the outcome would be also very accurate. It is as simple as checking the input and then finding the appropriate output for it from the given set of data. It would be provided with both input and output data which helps the model to learn quickly and efficiently to provide high accuracy. It is a type of x and y model. There are many purposes of this learning and is used in different real-world applications like any linear model, predicting, Image classification, and many more.

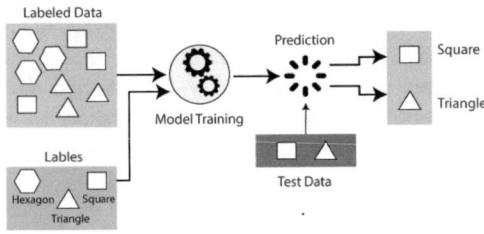

Fig. 1. A Sample Usecase diagram

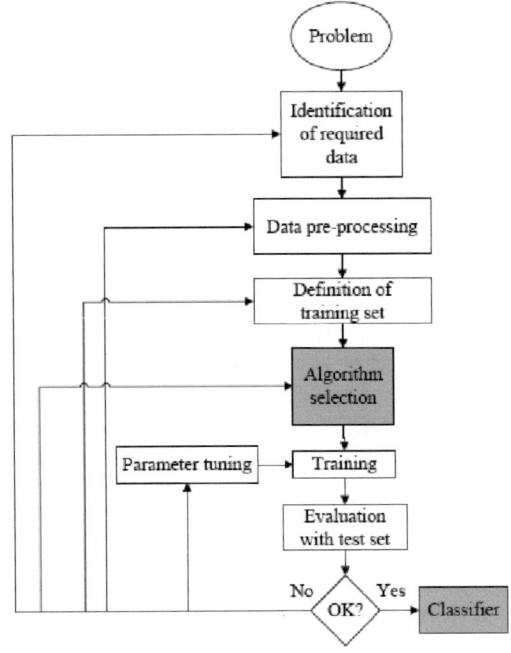

Fig. 2. The Process Flow of the Proposed Method

Image Preprocessing

i. In image pre-processing, at first, we will pre-process the images which are provided and it will improve the quality of the image, which further helps us to easily analyze them.

ii. This step is done before we are going to send our images for training the model and validating it.

iii. It helps to obtain great detail and quality of the image which we are going to work on.

iv. Image processing can also help in enhancing the features.

v. It converts the provided data into network-acceptable data. It also helps in improving the accuracy and efficiency of the prediction model.

vi. The images can be processed in many ways, some of which are mentioned below:

- Image Cropping and filtering.
- Intensity Adjustment.
- Histogram Equalization.
- Clearing noise.
- Detecting Edges.

Image Cropping and filtering

- Image cropping is nothing but removing unwanted and removing distorting parts of the image which affect the result of the prediction model.
- This is the foremost step in the Image pre-processing process.
- Tensor flow allows us to train the model using this pre-processing technique and all other steps involved in it.

Intensity Adjustment and Histogram Equalization

- It is used in managing the amount of intensity at each pixel of the images and helps to adjust that.
- Equalization techniques are used to manage the images to compare with the satellite observed images which are available at different resources and in contrast to them it will be equalizing.
- These are divided into many types and among those, these are managed locally and globally, according to their requirement they are used in helping out with the pre-processing of images.

Detecting Edges

- Every image we pass for processing is first classified by performing the detection of edges.
- As it is important to identify the coordinates of images to manage the pixels and also the breakage of the flow in them.
- Here image function plays an important role in describing the image edges and detecting those efficiently to describe partial derivatives which help in the growth of the image method.

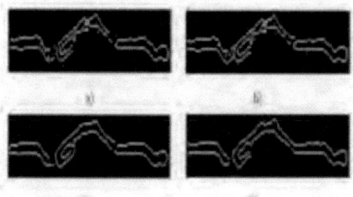

Fig. 3. The Sample Images Edges

Convolutional Neural Network

- As Neural networks have less power and not much accurate, we prefer using CNN.
- In CNN, every layer is independent of itself and they can automatically study or learn the different features from images and predict the output.
- It will not only work and think like the human brain but also includes some of the features of intelligent animals which have more accurate skills which are closer to truth.
- CNN also takes input in matrix form and work on direct images instead extracting features that are done by other neural networks.

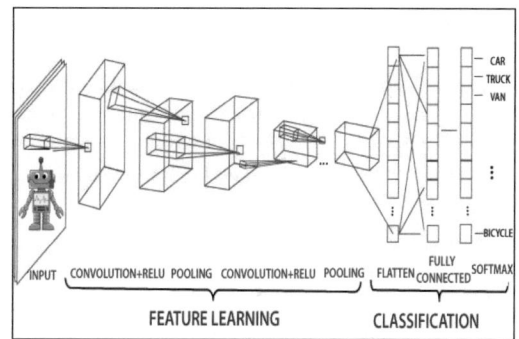

Fig. 4. The CNN Model

Output Screens

Accuracy Rate Is Calculated

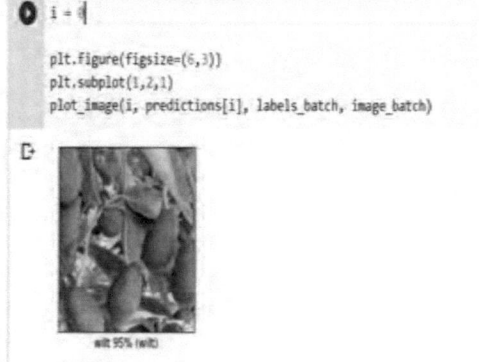

Output: Rings spot disease is identified in fruit with 95% accuracy.

Conclusion

As India is an agricultural nation and Indian farmers select a wide range of fruits and vegetable crops. The cultivation procedure can be improved with the help of our model which can detect the different defects that are present in different fruits and vegetables. If we can identify the defects prior that can help a lot of rural cultivators and also helps in saving a lot of resources. In our model, if we increase the epochs up to 20 we can achieve 80 to 100% of accuracy. The project's main aim is to reduce human effort and making especially farmers' life easier

References

Sonka, M., Hlavac, V., Boyle, R. (1998). Image Processing, Analysis and Machine Vision, PWS Publishing.

Prasad Babu, , Srinivasa Rao, B. (2007), Leaves Recognition Using Back Propagation Neural Network-Advice For Pest and Disease Control On Crops, IndiaKisan Net: Expert Advisory System.

Camargo, A., Smith, J. S. (2009). An image-processing based algorithm to automatically identify plant disease visual symptoms, Biosyst Engineering, 102(1).

Mukesh Kumar Tripathi, Dr. Dhananjay, D. Maktedar. Recent Machine Learning Based Approaches for Disease Detection and Classification of Agricultural Products, International Conference on Electrical, Electronics and Optimization Techniques (ICEEOT)-2016.

Bharathi, H., and Arulananth, T. S. (2017). A review of lung cancer prediction system using data mining techniques and self organizing map (SOM). International Journal of Applied Engineering Research, 12(10), 2190–2195.

Praveena, , Chinnasamy, P., Muneeswari, P., Ananthakumar, R., Bensujitha (2020). Detection and Categorization of Plant Leaf Diseases Using Neural Networks. European Journal of Molecular & Clinical Medicine, 7(4), 2438–2445.

Design of Remote Controlled Feeding and Smart Monitoring System for Aquaculture

Sudhakar Ajmera,[a,*] J. Jagan Babu,[b,*] S. V. S. Prasad,[c,*] Bittu Kumar,[d,*] Arulananth T S,[e,*] and Sangharsh snerla[f]

[a,c-f]Department of Electronics and Communication Engineering, MLR Institute of Technology, Hyderabad, India
[b]Department of Electronics and Communication Engineering, R.M.D Engineering College, Kavaraipettai, Chennai.
E-mail: *ajmera.sudhakar@gmail.com, jj.ece@rmd.ac.in, hodece@mlrinstitutions.ac.in, bittu.mlrit@gmail.com, arulananthece@mlrinstitutions.ac.in

Abstract

A device, that automatically feeds fish is known as an aquaculture automatic feeding system. With this ingenious approach, aqua farmers can feed aquatic organisms in their fisheries. A boat with a built-in container for loading fish food is part of the system. The autonomously driven boat can feed both hard and soft food. A farmer could simply sit on the lake's side and control the boat with a fly ski transmitter, with the receiver linked to the motor driver. The grade of fish and prawns is influenced by a number of factors, including water quality. As a result, a smart monitoring system keeps track of water quality measures and criteria to see if fish farming is a viable option. To monitor the water quality, pH and temperature sensors have been placed. These sensors will continuously take readings and display the results on an LCD screen.

Keywords: Boat, Arduino Atmega 2560, temperature sensor, pH Sensor, monster motor driver, LCD display.

Introduction

Aquaculture is a sort of farming in which diverse aquatic animals are provided with a proper living habitat, all of the essential circumstances, and healthy growth via food feeding. Aquafarmers primarily cultivate a variety of creatures like as crabs, prawns, fishes, and so on in this sort of agriculture. In Asian countries Such as Japan, India, the U.K., and others, this style of farming is more widespread (Hong-Jun Zhu, 2010, Changhui Deng et al., 2010, Sheetal Israni et al., 2015).

For them, this form of business has become a significant source of income. This form of aquaculture has various obstacles, particularly when feeding the fish, which entails entering the ponds in the middle. These concerns the pond's depth, i.e. the farmers had to be cautious because their lives were at risk. If the pond is particularly huge, the farmer will need to go to each end to spread the food evenly. The labor cost would be significant as well, due to the large amounts of food that would need to be handled. This Autonomous Feeding System would be extremely beneficial to aqua farmers in solving the aforementioned issues. The only thing left for the farmer to do was load the food into the boat and discharge it into the water. On the boat, both temperature and ph sensors are installed. This approach can be used to determine the pH and water temperature. The farmer can just sit and use a remote controller to operate the equipment (Kiran Patil et al., 2015).

A fly sky sender and receiver with a frequency of 2–4 miles are installed on the boat. Enabling the framer to operate the boat from a neighboring residence. The farmer will indeed be able to control the boat's movement thanks to the built-in RF transmitter and receiver. This system will assure consistent food distribution, save labor costs, and monitor water quality.

Methodology

Existing Problem

The present models of the food mechanism use a 200 RPM DC motor. The complete amount of food is initially loaded into a container in the upper section of the boat. The food will continue to push down to the bottom of the container, which will be conical in shape. So, once the DC motor has pumped some food out of the pond, the remaining food will settle to the bottom and be ready to be pumped out. However, food will not be given to the fish or prawns in a timely manner using this approach. While pumping food might damage the motor, it can also cause food to become caught in the motor. This method can only be used with soft food materials and not with hard things. Despite these limitations, we came up with a brilliant idea (Kayalvizhi, 2015).

Proposed Model

We want to create and execute an aquaculture automatic feeding and smart monitoring system, which will be a cutting-edge solution that allows aquaculturists to feed aquatic organisms in their fisheries and varied fishing habitats. The system basically comprises of a boat with an integrated food storage container into which the fish feed is loaded, as well as temperature and pH sensors installed in the boat to check water quality. The autonomous control boat is capable of feeding both soft and hard feed, as well as medicines. The farmer can just sit on the pond's edge and use an RF transmitter, such as fly sky, to control the boat. The pH and temperature readings are recorded and presented on the LCD, which helps with fisheries growth regulation.

Mechanism for Dumping Food and How It Works

To proceed, we are using Embedded Systems technology in this case, and a DC power supply is used to power the entire system. The motor driver is powered by a single LiPo (12v) battery. It is connected to high-load RPM motors (e.g., Johnson motor 500 rpm) and metallic servo motors via the motor driver. Arduino Atmeg The Ph sensor and temperature sensor are powered by the Atmega supply. The motor driver powers the LCD. The boat's motion is controlled by the Flysky, which connects four channels for the forward, backward, right, and left moments of the body. Two more channels are used to feed machinery. For the dropping of food, they are controlled by servo motors. For the moment being, two U-shaped 1-foot-long pipes are arranged on either side of the boat, with the servo motors at one end and a suspension nail at the other. At the top, these pipes are horizontally upright. As we turn the tuners in fly sky, the servo begins to rotate, allowing the pipes to rotate up to 90 degrees. Each pipe rotates in opposite directions. It can support a load of up to 5–6 kg. Following the dispensation of food, the tuners return to their normal position in fly sky (Bhatnagar, 2013).

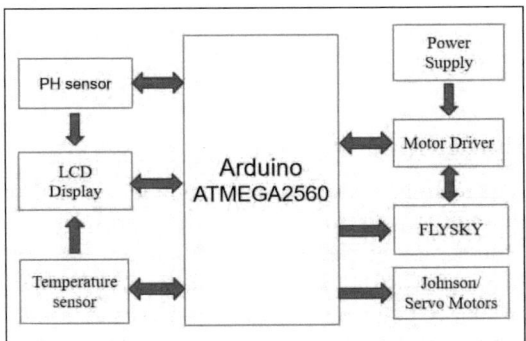

Fig. 1. Block Diagram Mechanism for dumping food

Components

Boat Design

A simple boat's function is to deliver food to the fish. The boat should be built with the center of gravity inside once the maximum feed has been loaded into it. To keep the boat balanced, the feed container needs to be placed evenly throughout. The boat should be constructed of a material that is both floatable and water resistant. Four simpler procedures can be used to complete the boat layout.

STEP-1 Arrange all the materials

STEP-2 Make a Sketch and Cut

Make a drawing of the pieces needed to create a boat on a Thermocol and then cut it out with a blade.

Fig. 2. Thermocol pieces

STEP-3 Join the pieces

Now join all the individual thermocol pieces (shown in Fig. 2) to form boat structure with the help of fevicol and razor needle.

STEP-4 Finishing and final design of boat as shown in Fig. 3.

Grab the shorty headers Thermocol and gradually give it a good polish with sandpaper.

Fig. 3. Final design of boat

ARDUINO ATmega 2560

A microcontroller board based on embedded systems is called the Arduino Mega 2560. On this board, there are four UARTs, a 16 MHz crystal oscillator, sixteen analog inputs, 54 digital I/O pins, a USB connector, a power jack, an ICSP header, and a reset button. To begin, just use a USB cable to connect it to a computer, or power it with an AC-to-DC adapter or battery. Figure 4 shows the ARDUINO ATmega 2560. The majority of shields created for the Uno and older boards are compatible with the Mega 2560. Arduino software is required to program the Mega 2560 board (IDE). The preprogrammed boot loader of the ATmega2560 enables you to upload fresh code to it without a computer. It has a boot loader that enables you to update its software without a computer.

Fig. 4. ARDUINO ATmega 2560

PH Sensor

A PH sensor (https://www.dfrobot.com/wiki/index.php /PH_meter(SKU_ SEN0161) can assess the pH of any solution and determine whether the substance is acidic, basic, or neutral. A pH scale is used to determine whether a water-based solution is acidic or basic. In comparison to basic solutions, which have a higher pH, acidic solutions have a lower pH. We can assess the water quality in fish and agricultural farms by knowing the Ph. A pH metre is a scientific tool that measures acidity or alkalinity by measuring hydrogen ion activity in water-based solutions. The electrical potential difference between a pH electrode and a reference electrode is what the pH metre (illustrated in Fig. 5) monitors.

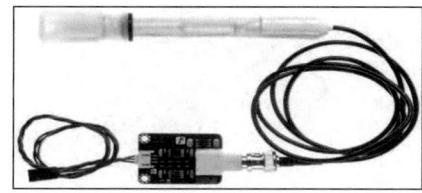

Fig.5. PH Sensor

Johnson Motor 500 RPM

A metal gearbox, an off-centered shaft, and a metal bushing for wear resistance are all features of the Johnson geared motor. It is a straightforward DC motor that utilizes a metal gearbox to move the motor shaft, making it a mechanically commutated DC electric motor. Johnson Geared Motors are well known for its small size and high torque-to-speed ratio. The motor will run smoothly between 6 and 18 V DC and provide 500 RPM when powered by a 12V supply. At 500 RPM, it produces 1 kg-cm of torque. Figure 6 shows the Johnson Motor 500 RPM.

Fig. 6. Johnson Motor 500 RPM

Metallic Servo Motor

The MG995 Servo Motor is powerful and dependable. It's a low-cost motor with a low-power output. The MG995 is a twin shock-proof ball-bearing servo with metal gear that is ideal for industrial operations. The motor responds quickly and rotates at a rapid rate. The VCC and GND power pins are used to energize the motor. The signal input pin on the MG995 Servo motor is used to rotate the motor. The pulse width modulation concept governs the operation of this motor. The Servo motor in question can only work at a frequency of 50 MHz. Figure 7 shows the Metallic Servo Motor.

Fig. 7. Servo Motor

16 × 2 Lcd Screen

Liquid crystal displays, or LCDs, are used in embedded system applications to show different system statuses and parameters. A 16-pin gadget called the LCD 16x2 has two rows of 16 characters each. Either 4-bit mode or 8-bit mode can be used with the LCD 16x2. Additionally, unique characters can be made. Eight data lines and three control lines are present for control. (Dupont C, 2017).

Monster Motor Shield (VNH2SP30)

One of the simplest ways to regulate a DC motor is with a VNH2SP30 is a full-bridge circuit. This method can be used to control both the motor direction and the speed. A pair of VNH2SP30 full-bridge motor drivers are included in this shield. In comparison to similar drivers, this driver has a higher current tolerance. This shield can tolerate a continuous current of up to 14A. Its power source should range from 5.5 to 16 volts. One of the control pins on the module is PWM. Longer duty cycles increase the voltage across the motor, which causes it to rotate more quickly. The longer the duty cycle, the more voltage there will be across the motor, causing it to rotate more quickly. This pin is connected to the MOSFET or transistor control pin (the gate or the base). internal electronics. Figure 8 shows the Monstor Motor Shield (Sridhar, B, 2017, Sridhar, B, 2018).

Fig. 8. Monster Motor Shield

Fly Sky (Fs-I6) Remote Control

The FS-i6 is a digital proportional radio control system that operates on the 2.4GHz global ISM band, making it globally usable. Each of the two sticks of a typical transmitter can move horizontally and vertically. As a result, each stick has two channels: one for horizontal and one for vertical movement. The Flysky FS-I6X comes with the FS-IA6B receiver, which is a 6-channel device. This device includes two tiny wire antennas, seven 3-pin connectors for input/output devices, and two more 3-pin connectors for iBUS transmission and receives connections.

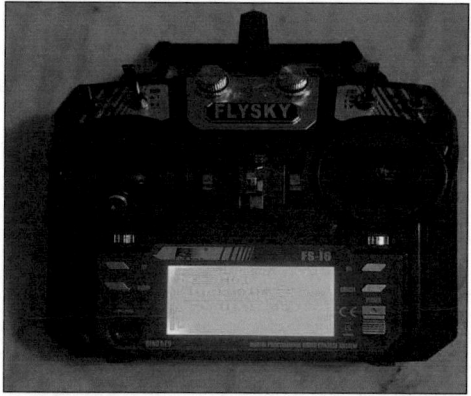

Fig. 9. Fly Sky (Fs-I6) Remote Control

Temperature Sensor(DS18B20)

The DS18B20 digital thermometer has a nonvolatile alarm function that monitors temperatures from 9 to 12 bits in Celsius and has user-programmable upper and lower trigger points. The DS18B20 connects to a central microprocessor via a 1-Wire bus, which by definition only needs one data line (and ground). The DS18B20 digital temperature sensor employs a single wire protocol to measure temperature with +-5 percent accuracy in the range of –67°F to +257°F (–55°C to +125°C). The data received via the 1-wire can range from 9 to 12 bits. This sensor can be operated with only one pin on the microcontroller because it employs the single-wire protocol(Bittu Kumar, 2015, Bittu Kumar, 2019).

Results

By using this Remote controlled Feeding and Smart Monitoring system, one can use this as a

replacement for the manual work of feeding the aquatics and can also monitor the required water pH and temperature content for the aquatics, it is displayed on LCD screen on top of the boat. Farmers can make necessary arrangements for water quality based on the readings from the LCD screen. The Figs. 10–12 show the different steps of a working model including the LCD screen.

Fig. 10. Working model

Fig. 11. Food deploying

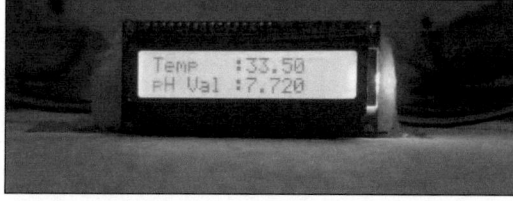

Fig. 12. Readings on the LCD screen

Conclusion

This system assists farmers in feeding the fishery and maintaining the pond's quality control, which is necessary for the fishery's continued expansion. South India has been identified as one of the top five aquatic industries in the world. Small farmers, who have a pond of less than an acre of land and are increasing the fishery, account for roughly 25% of the fishery. To feed them, the farmer must manually enter the pond and throw the food into it. It takes a significant amount of time and work. It has a significant impact on a person's health and, in certain cases, puts their life in jeopardy. When the quality of the water changes due to heavy rain or other factors, the growth of the fishery declines, resulting in fishery loss and death. In this case, they must immediately relocate the fishery to another pond and change the entire water in the pond. To avoid all of this, we've developed a remote-controlled feeding and smart monitoring system. This food can be ordered with a remote control. This requires very little manpower. It also features pH and temperature sensors that detect and display the data on the screen. These aid in the daily monitoring of the water, so that if the numbers are not as predicted, they may be quickly corrected by adding chemicals or other alternatives.

Future Scope

This research work can be further developed by adding some extra features to the present model. Additional features could include adding additional specialized sensors to the boat, which would help to ensure that the water quality content in all metrics is correct. We need to raise the boat's size in order to boost the feed. Fast deployment and increased load capacity can be achieved with high-power motors. It becomes more useful and trustworthy after these features are included.

References

Hong-Jun Zhu, (2010). Global Fisheries Development Status and Future Trend Analysis, Taiwan Economic Research Monthly, 33(3).

Changhui Deng, YanpingGao, Jun Gu, Xinying Miao (2010). Research on the Growth Model of Aquaculture Organisms Based on Neural Network Expert System, Sixth International Conference on Natural Computation (ICNC 2010); pp. 1812–1815, SEPTEMBER 2010.

Israni, Sheetal, Meharkure, Harshal, Yelore, Parag (2015). Application of IoT based System for Advance Agriculture in India, International Journal of Innovative research in Computer and Communication Engineering 3(11), November 2015.

Kayalvizhi, , Koushik Reddy, G., Vivek Kumar, P., VenkataPrasanth, N. (2015). Cyber Aqua Culture Monitoring System Using Ardunio and Raspberry Pi, International Journal of Advanced Research in Electrical, Electronics and Instrumentation Engineering, Vol. 4, Issue 5, pg:2320–3765; May 2015.

Patil, Kiran, Patil, Sachin, Patil, Mr. Sachin and Patil, Mr. Vikas, (2015). Monitoring of Turbidity, pH & Temperature of Water Based on GSM, International Journal for Research in Emerging Science and Technology, 2(3), March 2015.

Bhatnagar, Anita, and Devi, Pooja (2013). Water quality guidelines for the management of pond fish culture.International Journal of Environmental Sciences, 3(6), 1980.

Dupont C, Sheikhalishahi M, Biswas AR, Bures T. IoT, big data, and cloud platform for rural African needs. InIST-Africa Week Conference (IST-Africa), 2017 2017 May 30 (pp. 1–7).

Kumar, Bittu (2019). Real-time Performance Evaluation of Modified Cascaded Median based Noise Estimation for Speech Enhancement System' Fluctuation and Noise Letters, 18(04), Feb 2019.

Kumar, Bittu (2015). Spectral Subtraction using Modified Cascaded Median based Noise Estimation for Speech Enhancement. Sixth International Conference on Computer and Communication Technology, pp. 214–218. ACM, MNNIT, Allahabad, India, Sept. 2015

Sridhar, B. (2017). A blind image watermarking technique using most frequent wavelet coefficients. International Journal on Smart Sensing and Intelligent Systems, vol. 10 (4), pp. 863–878.

Sridhar, B. (2018). A wavelet based watermarking approach in concatenated square block image for high security. Journal of Automation, Mobile Robotics and Intelligent Systems, vol. 12 (3), pp. 68–72.

Hospital Data Security Using Blockchain and AI

Ajmeera Kiran,*,a B. Jency A Jebamani,b Ramesh Kumar Ayyasamy,c
V. Praveena,d and S. Arulananth T Se,*

aDepartment of Computer Science and Engineering, MLR Institute of Technology, Hyderabad, India
bDepartment of CSE, KPR Institute of Engineering and Technology, Coimbatore, India
cAssistant Professor, Department of Information Systems, Faculty of Information and Communication Technology, Universiti Tunku Abdul Rahman (UTAR), Malaysia, rameshkumar@utar.edu.my
dDepartment of CSE, Dr. N. G. P. Institute of Technology, Coimbatore, India
eDepartment of ECE, MLR Institute of Technology, Hyderabad, India,
E-mail: *kiranphd.jntuh@gmail.com, arulananthece@mlrinstitutions.ac.in

Abstract

Data security is crucial in the military, hospitals, government offices, colleges, and schools. In simple words, the process of securing data against illegal access and data corruption is described as data security. In order to achieve this security we can utilise key management, data encryption, hashing and tokenization. With the rising prevalence of data breaches, protecting your data is crucial. CIA-style data security. Confidentiality only offers access to authorized users. Access control lists, encryption, strong passwords, and two-factor authentication do this. Data integrity is preventing unauthorized changes that could corrupt it. Digital security enforces integrity. Availability is all for keeping software and security controls running smoothly. The blockchain is a secure chain of transaction blocks that cannot be hacked or modified after registration. Blockchain gives anonymity and decentralised identification. AI improves user security, analytical models, and data sets. SecNet secures data in cyberspace.

Keywords: Blockchain, Artificial Intelligence, Hashing, SecNet, cyberspace

Introduction

The amount of information that is being gathered has been constantly growing over the past several years. This data is used by a great number of businesses for both their business purposes and to conduct behavior analysis. We can generate tailored recommendations by performing in-depth analyses of the data and the patterns it exhibits. Sometimes this also results in unethical data sharing and the exploitation of data for personal use, which calls into question the availability of confidential information as well as its integrity and secrecy. Some businesses often exchange clients' personal information for monetary gain, which might damage their relationships with those customers because the information may be misappropriated or stolen. Privacy and security measures must be beefed up immediately in light of the recent spike in the number of data breaches and instances of digital surveillance. We are investigating the feasibility of securing data by utilizing blockchain technology (Chinnasamy et al., 2019) in conjunction with artificial intelligence. The decentralized system decides when, what, and how much data can be exchanged with whomever, as well as restricts the amount of data that can be shared. When combined, blockchain technology and artificial intelligence create details that are both more accurate and reliable. In addition to this, it strengthens the protection of the users and improves the analytical models. The blockchain has part of so many applications such as Education (Jain et al., 2021, Arunkumar et al., 2021) and smart cities.

Literature Survey

The research presented in Xia et al. (2017) suggests a system for the sharing of medical records that uses blockchain technology. This system would provide medical records with the highest possible level of security, while also making it easy to share data with several organizations. The research presented by Daniel et al. (2018) builds a foundational chain - a network that enables all companies to access trusted data without the risk of data manipulation and enables organizations to adapt to an environment that is continually subject to change. Because blockchain keeps a record of all changes, there is no way to alter or delete data without first receiving acknowledgment of the change. Shanthi et al. (2022) work based on Intrusion Detection System (IDS) and blockchain in depth. It then discusses the application of blockchain to IDS, delivers near to correct estimations of the threats, and does so in the following fashion. Please provide a detailed explanation of how AI is connected to big data and vice versa, as well as a few tips for improving data and information security through the use of AI. focuses on explaining the fact that the performance of AI can be improved by feeding a large amount of data to achieve a powerful model. More effective and enriched datasets can build a stronger model, which increases security. Arulananth et al. (2022) also explain that this can be done to improve the performance of 5G. focuses on launching a new market segment, one in which participants can switch between different machine-learning modes in exchange for rewards. This, in turn, makes AI believable

and accessible to everyone, and it also helps in providing better and more advanced AI algorithms for higher levels of data security.

The research presented by Praveena et al. (2021) establishes a framework in which Internet of Things (IoT) devices can collect real-time recordings that are then stored within the blockchain. Additionally, the majority of problems regarding insurance claims can be simply addressed, provided that fraudulent behaviors are avoided. The data that is saved in this manner is not only able to be protected, but also possesses transparency.

Methodology

The data security of these existing systems can be implemented in one of two ways: either by incorporating AI-based algorithms or by utilizing a blockchain that is compatible with certain applications. Unfortunately, neither one of these existing solutions addresses the issue from an architectural point of view. Blockchain and artificial intelligence were the two preexisting technologies that were included in the conventional network architecture that SecNet developed in order to address this shortcoming and find a solution. These are the data. In the modern world, everything is drawn from data, including the algorithms that power artificial intelligence, which rely on historical data to make predictions. Similarly, the healthcare infrastructure has been absorbing and adapting to the use of patients' previous medical information to stay up with their health, and in such instances, patients' confidentiality cannot be infringed upon and must be secured. Because everything is available on internet platforms, it has become common practice to use consumer data or sell it to companies. This is happening even though there is no way to regulate the interchange of data or the data that is saved on third-party servers. To combat these behaviors, private data centers have been developed. These data centers integrate artificial intelligence (AI) and blockchain technology with an intermediary architecture to produce a safe online environment. The primary goal of the SecNet architecture is to integrate its three primary components, which are as follows:

a. Data Sharing using Blockchain
b. Secure computing using artificial intelligence
c. Trust value-based exchange mechanism

Data Sharing Guaranteed by Blockchain: Blockchain-based data sharing with ownership guarantee permits information sharing among corporations without any fear. In this method clients can outline the access control and manage it, that is, which client is authorized to access the information and which client is unauthorized to access the information. Also, blockchain objects are going to be generated on it to access knowledge and permit solely those users to access information that has permissions. In the blockchain, object user can add data they want to share and provides permissions.

AI-Based Secure Computing: In this method, primarily the AI-based secure platform is employed to provide additional intelligent security rules, that facilitate to assembly of even better secure cyberspace. AI functions similarly to the human brain and it is responsible for executing the logic to check whether the string pattern matches and also if the requesting client has authorized permission to get entry to the shared information. If access is obtainable solely then AI will permit the blockchain to show the records in any other case it ignores the request. The SecNet model introduces the ASC component in the PDC to protect data.

Trust Value-Exchange Mechanism: In this method, all the customers who are sharing information will earn rewards every time their information is accessed. The trust value-exchange mechanism for purchasing security services is imparting a way for customers to gain economic rewards while giving out their information or service, which promotes information sharing and accordingly complements the overall performance of AI. To improve data security, all the information scattered throughout the web ought to be incorporated to supply higher and better security rules.

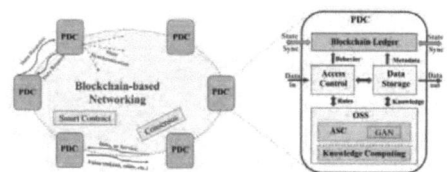

Fig. 1. Working of SecNet

Figure 1 below depicts the working of a SecNet. The nodes are linked with the help of a blockchain-based network. In this webwork, the nodes communicate with one another and attain an agreement primarily based on blockchain techniques. During this process, they collaborate via the execution of smart contracts. All the nodes contain a blockchain ledger to synchronize the state with different nodes so that we can attain an agreement, both on node state and smart contract execution results. In terms of data, the nodes are equipped with a data storage module and an access control module for providing data security. SecNet nodes additionally have an Operation Support System (OSS) module which allows AI-based secure computing (ASC) for generating expertise and security rules from data.

Patient Profile Creation: The first step for every patient is to create a profile, this module consists of all the patient details. There's an option to select which hospital the client/user wants to share the information with. Once the user clicks on create, a blockchain object is generated, that solely provides permission to particular hospitals that are authorized.

Patient Login: Patient login is done using the patient id generated after creating the profile. This module displays that particular patient's information along with the blockchain hash code and the economic rewards earned.

Hospital Login: Patient data can be shared with the two hospitals that are using this application as two organizations (namely Hospital 1 and Hospital 2). If you give "Hospital1" as the username and password then we'll log in to Hospital1, similarly for Hospital2. This module also consists of accessing patient data through which we'll get the desired information.

Access Patient Data: In order to fetch certain information from the hospital database we use an AI algorithm. This particular algorithm takes the disease as input and matches it with the patient diseases in the database. Once the algorithm finds a match, the records only in that particular hospital that we have logged into will be displayed

Results

To run the development server, navigate to the project directory and open it in a terminal. Now to start the server use the command - "python manage.py run server"

Fig. 2. Connecting to the server

Copy the URL generated after starting the server to a browser and add the HTML file name. The following screen will appear once we run the browser.

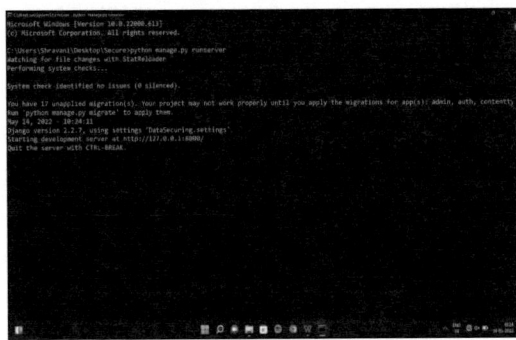

Fig. 3. Home Screen

Click on create profile and enter patient details.

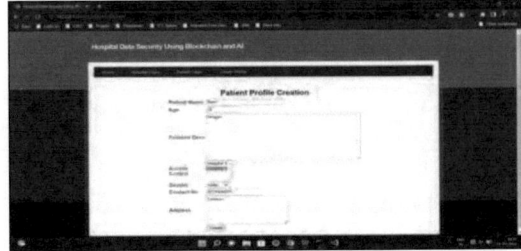

Fig. 4. Profile Creation

After clicking on create, the patient id is generated.

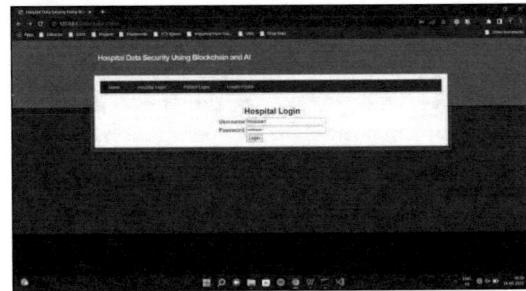

Fig. 5. Patient ID

Now we click on Hospital Login, we can check the data shared to a particular hospital by giving the username and password.

For example - If we give the username as Hospital1 and password as Hospital1, we can access all the patient records that have been given access to Hopital1.

Similarly, we can access all the patient records that have been given access to Hospital2 by giving that as username and password.

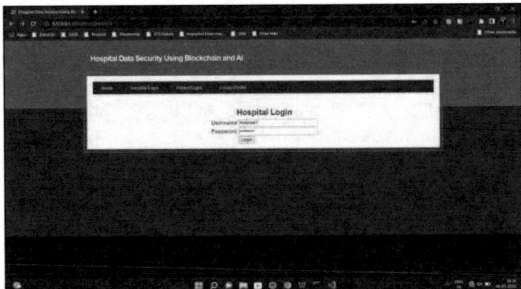

Fig. 6. Hospital Login Screen

Now we have access to Hospital1.

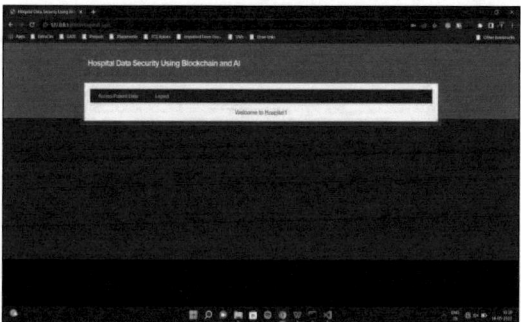

Fig. 7. Logged into Hopital1

By clicking on access patient data we enter a string such that it provides records of all the patients that match with it.

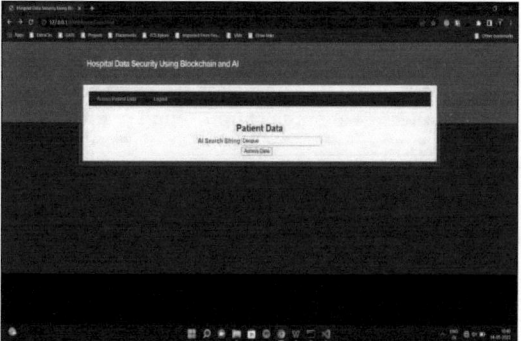

Fig. 8. Access Patient Data

Since we have no patient records that match the disease string, no records will be displayed.

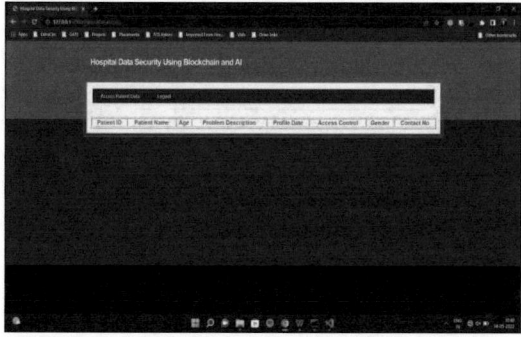

Fig. 9. Records in Hospital1 that match with 'Dengue'

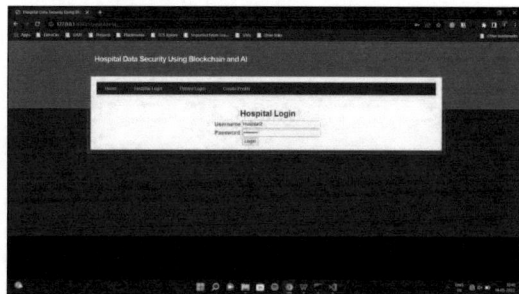

Fig. 10. Hospital Login Screen

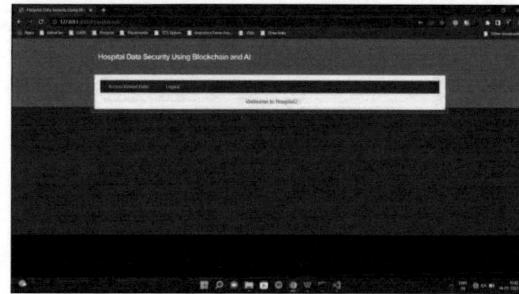

Fig. 11. Logged into Hospital2

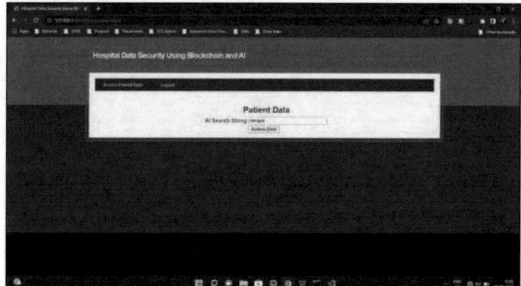

Fig. 12. Access Patient Data

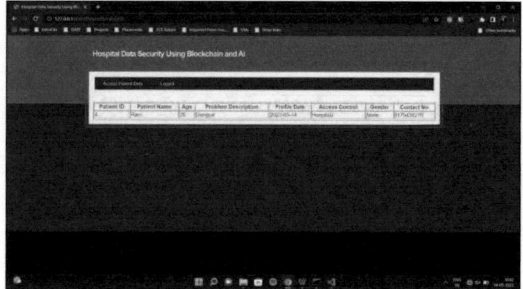

Fig. 13. Records in Hospital2 that match with "dengue"Using the patient id generated after creating the profile we can check the details.

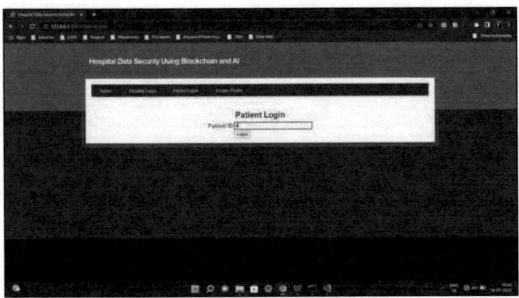

Fig. 14. Patient Login Screen

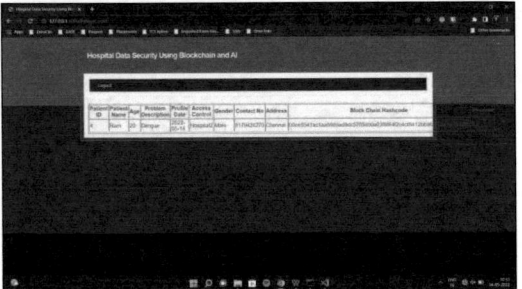

Fig. 15. Hashcode and Revenue generated for every patient

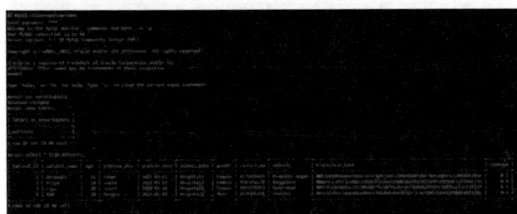

Fig. 16. Patient records stored in database

Conclusion

We proposed the SecNet architecture, as a networking model that only focuses on data security. This architecture integrates blockchain and AI to its maximum use so that unauthorized modifications cannot be done to the data, it also protects the data from being misused. SecNet provides data ownership guarantee using blockchain technologies, an AI-based secure computing platform, and a trust value exchange system. Further, blockchain technology can be used to our benefit by accepting requests based on the authorization of data, unique smart contracts for sharing information and AI based secure computing provider in SecNet. We can also enhance SecNet on advanced structures by integrating IPFS and Ethereum.

References

Chinnasamy, , Deepalakshmi, P., Praveena, V., Rajakumari, K., Hamsagayathri, P., (2019). Blockchain Technology: A Step Towards Sustainable Development. International Journal of Innovative Technology and Exploring Engineering (IJITEE)

Chinnasamy, P., Vinothini, C., Arun Kumar, S., Allwyn Sundarraj, A., Annlin Jeba, S. V., Praveena, V. (2021) Blockchain Technology in Smart-Cities. In: Panda S. K., Jena A. K., Swain S. K., Satapathy S. C. (eds) Blockchain Technology: Applications and Challenges. Intelligent Systems Reference Library, vol. 203. Springer, Cham. https://doi.org/10.1007/978-3-030-69395-4_11

Jain, , Kumar Tripathi, A., Chandra, N. and Chinnasamy, P. (2021). Smart Contract enabled Online Examination System Based in Blockchain Network, 2021 International Conference on Computer Communication and Informatics (ICCCI).

Xia, , Sifah, , Asamoah, , Gao, J., Du, X., and Guizani, M. (2017). MeDShare: Trust-less medical data sharing among cloud service providers via blockchain.

Kurtulmus, , and Daniel, K. (2018). Trustless machine learning contracts; evaluating and exchanging machine learning models on the Ethereum blockchain,

Shanthi, D., Swapna, N., Kiran, Ajmeera (2022). A Anoosha, Ensemble Approach Of GP,ACOT,PSO, and SNN For Predicting Software Reliability, International Journal of Engineering Systems Modelling And Simulation.

Arulananth, T. S., Baskar, M., Ramkumar, J., and Rao, K. S. (2022). Enhanced blockchain technology associated with IoT for secure and privacy communications in 5G. Blockchain for 5G Healthcare Applications: Security and privacy solutions, 83–113.

Chinnasamy, P., and Vinodhini, B., and Praveena, V., and Vinothini, C., and Ben Sujitha, B. (2021). Blockchain based Access Control and Data Sharing Systems for Smart Devices, Journal of Physics: Conference Series, 1767(1), pp. 012056, doi.10.1088/1742-6596/1767/1/012056.

MIMO Antenna for 5G Communication Using HFSS

B. Sridhar[*,a], *M. Koteswara Rao*[b], *K. Naveen*[c], *Kiran Dasari*[d] and *Bittukumar*[e]

[a,c,d&e]Department of Electronics and Communication Engineering, MLR Institute of Technology, Hyderabad, India.
[b]Department of Electronics and Communication Engineering, Sri Vasavi Engineering College, Tadepalligudem, India.
E-mail: [*]sridharbece@gmail.com

Abstract

A MIMO antenna with an increased gain is proposed, for 5G applications, a single patch antenna has limited gain and a poor data rate. To boost the upload/download data throughput for 5G applications/communications, high-gain antennas are necessary. This paper creates a dual-band mimo antenna for 5G connectivity utilizing HFSS. This dual-band mimo antenna is utilized for smart phone 5G communication. This MIMO antenna is made up of two patches connected by feed lines. The suggested antenna contains four patches that provide the necessary bandwidth for 5G connectivity. We devised and met the criteria of 5G applications by using only two patch antennas instead of four. As a result, the necessary bandwidth is obtained by decreasing the design process. In terms of reflection coefficients, gains, surface current distributions, and radiation patterns, the simulated antenna performances are reported and analyzed. The reported gains and reflection coefficient suggest that the antenna might be used for 5G operations/applications.

Keywords: Antenna gain, Radiation Efficiency, Directivity, Return loss

Introduction

The issue of high gain in planar antennas is becoming a major topic in microwave antenna design. High-gain antennas are required for applications in portable devices to fulfill multiple wireless device characteristics such as small size, efficiency, and high radiation power (Murugan et al., 2021). Size is one of the major design concerns. A piece of technology known as an antenna transforms an information stream into electromagnetic waves which converts electrical energy into electromagnetic waves and back again. Signals may be transmitted and received using an antenna. An antenna that transmits electromagnetic waves by converting electrical impulses into them is known as a transmitting antenna (Omar et al., 2021). One type of reception antenna converts the electromagnetic waves from a beam into electrical impulses. The same antenna can be used for both transmitting and receiving during two-way communication. Wireless local area network (WLAN) and 5G applications make extensive use of these antennas. The term 5G refers to the next generation of mobile networks following the 4G LTE network. It will be available for purchase in the year 2020. Cell phone operators throughout the world debuted 5G, the fifth-generation technological standard for broadband cellular networks, in 2019. It is intended to succeed 4G networks, which currently connect the vast majority of existing cell phones.

All 5G wireless devices in a cell are linked to the Internet, and because 5G networks have greater bandwidth than earlier networks, it can connect more devices and improve Internet service quality in crowded. locations. The rest of the paper is organized as follows: literature survey, methodology, Construction, Results and conclusion.

Literature Survey

The numerous Multiple-input-multiple-output (MIMO) antennas are described in this survey. The MIMO antenna is crucial in today's wireless networking environment. MIMO antennas have received a lot of attention during the previous decade (Hasan et al., 2019). The parameters of several MIMO antennas are discussed on this page. This research also examines many MIMO antennas with various components and how they affect MIMO antenna difficulties such as cross-polarization and reciprocal coupling across patch arrays (Pramod et al., 2020)

In this work, a compact dual-band MIMO antenna for 5G mobile communications. The primary component is a dual-band monopole operating at 3.4 and 4.9 GHz. To make this site band-notch, an inverted L-shaped strip is added to the monopole's ground (Parvathi et al., 2021). The sub-6 GHz commercial 5G communication frequency is covered by a quasi-directional emission pattern and S11 at 3.4 GHz (3.3–3.6 GHz and 4.8–5.0 GHz). The suggested MIMO antenna is 50*50*0.8mm3 and is made up of four upgraded sections (Wang et al., 2019).

A dual-band MIMO antenna for 5G communication is suggested in this work. The proposed antenna consists of four antennas that operate at frequencies of 3300–3600 MHz and 4800–5000 MHz. The antenna proposed in this letter differs from customary 5G antennas in that it is perpendicular to the edge of the system circuit board and may be put on a standard full-screen mobile phone (Singh et al., 2022).

The internet of things firms are providing various innovations and solutions to the market, and tiny chip manufacturing industries are investing more money in the industry, which is rapidly developing. It is not without its challenges. One of the most challenging aspects of figuring out the IoT is ensuring that those "items" or end hubs are really ready to communicate with the web (Saad et al., 2019).

This work presents novel, small, microstrip line-fed dual-band printed MIMO antennas that resonate at 28 and 38 GHz and are suitable for 5G mobile communications. The first design for the 28 GHz and 38 GHz bands consists of a rectangular microstrip patch antenna with two elements and an inset feed. Although a number of substrate types may be employed to create microstrip patch antennas, the dielectric constant of the substrates must be taken into account for antenna performance (power efficiency, bandwidth, radiation pattern).

Pattern division multiple access (PDMA), a revolutionary non-orthogonal multiple access technique for fifth-generation mobile networks, can enable greater connectivity and significantly better spectral efficiency as compared to conventional orthogonal multiple access (OMA).

This paper proposes a novel PDMA framework paired with multiple-input multiple-output (MIMO) technology for downlink transmission utilizing a space-frequency block coding method in order to increase system transmission reliability and spectrum efficiency for PDMA.

Methodology

MIMO antennas have gained more attention in the real time communication system. It can use multi-paths to transmit or receive data, and result we can increase the range and output performance. The multiple-antenna technology is the most excellence technologies for futuristic wireless devices. This paper discusses the project's block diagram as well as the creation of separate modules. And how these modules will aid you in your project. To create antennas, the HFSS (High-Frequency Structured Simulator) program is used, and the flow chart for each antenna design is as follows: displayed in the figure below, which may be utilized for the basic design of any antenna designed in the HFSS software of ANSYS.

Other software are used to design antennas like Microwave Studio CST, Antenna Magus,

FEKO, and ZELAND IE3D. However, this project was created with ANSYS HFSS One of the most advanced 3D EM software packages for antenna design and complex RF electrical circuit design is software, which is one of the most advanced 3D EM software packages for antenna design and complex RF electrical circuit design. The software also includes the necessary features and tools for filter design, transmission line design, and modeling. The Finite Element Method is the foundation for IT's design and analysis methods. It has an automatic design process that allows the user to just indicate what they want. The program builds a mesh that meets these parameters based on needed geometry, design material qualities, and so on.

It supports the design of linear circuits, Figure 1 shows the proposed flow diagram of antenna design.

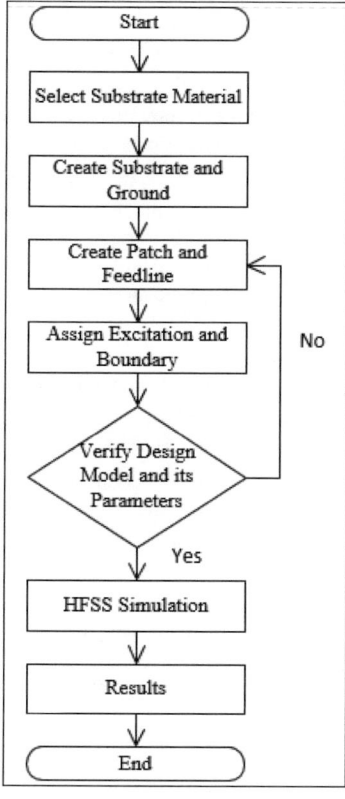

Fig. 1. Proposed flow diagram of Antenna design

Construction

The figure 2 depicts the structural arrangement and geometry of the patch antenna in this part. The most current wireless communication trends and improvements call for high-performance antenna designs with low profile qualities, such as cheap cost, simplicity of fabrication, handling, maintenance, simple design, lightweight, etc. Due to its compact size, a microstrip patch antenna is ideal for these uses. The patch's impedance is highest along the edges and lowest in the centre. Corporate feed networks with quarter wave transformers are employed to match the high edge impedance. The antenna feed can be considered a component of the antenna since it provides an efficient flow of EM energy, correct antenna excitation, and interaction with the rest of the communication chain. These goals guide the selection of the antenna feed (Kumar 2019, 2021).

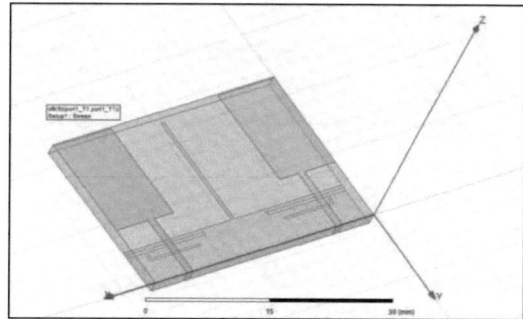

Fig. 2. Design of MIMO antenna

Results

The proposed antenna is developed and modeled in the ANSYS HFSS software. Examining the gain, S parameter, EM radiation, and radiation pattern of the proposed antenna. The results of the gain-enhanced MIMO antenna for 5G applications are obtained, and a variety of factors, including radiation efficiency, gain, radiation pattern, and power loss are also obtained and analyzed for the MIMO antennas. As a result, we can also demonstrate how the design has improved gain in Figure 3.

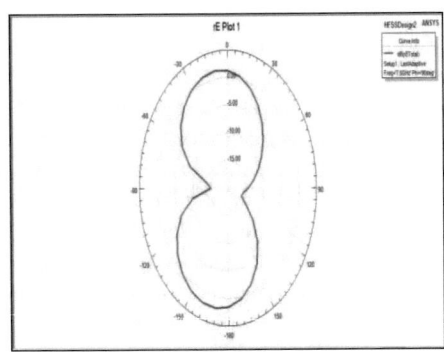

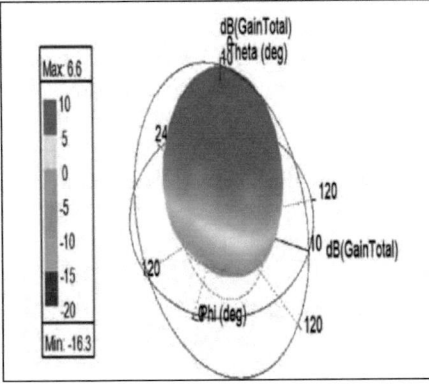

Fig. 3. Gain plot

Conclusion

We have constructed the dual band antenna, and we have a total gain plot of 6.6dB for 5G connection, with a maximum gain band of roughly 8dB. The ground's substrate material was FR4 epoxy. The gain has risen, and the requisite radiation pattern and gain for the MIMO antenna have been met. The antenna system was designed and simulated using ANSYS HFSS. The antenna design figure shows the top view of the completed antenna system. The recommended antenna construction is modeled using ground plane and FR4 epoxy as base materials. Two patches on the substrate are given in the spaces between the radiation patches. Modeling is done using the High-Frequency Structure Simulator (Ansys HFSS Version-18) software.

References

Murugan, S. (2021). Compact MIMO Shorted Microstrip Antenna for 5G Applications. International Journal of Wireless and Microwave Technologies (IJWMT), 11(1), pp.22–27.

Omar, A., Hussein, M., Rajmohan, I. J., and Bathich, K. (2021). Dual-band MIMO coplanar waveguide-fed-slot antenna for 5G communications. Heliyon, 7(4), p.e06779.

Hasan, M. N., Bashir, S. and Chu, S. (2019). Dual band omnidirectional millimeter wave antenna for 5G communications. Journal of Electromagnetic Waves and Applications, 33(12), pp. 1581–1590.

PramodKumar, A., Suman, N., Koteswararao, Y. V., and Venkatachari, D. (2020). Design and analysis of high beam forming MIMO antenna for 5G applications. Materials Today: Proceedings.

Parvathi, K. S., and Gupta, S. R. (2021). Novel dual-band EBG structure to reduce mutual coupling of air gap-based MIMO antenna for 5G application. AEU-International Journal of Electronics and Communications, 138, p.153902.

Wang, F., Duan, Z., Wang, X., Zhou, Q. and Gong, Y., 2019. High isolation millimeter-wave wideband MIMO antenna for 5G communication. International Journal of Antennas and Propagation, 2019.

Singh, A. K., Dwivedi, A. K., Jha, C., Singh, S., Singh, V. and Yadav, R. S., 2022. A compact MIMO antenna for 5G NR frequency bands n257/n258/n261 under millimeter-wave communication. IETE Journal of Research, pp.1–13.

Saad, A. A. R., and Mohamed, H. A. (2019). Printed millimeter-wave MIMO-based slot antenna arrays for 5G networks. AEU-International Journal of Electronics and Communications, 99, pp. 59–69.

Kumar, B. (2021). Comparative performance evaluation of greedy algorithms for speech enhancement system. Fluctuation and Noise Letters, 20(02), p.2150017.

Kumar, B. (2019). Real-time performance evaluation of modified cascaded median-based noise estimation for speech enhancement system. Fluctuation and Noise Letters, 18(04), p.1950020.

Multiple Biometric Authentication Through Image Assessment Using Machine Learning

A. Sangeetha,[a] P. Purusotham,[b] T. Vinod,[c] Ajmeera Kiran,[d] S. Sinduja,[e] and P. Chinnasamy[f]

[a,b,c,d,f]Department of Computer Science and Engineering, MLR Institute of Technology, Hyderabad, India
[e]Department of Computer Science and Engineering, Vivekanandha College of Engineering for Women,
E-mail: *sangeetha.a@mlrinstitutions.ac.in, chinnasamyponnusamy@gmail.com

Abstract

It is proposed to use a combination of three biometric identities: iris, palm print, and face. Firstly, Input images that contain noise undergo pre-processing from which key features are extracted. These features are compared with the existing images present in the database that are considered to be Authenticated. In this stage, we obtain Matching scores. A module fuses these individual scores obtained. The final fusion is capable of classifying a person to be authenticated or unauthenticated.

Keywords: Image Quality Assessment, Pre-processing, Feature extraction, Biometrics, DWT Segmentation, Image Fusion

Introduction

The process begins with image pre-processing, which includes noise removal and then feature extraction. We have to interpret the shapes of Iris, palm prints, and face properly because we need to train the computer which cannot distinguish the shapes as humans do. Another issue is that an image can be influenced by a variety of elements such as the angle at which we take the picture or the lighting conditions in the area (Valli, 2018; Chinnasamy et al., 2020; Chinnasamy et al., 2020).

Overview

The iris is a thin layer, circular component in the eye, and is responsible for the control of the diameter and size of the pupil which is proportional to the amount of light reaching the retina. The color of the iris is often referred to as "eye color." Iris recognition refers to the capability of verifying the identity of a person through his eyes i.e., Iris. The human iris is the space located around the pupil of the eye and is covered by the cornea layer which comes under the unique information of a person (Kavati, 2017).

For each subject, we collect palm print images from both left and right hands. These images are 8-bit gray-level JPEG files by our self-developed face recognition device. Subjects are required to put their face near the capturing hardware and make sure that he is in an evenly colored background. The device takes the biological features of a person.

Literature Survey

During a literature survey, we collected some information about the mechanisms of detecting fake biometrics. on the vulnerability of face verification systems to hill-climbing attacks using Bayesian we try to test the weaknesses of two face recognition systems via adaptation.

The hill-climbing assault algorithm is used here. This attack considers distinct properties such as eigenface-based and parts-based verification systems. We may conclude that the proposed method has successfully bi-passed over the bulk of the accounts that have been attacked [5]. A new software-based liveness detection method is proposed that takes image quality parameters into account. This scenario is tested on many of the databases that are comprised of both real and duplicate fingerprints. The proposed solution is proved to be efficient to the multi-scenario dataset and presents accurately classified samples that are nearly 90%. This method added the proof for previous technologies which propose that images are required for authentication but in a more secure format (Galbally et al., 2011).

All the possible test cases that are given to fingerprint verification systems that include and exclude the user's knowledge are studied. Considering the quality of the images, and the observations that are obtained after different operations, we conclude the robustness of the systems to various direct attacks (Chinnasamy et al., 2020)

Existing System

Existing systems include techniques such as Edge Detection, Segmentation, and Feature Vector. Presently biometrics are done using Fingerprinting. But it is not much flexible because we can make duplicates of fingers and bluff people. Hence making it less efficient. We will be using PCA i.e. Principal Component Analysis algorithm for finding the covariance and variance. Only spatial domain is calculated.

Proposed System

We can implement the proposed method, On applying downscaling and color level transform, we pre-process the image. Biometrics based on the consideration of iris, palm print, and face features for a person's authentication using Sequential modified hear wavelet and Energy feature extraction using key generation analysis (Chinnasamy et al., 2020).

System Architecture

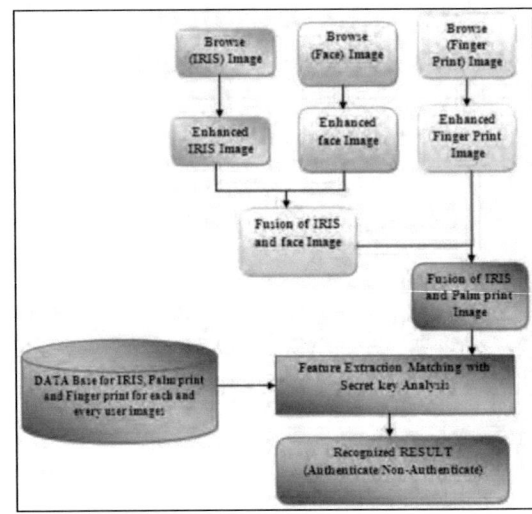

Fig. 1. System Architecture

Image Acquisition

It is the first step in any vision system, because, without an image, no processing is possible. In this process, we retrieve our required images from the source i.e. a human being is used in this case. These retrieved images are completely unprocessed which means they consist of so much unnecessary data like noise etc. Hence these images are now sent for dismissal of all such weird data.

Pre-Processing

For Pre-processing the images, here we use Median filtering method. The main use of inculcating this method is that it replaces each pixel value with the median value of the pixels that are in the neighborhood of that particular pixel. Previously, that was not the case when we considered the central value of the pixel. Also, it is capable of obtaining filtered images with less blurring rather than traditional methods. Hence, it is an effective method in Data cleaning process.

Discrete Wavelet Transform

A signal is decomposed into a set of bases by Wavelet transform which is known as wavelets.

$$\psi_{a,b}(t) = \frac{1}{\sqrt{a}}\psi\left(\frac{t-b}{a}\right)$$

Where a-> scaling parameter

b -> shifting parameter

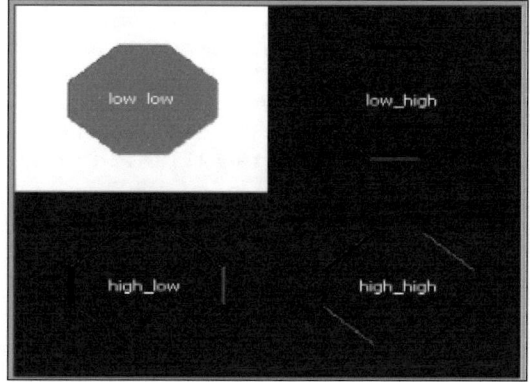

Fig. 2. 1D- Discrete Wavelet Transform

Discrete wavelet transform decomposes the input image into four levels which are LL, HL, LH, and HH.

Features Extraction Process

In this process, we obtain a matrix named the Co-occurrence Matrix (CCM). Each element of the matrix is comprised of the values i and j. It calculates the occurrence of i-valued pixel intensity in relationship with j valued pixel. The size of the CCM is determined by the number of gray levels in the image. The levels that are obtained from DWT Segmentation are used for considering the data energy. Based on the orientation and distance between image pixels, a co-occurrence matrix is constructed.

Evaluation and Results

Firstly, images are acquired through the Image Acquisition process and then these are pre-processed to remove the unnecessary data. Now, we apply DWT transformation for decomposition. Features are extracted and compared to that of

database images obtaining the matching scores that are put into CCM matrix.

i. GUI of the Application:

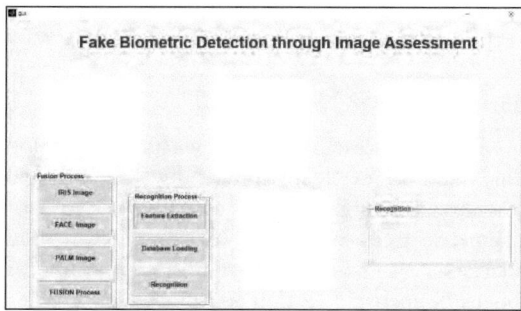

Fig. 3. Application Interface

ii. Fused Images formed out of input Images:

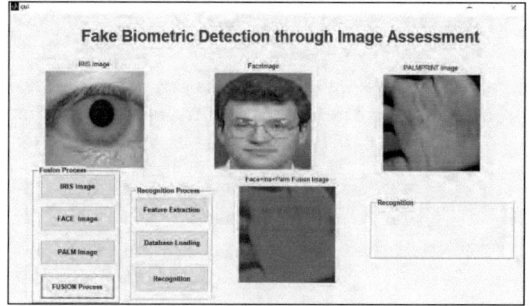

Fig. 4. Fused Image is formed

iii. Output classifying Authenticate or not:

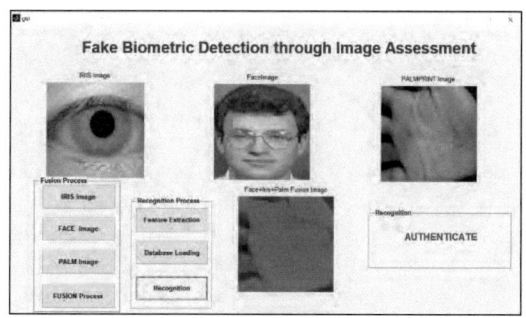

Fig. 5. POC showing that the person is Authenticate

Conclusion

Our existing database consists of 4 different iris, face, and fingerprint images of 4 persons, and a code matrix is formed. When the photographs of

Iris, palm, and face of the individual who needs to be classified are received, a code matrix is produced, similar to the prior technique. Once the matching of patterns is done, then we can classify if the person is authenticated or not.

Future Work

As part of future enhancement, we have to develop this method even for physically handicapped people who might be in loss any one of the biometric traits which are Iris, palm Print, and face. As a result, an effective real-time solution for bogus biometric detection is provided.

References

Susmitha Valli, (2018). Scalar Vs Vector Color Image Processing: An analysis, JARDCS, 7 Special ISSN No:1943–023X 2018.

Srinivas Rao, K. (2018). Fast finger print retrieval using minutiae neighbor structure. Springer volume 705 ISBN978-981-10-8569-7.

Dr. Illaiah Kavati (2017). Score based indexing and retrieval technique for biometric databases. IJPRAI Vol. 31, No. 06, 1756009 (2017) 2017.

Galbally, , Fierrez, J., Alonso-Fernandez, F., and Martinez-Diaz, M. (2011). Evaluation of direct attacks to fingerprint verification systems, J. Telecommun. Syst., 47(3–4), pp. 243–254. Chinnasamy, , Deepalakshmi, P., Shankar, K., Chapter 6 - An analysis of security access control on healthcare records in the cloud, Editor(s): Amit Kumar Singh, Mohamed Elhoseny, In Intelligent Data-Centric Systems, Intelligent Data Security Solutions for e-Health Applications, Academic Press, 2020, Pages 113–130, ISBN 9780128195116, https://doi.org/10.1016/B978-0-12-819511-6.00006-6.

Chinnasamy, , Ganesan, A., Prasathkumar, V., Praveena, V. (2020). A Trusted-Role Based Access Control Model for Secure Cloud Storage. International Journal of Advanced Science and Technology, 29(9s), 3253–3259.

Skin Cancer Detection Using CNN

Arulananth T S,[a,*] Jayanthi D,[b,*] S. Monika,[c,*] Pulkit Singh,[d,*]
Bathula Jeevan Chaitanya,[e] and Jayasree Singi Reddy[f]

[a,c-f]Department of Electronics and Communication Engineering, MLR Institute of Technology,
Hyderabad, India
[b]Department of Electronics and Communication Engineering, R.M.D Engineering College
Kavaraipettai, Chennai,India
E-mail: *arulananthece@mlrinstitutions.ac.in, slvjayanthi@gmail.com,
monika.sirigiri@mlrinstitutions.ac.in, pulkitsinghp24@gmail.com

Abstract

Melanoma has the lowest survival rate of the three styles of pores and pores and skin maximum cancers: BCC, SCC, and Melanoma. Melanoma is more likely to be decided early, developing the opportunities of survival. Hair removal, de-noising, sharpening, and scaling of the given pores and pores and skin image, further to segmentation, that is used to interrupt up off the area of interest from the given image, are the four maximum vital components of pores and pores and skin maximum cancers detection methods. Machine analyzing and deep analyzing-based totally definitely algorithms are used for kind in cutting-edge day pores and pores skin maximum cancers detection technology. A deep convolutional neural network is used in this method.

Keywords: Convolutional neural network, image processing, image segmentation

Introduction

Squamous Cell Carcinoma (SCC), Melanoma, and Basal Cell Carcinoma are the three varieties of pores and pores and skin maximum cancers (BCC). Melanoma is the most dangerous, with a completely low survival rate. Melanoma detection early improves a character's opportunities for survival (Ho Take Lau et al., 2009). The following facts on melanomas were accumulated withinside the United States. In the United States, one character is killed through the manner of approach of most cancers every hour. In 2018, there were 87, 110 new most cancers cases, in line with a survey. Melanoma is predicted to claim the lives of nine, 730 people each year. Melanoma is the most commonplace vicinity cause of pores and pores and skin maximum cancers-related death, accounting for approximately 1% of all cases. The sun is the most commonplace vicinity cause of melanomas. The chance of developing pores and pores and skin maximum cancers will grow with age.

Regularly using sunscreen with an SPF of 15 or higher can reduce the threat of most cancers through manner of approach of 50% and squamous molecular carcinoma through manner of approach of 40%. A variety of instructors have said photograph processing is based on techniques for recognizing various varieties of pores and pores and skin problems. A few techniques based mostly on research may be described. Color pics are used to diagnose pores and pores and skin problems without the help of a doctor. The tool is broken up into stages: the number one detects and classifies illness kinds the usage of coloration photograph processing techniques, k-way clustering, and coloration gradient algorithms, and the second classifies illness kinds through the usage of artificial neural networks (Ali AI-Haj et al., 2008).

The techniques grow to be tested on six unique varieties of pores and pores and skin problems and yielded first-diploma accuracy of 95, 99%, and second-diploma accuracy of 94.016%. The first step in diagnosing pores and pores and skin troubles is to use a photograph-characteristic extraction technique. As more attributes are gathered from a photograph, the set of guidelines becomes more accurate in this regard. The techniques grow to be used to diagnose nine unique varieties of pores and pores and skin troubles with a 90 curacy rate. Melanoma is a form of pores and pores and skin maximum cancer that can be lethal if now not detected and treated early.

The creator investigated a number of photograph processing segmentation algorithms that are probably used to find out most cancers in his study. It is hooked up in a manner to split sick spot obstacles in a manner to extract more features. According to the findings, a Melanoma detection tool for dark pores and pores and skin has to be advanced with the usage of a custom set of guidelines databases containing pics from various Melanoma resources. The manual vector system techniques grow to be used to classify most cancers, basal molecular carcinoma (BCC), nevus, and seborrheic keratosis (SK) (SVM). It outperforms masses of strategies in terms of accuracy. On the opportunity hand, the cross-border spread of persistent pores and pores and skin diseases have to have disastrous consequences (Agrawal et al., 2010).

As a result, he hooked up a laptop tool that could find out and test the severity of eczema automatically. The technique is broken up into three stages: the number one recognizes the pores and pores and skin, the second extract a hard and fast of features that incorporate coloration, texture, and borders, and the 1/three employs a Support Vector Machine to assess the severity of the eczema (SVM). A new technique for detecting pores and pores and skin problems that combines laptop vision and system learning has been advanced. Machine learning recognizes pores and pores and skin troubles, on the equal time as laptop vision extracts facts from photos. With a ninety- five-percent accuracy rate, the techniques grow to be tested on six unique varieties of pores and pores and skin problems.

Software Components

MATLAB

MATLAB, which stands for "matrix laboratory," is a software program software platform used to remedy mathematical and scientific problems. Matrix manipulation, function, and facts visualization, set of policies implementation, patron interface development, and integration with programs written in programming languages which consist of C, C++, Java, and others are all possible with Math Works" proprietary programming language. Figure 1 shows the logo of MATLAB. The IPT is hard and fast of MATLAB capability that extends the skills of the numeric computing environment (Kumar, 2016, 2018, 2019, 2021). It includes many reference necessities for fact processing, analysis, visualization, and a set of policy development. Algorithms and software program software for workflow manipulation Image segmentation, enhancement, noise reduction; geometric transformations, image registration, and 3-D image processing (Tanaka, T, et al 2004) are all possible with it. Many IPT sports can be used to generate C/C++ code for pc prototypes and embedded vision systems.

Digital Image

A digital photograph is a two-dimensional feature f (x, y), in which x and s are spatial coordinates and photograph intensity equals the amplitude off at that point. In a digital photograph, the values of "x," "y," and the amplitude values of "f" are all finite discrete numbers. Sampling and quantization are terms used to explain the digitalization of coordinate and amplitude data, respectively. A matrix of real numbers is the stop-end result of sampling and quantization (Prasad et al., 2021, Haribabu et al., 2018, Prasad et al., 2017).

Image Processing

Image processing (Lee et al, 1997) is multi-disciplinary vicinity that includes mathematics, physics, optics, and electric-powered engineering. This vicinity combines pattern recognition, machine learning, artificial intelligence, and human vision research. The picture processing machine includes importing a picture from an optical scanner or a digital camera, evaluating and changing the picture (statistics compression, picture enhancement, and filtering), and generating the favored output picture. Image processing has advanced because of the need to extract data from pox and comprehend its content. Image processing is applied in an entire lot of fields, consisting of medicine, industry, the military, and consumer electronics. Digital radiography, positron emission tomography (PET), computerized axial tomography (CAT), magnetic resonance imaging (MRI) (Sridhar, 2019), and sensible magnetic resonance imaging are all diagnostic imaging modalities (FMRI). Industrial packages embody safety systems, first-class control, and self-maintaining guided vehicle control, to name a few. Complex picture processing strategies are applied in packages beginning from recognizing infantrymen or automobiles to missile guidance, object recognition, and reconnaissance. Fingerprinting, facial recognition, iris recognition, and hand recognition are all commonplace area biometric strategies applied in law enforcement and security. Consumer electronics inclusive of digital cameras and camcorders, high-definition televisions, monitors, DVD players, private video recorders, and molecular phones all require picture processing (Sridhar, 2017).

Methodology

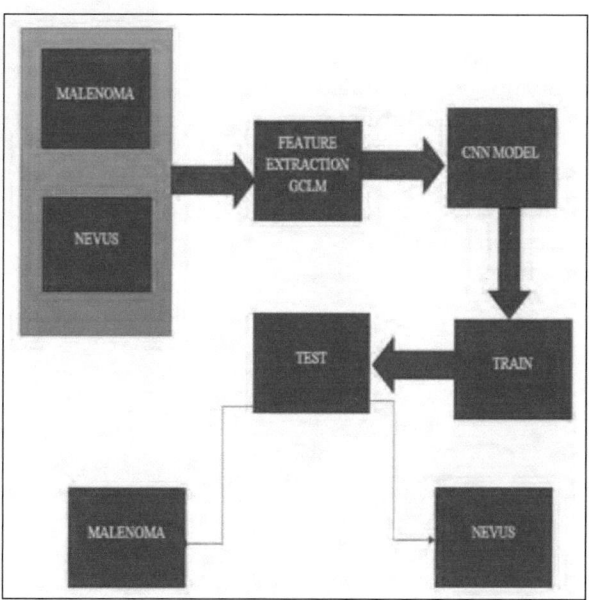

Fig. 1. Methodology

The approach (as shown in Fig. 1) of digitizing information from our surroundings and presenting, analyzing, and saving it in a laptop is called information acquisition. Data transformation is the approach of converting information from one format to another, which encompass a database file, XML document, or Excel spreadsheet. The most commonplace vicinity shape of transformation is converting raw information deliver proper right into a cleansed, validated, and ready-to- use format.

Data extraction is the approach of extracting information from a specific deliver and moving it to a brand-new surroundings, whether or not or now no longer on-premises, with inside the cloud, or a hybrid of every. This intention can be achieved in a number of ways, all of which may be difficult and generally accomplished with the resource of the usage of hand. It can enhance every supervised and unsupervised mastering with the resource of the usage of making information adjustments much less hard and faster on the equal time as moreover improving model validity. Feature implementation is critical even as running with tool mastering models. A broken feature may have an immediate impact on your model, irrespective of the information or design. Feature desire reduces information dimensionality with the resource of the usage of specializing in a subset of quantifiable abilities in case you need to create a further understandable dataset model (predictor variables) (Rioul, et al., 1991). Feature desire algorithms are in search of a subset of predictors that best anticipate placed responses on the equal time as accounting for constraints that encompass required and excluded attributes, similarly to the scale of the available subset of predictors. Feature desire blessings encompass superior prediction performance, faster and plenty much less luxurious predictors, and a better knowledge of the information-generating approach. Even if all of the abilities are massive and provide information on the response variable, having too many can reduce the accuracy of prediction (Sheha et al., 2012).

Results and Discussions

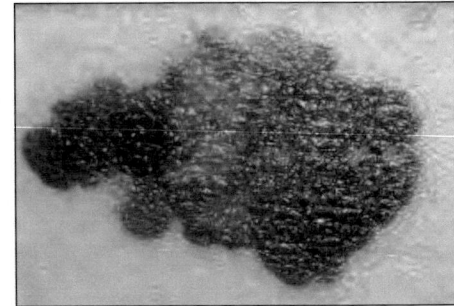

Fig. 2. Input image

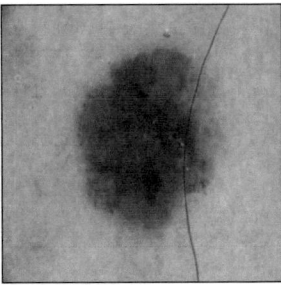

Fig. 3. Output image

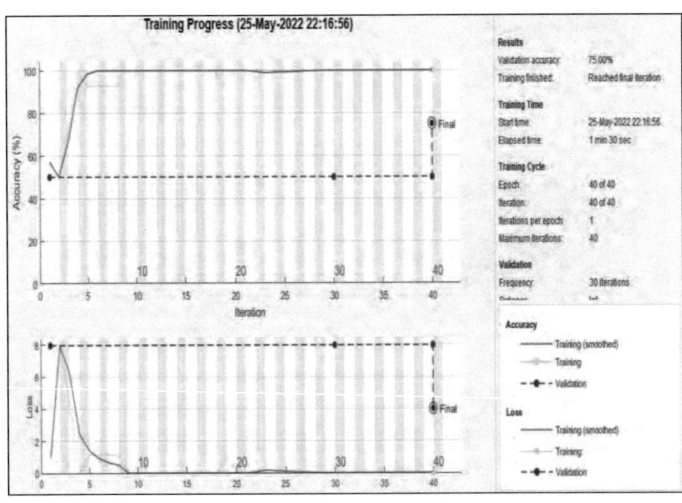

Fig. 4. Training progress

The proposed approach may also need to find out the presence of a dermatological scenario in a photo. It can be used to help human beings all over the worldwide similarly to perform useful tasks. The system can be deployed for free of charge because of the reality the equipment is unfastened to use and available to the man or woman. Despite the regulations of the gadget gaining knowledge of the facts set, the software program software efficiently identified the scenario. The finished product is small and light, making it quality for devices with low system requirements. Figures 2 and 3 show the input image and output image respectively. Figure 4 shows the training progress stages. It moreover has a clean man or woman interface for the man or woman's convenience. Machine-gaining knowledge of photo-processing techniques was successfully developed as presented in Fig. 5.

	1	2	3
10	malenoma		
11	malenoma		
12	nevus		
13	nevus		
14	malenoma		
15	malenoma		
16	nevus		

Fig. 5. Results

Conclusion and Future Scope

Traditional biopsy strategies are a lot much less effective in detecting pores and pores and skin maximum cancers early than computer-based tools. The recommended technique reduces the cost and the time required to find something. The approach employs Artificial Intelligence and Digital Image Processing to stumble on pores and pores and skin maximum cancers. An ANN-based total absolutely classifier is compared to a CNN AVG Filter for photo reputation from taking a look at and training samples. Traditional techniques have an accuracy of 74% for ANN and 87.3% for CNN, the same time as the proposed technique has an accuracy of 74%

for ANN and 87.3% for CNN. In the future, present-day practices with inside subsequent instructions can be improved: 1. a standardized approach ought to be used to stumble on all forms of pores and pores and skin diseases.

References

Lau, H. T., and Al-Jumaily, A. (2009, December). Automatically early detection of skin cancer: Study based on neural netwok classification. In 2009 International Conference of Soft Computing and Pattern Recognition (pp. 375–380). IEEE.

Al-Haj, A. (2008, July). Wavelets pre-processing of Artificial Neural Networks classifiers. In 2008 5th International Multi-Conference on Systems, Signals and Devices (pp. 1–5). IEEE..

Agrawal, P., Shriwastava, S.K. and Limaye, S. S. (2010), July. MATLAB implementation of image segmentation algorithms. In 2010 3rd International Conference on Computer Science and Information Technology (Vol. 3, pp. 427–431). IEEE.

Tanaka, T., Yamada, R., Tanaka, M., Shimizu, K. and Oka, H. (2004, September). A study on the image diagnosis of melanoma. In The 26th Annual International Conference of the IEEE Engineering in Medicine and Biology Society (Vol. 1, pp. 1597–1600). IEEE.

Kumar, B. (2021). Comparative performance evaluation of greedy algorithms for speech enhancement system. Fluctuation and Noise Letters, 20(02), p.2150017.

Kumar, B. (2019). Real-time performance evaluation of modified cascaded median-based noise estimation for speech enhancement system. Fluctuation and Noise Letters, 18(04), p.1950020.

Kumar, B. (2018). Comparative performance evaluation of MMSE-based speech enhancement techniques through simulation and real-time implementation. International Journal of Speech Technology, 21(4), pp. 1033–1044.

Kumar, B. (2016). Mean-median based noise estimation method using spectral subtraction for speech enhancement technique. Indian J. Sci. Technol, 9, pp. 1–6.

Prasad, S. V. S., Pittala, C. S., Vijay, V., Vallabhuni, R. R. (2021). Complex Filter Design for Bluetooth Receiver Application. Proceedings of the 6th International Conference on Communication and

Electronics Systems, ICCES 2021, art. no. 9489020, pp. 442–446.

Haribabu, K., Prasad, S.V.S., Satish Kumar, M. (2018). An IOT based smart home automation using LabVIEW. Journal of Engineering and Applied Sciences, 13(6), pp. 1421–1424.

Prasad, S.V.S., Savithri, T.S., Krishna, I.V.M. performance evaluation of svm kernels on multispectral liss iii data for object classification (2017) International Journal on Smart Sensing and Intelligent Systems, 10(4), pp. 863-878.

Sridhar, B. (2019). Applications of digital image processing in real time world. International Journal of Scientific and Technology Research, 8(12), pp. 3354–3357.

Sridhar, B. (2017). Joint encryption and watermarking technique using block cipher and wavelet. Journal of Theoretical and Applied Information Technology, 95(7), pp. 1479–1484.

Sheha, M.A., Mabrouk, M.S. and Sharawy, A., 2012. Automatic detection of melanoma skin cancer using texture analysis. International Journal of Computer Applications, 42(20), pp. 22–26.

Rioul, O. and Vetterli, M. (1991). Wavelets and signal processing. IEEE signal processing magazine, 8(4), pp.14–38.

Lee, T., Ng, V., Gallagher, R., Coldman, A. and McLean, D. (1997). Dullrazor®: A software approach to hair removal from images. Computers in biology and medicine, 27(6), pp. 533–543.